Membrane Trafficking

METHODS IN MOLECULAR BIOLOGY™

John M. Walker, SERIES EDITOR

484. **Functional Proteomics:** *Methods and Protocols*, edited by *Julie D. Thompson, Christine Schaeffer-Reiss, and Marius Ueffing, 2008*
483. **Recombinant Proteins From Plants:** *Methods and Protocols*, edited by *Loïc Faye and Veronique Gomord, 2008*
482. **Stem Cells in Regenerative Medicine:** *Methods and Protocols*, edited by *Julie Audet and William L. Stanford, 2008*
481. **Hepatocyte Transplantation:** *Methods and Protocols*, edited by *Anil Dhawan and Robin D. Hughes, 2008*
480. **Macromolecular Drug Delivery:** *Methods and Protocols*, edited by *Mattias Belting, 2008*
479. **Plant Signal Transduction:** *Methods and Protocols*, edited by *Thomas Pfannschmidt, 2008*
478. **Transgenic Wheat, Barley and Oats: Production and Characterization Protocols**, edited by *Huw D. Jones and Peter R. Shewry, 2008*
477. **Advanced Protocols in Oxidative Stress I**, edited by *Donald Armstrong, 2008*
476. **Redox-Mediated Signal Transduction:** *Methods and Protocols*, edited by *John T. Hancock, 2008*
475. **Cell Fusion:** *Overviews and Methods*, edited by *Elizabeth H. Chen, 2008*
474. **Nanostructure Design:** *Methods and Protocols*, edited by *Ehud Gazit and Ruth Nussinov, 2008*
473. **Clinical Epidemiology:** *Practice and Methods*, edited by *Patrick Parfrey and Brendon Barrett, 2008*
472. **Cancer Epidemiology, Volume 2:** *Modifiable Factors*, edited by *Mukesh Verma, 2008*
471. **Cancer Epidemiology, Volume 1:** *Host Susceptibility Factors*, edited by *Mukesh Verma, 2008*
470. **Host-Pathogen Interactions:** *Methods and Protocols*, edited by *Steffen Rupp and Kai Sohn, 2008*
469. **Wnt Signaling, Volume 2:** *Pathway Models*, edited by *Elizabeth Vincan, 2008*
468. **Wnt Signaling, Volume 1:** *Pathway Methods and Mammalian Models*, edited by *Elizabeth Vincan, 2008*
467. **Angiogenesis Protocols:** *Second Edition*, edited by *Stewart Martin and Cliff Murray, 2008*
466. **Kidney Research:** *Experimental Protocols*, edited by *Tim D. Hewitson and Gavin J. Becker, 2008.*
465. **Mycobacteria**, *Second Edition*, edited by *Tanya Parish and Amanda Claire Brown, 2008*
464. **The Nucleus, Volume 2:** *Physical Properties and Imaging Methods*, edited by *Ronald Hancock, 2008*
463. **The Nucleus, Volume 1:** *Nuclei and Subnuclear Components*, edited by *Ronald Hancock, 2008*
462. **Lipid Signaling Protocols**, edited by *Banafshe Larijani, Rudiger Woscholski, and Colin A. Rosser, 2008*

461. **Molecular Embryology:** *Methods and Protocols, Second Edition*, edited by *Paul Sharpe and Ivor Mason, 2008*
460. **Essential Concepts in Toxicogenomics**, edited by *Donna L. Mendrick and William B. Mattes, 2008*
459. **Prion Protein Protocols**, edited by *Andrew F. Hill, 2008*
458. **Artificial Neural Networks:** *Methods and Applications*, edited by *David S. Livingstone, 2008*
457. **Membrane Trafficking**, edited by *Ales Vancura, 2008*
456. **Adipose Tissue Protocols**, *Second Edition*, edited by *Kaiping Yang, 2008*
455. **Osteoporosis**, edited by *Jennifer J. Westendorf, 2008*
454. **SARS- and Other Coronaviruses:** *Laboratory Protocols*, edited by *Dave Cavanagh, 2008*
453. **Bioinformatics, Volume 2:** *Structure, Function, and Applications*, edited by *Jonathan M. Keith, 2008*
452. **Bioinformatics, Volume 1:** *Data, Sequence Analysis, and Evolution*, edited by *Jonathan M. Keith, 2008*
451. **Plant Virology Protocols:** *From Viral Sequence to Protein Function*, edited by *Gary Foster, Elisabeth Johansen, Yiguo Hong, and Peter Nagy, 2008*
450. **Germline Stem Cells**, edited by *Steven X. Hou and Shree Ram Singh, 2008*
449. **Mesenchymal Stem Cells:** *Methods and Protocols*, edited by *Darwin J. Prockop, Douglas G. Phinney, and Bruce A. Brunnell, 2008*
448. **Pharmacogenomics in Drug Discovery and Development**, edited by *Qing Yan, 2008.*
447. **Alcohol:** *Methods and Protocols*, edited by *Laura E. Nagy, 2008*
446. **Post-translational Modifications of Proteins:** *Tools for Functional Proteomics, Second Edition*, edited by *Christoph Kannicht, 2008.*
445. **Autophagosome and Phagosome**, edited by *Vojo Deretic, 2008*
444. **Prenatal Diagnosis**, edited by *Sinhue Hahn and Laird G. Jackson, 2008.*
443. **Molecular Modeling of Proteins**, edited by *Andreas Kukol, 2008.*
442. **RNAi:** *Design and Application*, edited by *Sailen Barik, 2008.*
441. **Tissue Proteomics:** *Pathways, Biomarkers, and Drug Discovery*, edited by *Brian Liu, 2008*
440. **Exocytosis and Endocytosis**, edited by *Andrei I. Ivanov, 2008*
439. **Genomics Protocols**, *Second Edition*, edited by *Mike Starkey and Ramnanth Elaswarapu, 2008*
438. **Neural Stem Cells:** *Methods and Protocols, Second Edition*, edited by *Leslie P. Weiner, 2008*
437. **Drug Delivery Systems**, edited by *Kewal K. Jain, 2008*

METHODS IN MOLECULAR BIOLOGY™

Membrane Trafficking

Edited by

Ales Vancura

St. John's University, Queens, NY, USA

 Humana Press

Editor
Ales Vancura
Department of Biological Sciences
St. John's University
Queens, NY
USA
vancuraa@stjohns.edu

Series Editor
John M. Walker
University of Hertfordshire
Hatfield, Hertfordshire
UK

ISBN: 978-1-58829-925-3   e-ISBN: 978-1-59745-261-8
DOI: 10.1007/978-1-59745-261-8

Library of Congress Control Number: 2008933671

© 2008 Humana Press, a part of Springer Science+Business Media, LLC
All rights reserved. This work may not be translated or copied in whole or in part without the written permission of the publisher (Humana Press, 999 Riverview Drive, Suite 208, Totowa, NJ 07512 USA), except for brief excerpts in connection with reviews or scholarly analysis. Use in connection with any form of information storage and retrieval, electronic adaptation, computer software, or by similar or dissimilar methodology now known or hereafter developed is forbidden.
The use in this publication of trade names, trademarks, service marks, and similar terms, even if they are not identified as such, is not to be taken as an expression of opinion as to whether or not they are subject to proprietary rights.
While the advice and information in this book are believed to be true and accurate at the date of going to press, neither the authors nor the editors nor the publisher can accept any legal responsibility for any errors or omissions that may be made. The publisher makes no warranty, express or implied, with respect to the material contained herein.

Printed on acid-free paper

9 8 7 6 5 4 3 2 1

springer.com

Preface

Over the past 30 years, there has been a massive expansion in membrane trafficking research. The growth in this field has been facilitated primarily by the advent of molecular cloning of the 1970s and 1980s, and by the development and technical improvements of new cell-imaging and fractionation techniques. The field of membrane trafficking has its beginnings most likely in the work of George Palade (1974 Nobel Prize in Physiology and Medicine) and his co-workers, who used electron microscopy to lay down a morphological map of the secretory pathway. Another milestone in membrane trafficking was the signal hypothesis, proposed by Günther Blobel (1999 Nobel Prize in Physiology and Medicine) and his co-workers. In subsequent years, membrane trafficking research witnessed a remarkable convergence of information gained using genetic approaches in yeast cells with biochemical approaches in mammalian cells.

Membrane trafficking is responsible for selective transfer of proteins and lipids between membrane-delimited organelles and includes two fundamentally different transport processes: (a) translocation of proteins across or into membranes of the endoplasmic reticulum (ER), mitochondria, nucleus, or peroxisomes; and (b) vesicle-mediated transport between ER, Golgi, lysosomes, peroxisomes, and plasma membrane. The aim of *Membrane Trafficking* is to bring together a series of articles, each describing commonly used methods to elucidate the specific aspects of membrane trafficking. This volume of *Methods in Molecular Biology*™ is divided into two parts that reflect contributions of yeast and mammalian cells as the two major experimental models in membrane trafficking research. The methods in both parts have been chosen to represent both classic and cutting-edge techniques to study macromolecular transport across the membranes. The reliability of all the protocols has been tested in laboratories around the world. Each chapter is appended by notes that navigate through the protocol and serve as a troubleshooting guide. It is our hope that *Membrane Trafficking* will be useful not only to senior researchers and scientists who are experienced in cellular membranes and membrane trafficking, but also to graduate students and scientists who want to study membrane trafficking for the first time.

I would like to thank all the authors for their enthusiastic help and support in assembling this volume; I fully realize that in the highly competitive environment of academic research, many scientists are reluctant to commit time to writing book chapters and review articles. I would also like to express my gratitude to the series editor, Dr. John Walker, and the outstanding staff of Humana Press for their support, help, and encouragement.

Ales Vancura

Contents

Preface v
Contributors xi

PART I YEAST CELLS

1. Freight Management in the Cell: *Current Aspects of Intracellular Membrane Trafficking* 3
   Johannes M. Herrmann and Anne Spang

2. High-Throughput Protein Extraction and Immunoblotting Analysis in *Saccharomyces cerevisiae* 13
   Todd C. Lorenz, Vikram C. Anand, and Gregory S. Payne

3. Genome-Wide Analysis of Membrane Transport Using Yeast Knockout Arrays 29
   Helen E. Burston, Michael Davey, and Elizabeth Conibear

4. A Cell-Free System for Reconstitution of Transport Between Prevacuolar Compartments and Vacuoles in *Saccharomyces cerevisiae* 41
   Thomas A. Vida

5. *In vitro* Analysis of the Mitochondrial Preprotein Import Machinery Using Recombinant Precursor Polypeptides 59
   Dorothea Becker, Martin Krayl, and Wolfgang Voos

6. *In Vitro* Import of Proteins Into Isolated Mitochondria 85
   Karl Bihlmaier, Melanie Bien, and Johannes M. Herrmann

7. Synthesis and Sorting of Mitochondrial Translation Products 95
   Heike Bauerschmitt, Soledad Funes, and Johannes M. Herrmann

8. *In Vivo* Labeling and Analysis of Mitochondrial Translation Products in Budding and in Fission Yeasts 113
   Karine Gouget, Fulvia Verde, and Antoni Barrientos

9. Exploring Protein–Protein Interactions Involving Newly Synthesized Mitochondrial DNA-Encoded Proteins 125
   Darryl Horn, Flavia Fontanesi, and Antoni Barrientos

10. Purification of Yeast Membranes and Organelles by Sucrose Density Gradient Centrifugation 141
    Jennifer Chang, Victoria Ruiz, and Ales Vancura

11 Microscopic Analysis of Lipid Droplet Metabolism and Dynamics
    in Yeast   151
    Heimo Wolinski and Sepp D. Kohlwein

12 Use of Bimolecular Fluorescence Complementation
    in Yeast *Saccharomyces cerevisiae*   165
    *Kari-Pekka Skarp, Xueqiang Zhao, Marion Weber,
    and Jussi Jäntti*

PART II MAMMALIAN CELLS

13 Using Quantitative Fluorescence Microscopy to Probe Organelle
    Assembly and Membrane Trafficking   179
    Brian Storrie, Tregei Starr, and Kimberly Forsten-Williams

14 Measuring Secretory Membrane Traffic   193
    *Vytaute Starkuviene, Arne Seitz, Holger Erfle,
    and Rainer Pepperkok*

15 A Correlative Light and Electron Microscopy Method Based on
    Laser Micropatterning and Etching   203
    *Julien Colombelli, Carolina Tängemo, Uta Haselman,
    Claude Antony, Ernst H.K. Stelzer, Rainer Pepperkok,
    and Emmanuel G. Reynaud*

16 Approaches to Investigate the Role of Signaling in ER-to-Golgi
    Transport   215
    Laura J. Sharpe, Ximing Du, and Andrew J. Brown

17 Recruitment of Coat Proteins to Peptidoliposomes   227
    Gregor Suri, Martin Spiess, and Pascal Crottet

18 SNARE-Mediated Fusion of Liposomes   241
    Jérôme Vicogne and Jeffrey E. Pessin

19 Use of Polarized PC12 Cells to Monitor Protein Localization in
    the Early Biosynthetic Pathway   253
    *Ragna Sannerud, Michaël Marie, Bodil Berger Hansen,
    and Jaakko Saraste*

20 Tracking the Transport of E-Cadherin
    to and From the Plasma Membrane   267
    Matthew P. Wagoner, Kun Ling, and Richard A. Anderson

21 Analysis of Nucleocytoplasmic Shuttling of NFκB Proteins in
    Human Leukocytes   279
    *Chandra C. Ghosh, Hai-Yen Vu, Tomas Mujo,
    and Ivana Vancurova*

22 Imaging pHluorin-Based Probes
at Hippocampal Synapses 293
**Stephen J. Royle, Björn Granseth, Benjamin Odermatt,
Aude Derevier, and Leon Lagnado**

23 Analysis of Receptor Tyrosine Kinase Internalization Using Flow
Cytometry 305
Ning Li, Kristen S. Hill, and Lisa A. Elferink

24 Measurement of Receptor Endocytosis and Recycling 319
Jane M. Knisely, Jiyeon Lee, and Guojun Bu

25 Styryl Dye-Based Synaptic Vesicle Recycling Assay in Cultured
Cerebellar Granule Neurons 333
Victor Anggono, Michael A. Cousin, and Phillip J. Robinson

26 Use of Quantitative Immunofluorescence Microscopy to Study
Intracellular Trafficking: *Studies of the GLUT4
Glucose Transporter* 347
Vincent Blot and Timothy E. McGraw

27 Dissecting GLUT4 Traffic Components in L6 Myocytes
by Fluorescence-Based, Single-Cell Assays 367
Costin N. Antonescu, Varinder K. Randhawa, and Amira Klip

28 Functional Genetic Analysis of the Mammalian Mitochondrial
DNA Encoded Peptides 379
**María Pilar Bayona-Bafaluy, Nieves Movilla,
Acisclo Pérez-Martos, Patricio Fernández-Silva, and
José Antonio Enriquez**

Index 391

Contributors

VIKRAM C. ANAND • *Department of Biological Chemistry, David Geffen School of Medicine, UCLA, Los Angeles, CA*
RICHARD A. ANDERSON • *Department of Pharmacology, University of Wisconsin Medical School, Madison, WI*
VICTOR ANGGONO • *Children's Medical Research Institute, Wentworthville, New South Wales, Australia*
COSTIN N. ANTONESCU • *Hospital for Sick Children and Department of Biochemistry, University of Toronto, Toronto, Ontario, Canada*
CLAUDE ANTONY • *Cell Biology and Cell Biophysics Unit, EMBL, Heidelberg, Germany*
ANTONI BARRIENTOS • *Department of Neurology, Miller School of Medicine, University of Miami, Miami, FL*
HEIKE BAUERSCHMITT • *Adolf-Butenandt-Institute of Physiological Chemistry, University of Munich, Munich, Germany*
MARÍA PILAR BAYONA-BAFALUY • *Departamento de Bioquímica y Biología Molecular y Celular, Universidad de Zaragoza, Zaragoza, Spain*
DOROTHEA BECKER • *Institute for Biochemistry and Molecular Biology, University of Freiburg, Freiburg, Germany*
MELANIE BIEN • *Cell Biology, University of Kaiserslautern, Kaiserslautern, Germany*
KARL BIHLMAIER • *Cell Biology, University of Kaiserslautern, Kaiserslautern, Germany*
VINCENT BLOT • *Department of Biochemistry, Weill Medical College of Cornell University, New York, NY*
ANDREW J. BROWN • *School of Biotechnology and Biomolecular Sciences, University of New South Wales, Sydney, New South Wales, Australia*
GUOJUN BU • *Departments of Pediatrics, and Cell Biology and Physiology, Washington University School of Medicine, St. Louis, MO*
HELEN E. BURSTON • *Centre for Molecular Medicine and Therapeutics, University of British Columbia, Vancouver, British Columbia, Canada*
JENNIFER CHANG • *Department of Biological Sciences, St. John's University, Queens, NY*
JULIEN COLOMBELLI • *Cell Biology and Cell Biophysics Unit, EMBL, Heidelberg, Germany*

ELIZABETH CONIBEAR • *Centre for Molecular Medicine and Therapeutics, University of British Columbia, Vancouver, British Columbia, Canada*
MICHAEL A. COUSIN • *Centre for Integrative Physiology, University of Edinburgh, Edinburgh, UK*
PASCAL CROTTET • *Biozentrum, University of Basel, Basel, Switzerland*
MICHAEL DAVEY • *Centre for Molecular Medicine and Therapeutics, University of British Columbia, Vancouver, British Columbia, Canada*
AUDE DEREVIER • *MRC Laboratory of Molecular Biology, Cambridge, UK*
XIMING DU • *School of Biotechnology and Biomolecular Sciences, University of New South Wales, Sydney, New South Wales, Australia*
LISA A. ELFERINK • *Department of Neuroscience and Cell Biology, University of Texas Medical Branch, Galveston, TX*
JOSÉ ANTONIO ENRIQUEZ • *Departamento de Bioquímica y Biología Molecular y Celular, Universidad de Zaragoza, Zaragoza, Spain*
HOLGER ERFLE • *Bioquant University of Heidelberg, Heidelberg, Germany*
ATRICIO FERNÁNDEZ-SILVA • *Departamento de Bioquímica y Biología Molecular y Celular, Universidad de Zaragoza, Zaragoza, Spain*
FLAVIA FONTANESI • *Department of Neurology, Miller School of Medicine, University of Miami, Miami, FL*
KIMBERLY FORSTEN-WILLIAMS • *Department of Chemical Engineering, Virginia Polytechnic Institute and State University, Blacksburg, VA*
SOLEDAD FUNES • *Adolf-Butenandt-Institute of Physiological Chemistry, University of Munich, Munich, Germany*
CHANDRA C. GHOSH • *Department of Biological Sciences, St. John's University, Queens, NY*
KARINE GOUGET • *Department of Neurology, Miller School of Medicine, University of Miami, Miami, FL*
BJÖRN GRANSETH • *MRC Laboratory of Molecular Biology, Cambridge, UK*
UTA HASELMAN • *Cell Biology and Cell Biophysics Unit, EMBL, Heidelberg, Germany*
BODIL BERGER HANSEN • *Department of Biomedicine, University of Bergen, Bergen, Norway*
JOHANNES M. HERRMANN • *Cell Biology, University of Kaiserslautern, Kaiserslautern, Germany*
KRISTEN S. HILL • *Department of Neuroscience and Cell Biology, University of Texas Medical Branch, Galveston, TX*
DARRYL HORN • *Department of Biochemistry and Molecular Biology, Miller School of Medicine, University of Miami, Miami, FL*
JUSSI JÄNTTI • *Institute of Biotechnology, University of Helsinki, Finland*

Contributors

AMIRA KLIP • *Hospital for Sick Children and Department of Biochemistry, University of Toronto, Toronto, Ontario, Canada*
JANE M. KNISELY • *Departments of Pediatrics, and Cell Biology and Physiology, Washington University School of Medicine, St. Louis, MO*
SEPP D. KOHLWEIN • *Institute of Molecular Biosciences, University of Graz, Graz, Austria*
MARTIN KRAYL • *Institute for Biochemistry and Molecular Biology, University of Freiburg, Freiburg, Germany*
LEON LAGNADO • *MRC Laboratory of Molecular Biology, Cambridge, UK*
JIYEON LEE • *Departments of Pediatrics, and Cell Biology and Physiology, Washington University School of Medicine, St. Louis, MO*
NING LI • *Department of Neuroscience and Cell Biology, University of Texas Medical Branch, Galveston, TX*
KUN LING • *Department of Pharmacology, University of Wisconsin Medical School, Madison, WI*
TODD C. LORENZ • *Department of Biological Chemistry, David Geffen School of Medicine, UCLA, Los Angeles, CA*
TIMOTHY E. MCGRAW • *Department of Biochemistry, Weill Medical College of Cornell University, New York, NY*
MICHAËL MARIE • *Department of Biomedicine, University of Bergen, Bergen, Norway*
NIEVES MOVILLA • *Departamento de Bioquímica y Biología Molecular y Celular, Universidad de Zaragoza, Zaragoza, Spain*
TOMAS MUJO • *Department of Biological Sciences, St. John's University, Queens, NY*
BENJAMIN ODERMATT • *MRC Laboratory of Molecular Biology, Cambridge, UK*
GREGORY S. PAYNE • *Department of Biological Chemistry, David Geffen School of Medicine, UCLA, Los Angeles, CA*
RAINER PEPPERKOK • *Cell Biology and Cell Biophysics Unit, EMBL, Heidelberg, Germany*
ACISCLO PÉREZ-MARTOS • *Departamento de Bioquímica y Biología Molecular y Celular, Universidad de Zaragoza, Zaragoza, Spain*
JEFFREY E. PESSIN • *Department of Pharmacological Sciences, Stony Brook University, Stony Brook, NY*
VARINDER K. RANDHAWA • *Hospital for Sick Children and Department of Biochemistry, University of Toronto, Toronto, Ontario, Canada*
EMMANUEL G. REYNAUD • *Cell Biology and Cell Biophysics Unit, EMBL, Heidelberg, Germany*
PHILLIP J. ROBINSON • *Children's Medical Research Institute, Wentworthville, New South Wales, Australia*

STEPHEN J. ROYLE • *School of Biomedical Sciences, University of Liverpool, Liverpool, UK*
VICTORIA RUIZ • *Department of Biological Sciences, St. John's University, Queens, NY*
RAGNA SANNERUD • *Department of Biomedicine, University of Bergen, Bergen, Norway*
JAAKKO SARASTE • *Department of Biomedicine, University of Bergen, Bergen, Norway*
ARNE SEITZ • *Cell Biology and Cell Biophysics Unit, EMBL, Heidelberg, Germany*
LAURA J. SHARPE • *School of Biotechnology and Biomolecular Sciences, University of New South Wales, Sydney, New South Wales, Australia*
KARI-PEKKA SKARP • *Institute of Biotechnology, University of Helsinki, Finland*
ANNE SPANG • *Biozentrum, University of Basel, Basel, Switzerland*
MARTIN SPIESS • *Biozentrum, University of Basel, Basel, Switzerland*
VYTAUTE STARKUVIENE • *Bioquant, University of Heidelberg, Heidelberg, Germany*
TREGEI STARR • *Department of Physiology and Biophysics, University of Arkansas for Medical Sciences, Little Rock, AR*
ERNST H.K. STELZER • *Cell Biology and Cell Biophysics Unit, EMBL, Heidelberg, Germany*
BRIAN STORRIE • *Department of Physiology and Biophysics, University of Arkansas for Medical Sciences, Little Rock, AR*
GREGOR SURI • *Biozentrum, University of Basel, Basel, Switzerland*
CAROLINA TÄNGEMO • *Cell Biology and Cell Biophysics Unit, EMBL, Heidelberg, Germany*
ALES VANCURA • *Department of Biological Sciences, St. John's University, Queens, NY*
IVANA VANCUROVA • *Department of Biological Sciences, St. John's University, Queens, NY*
FULVIA VERDE • *Department of Pharmacology, Miller School of Medicine, University of Miami, Miami, FL*
JÉRÔME VICOGNE • *Department of Pharmacological Sciences, Stony Brook University, Stony Brook, NY*
THOMAS A. VIDA • *Department of Microbiology and Molecular Genetics, University of Texas Health Science Center, Houston, TX*
WOLFGANG VOOS • *Institute for Biochemistry and Molecular Biology, University of Bonn, Bonn, Germany*
HAI-YEN VU • *Department of Biological Sciences, St. John's University, Queens, NY*

MATTHEW P. WAGONER • *Department of Pharmacology, University of Wisconsin Medical School, Madison, WI*
MARION WEBER • *Institute of Biotechnology, University of Helsinki, Finland*
HEIMO WOLINSKI • *Institute of Molecular Biosciences, University of Graz, Graz, Austria*
XUEQIANG ZHAO • *Institute of Biotechnology, University of Helsinki, Finland*

I

YEAST CELLS

1

Freight Management in the Cell
Current Aspects of Intracellular Membrane Trafficking

Johannes M. Herrmann and Anne Spang

Summary

The correct transport of each protein to its respective location in the cell is a major task of eukaryotic cells. This problem is tackled by the employment of a large variety of cellular factors which, in a concerted action, recognize cargo proteins, translocate them across or insert them into specific membranes, sort them into transport vesicles, and confer the correct delivery to their final destinations. Like in an airport's cargo sorting system, the freight typically passes through several sequential sorting stations until it finally reaches the location that is specified by its individual address label. Although each membrane system employs its specific set of factors, the transport processes typically rely on common principles. Over the last years, many of these principles and the players involved were discovered. These findings allowed a rather detailed understanding of several transport processes like those into the endoplasmic reticulum (ER), the mitochondria, and the nucleus, or of the process of vesicular trafficking between the ER and the Golgi apparatus. With the advent of genome- and proteome-wide screens, a better understanding of the structures of the components involved and detailed studies of their dynamics, the future promises a broad comprehensive picture of the processes by which eukaryotic cells sort their proteins.

Key words: Endoplasmic reticulum; membrane biology; mitochondria; protein trafficking; secretion; translocation; vesicular transport.

1. Introduction

Textbooks on cell biology typically start with an illustration showing a section through a eukaryotic cell in which the different cellular compartments are neatly ordered and composed in a well-arranged and static coexistence. In contrast, the recent advances in high-resolution microscopy allowed exciting insights into the spatial, three-dimensional orientation of cellular membranes showing that the cell is tightly stuffed with a multitude of small membrane structures (for examples, *see* **refs.** *1–4*). At least on the basis of their structural appearance, it is almost impossible to attribute most of these membrane structures to specific cellular compartments. Moreover, these membrane structures are highly dynamic, merge by fusion or divide by fission, and are actively transported by motor proteins through the cellular interior. Despite all this seeming chaos, the distribution and dynamics of cellular compounds is highly organized and regulated. As is true for moving companies, the logistics of material transport is of outstanding importance for eukaryotic cells. Hundreds of factors take care that each protein is safely delivered to its appropriate position. Over the last two decades, the transport of proteins and intracellular membranes has been one of the major interests of molecular cell biologists and many of these processes were studied in great detail.

Despite all current knowledge on membrane-trafficking processes, many fundamental questions still remain to be addressed. Examples of such issues are: How are targeting signals recognized by chaperones and translocases? what decides co- or posttranslational targeting of proteins? what the energetic basis of vectorial protein translocation through membranesis? How are integral membrane proteins folded or assembled, and do chaperones exist which are dedicated for the folding and assembly of membrane proteins? How are proteins specifically recruited into transport vesicles? How do cargo proteins regulate the flow of vesicles? Do vesicles reach their target membranes randomly or are actively guided from donor to target compartments? All these questions are currently addressed by different experimental approaches and first insights reveal a large variety and complexity of the mechanisms underlying these processes.

2. Protein Sorting

In eukaryotic cells there exists a huge diversity of signals, components, and energy requirements by which proteins are translocated into the ER, the nucleus, mitochondria, chloroplasts, and peroxisomes, or by which vesicles are sorted between the membranes of the ER, the Golgi, lysosomes, endosomes, and the plasma membrane. Nevertheless, all these processes underlie some general principles *(5,6)*, which are depicted in **Fig. 1**. Proteins contain specific signals in their sequences that specify them for their individual targeting membrane. These signals (and in some cases also the rest of the protein sequences) are typically recognized by targeting factors and chaperones that usher their cargo to appropriate translocases in the targeting membrane. The translocases mediate the transfer of the polypeptides across the lipid bilayer and also the insertion into the

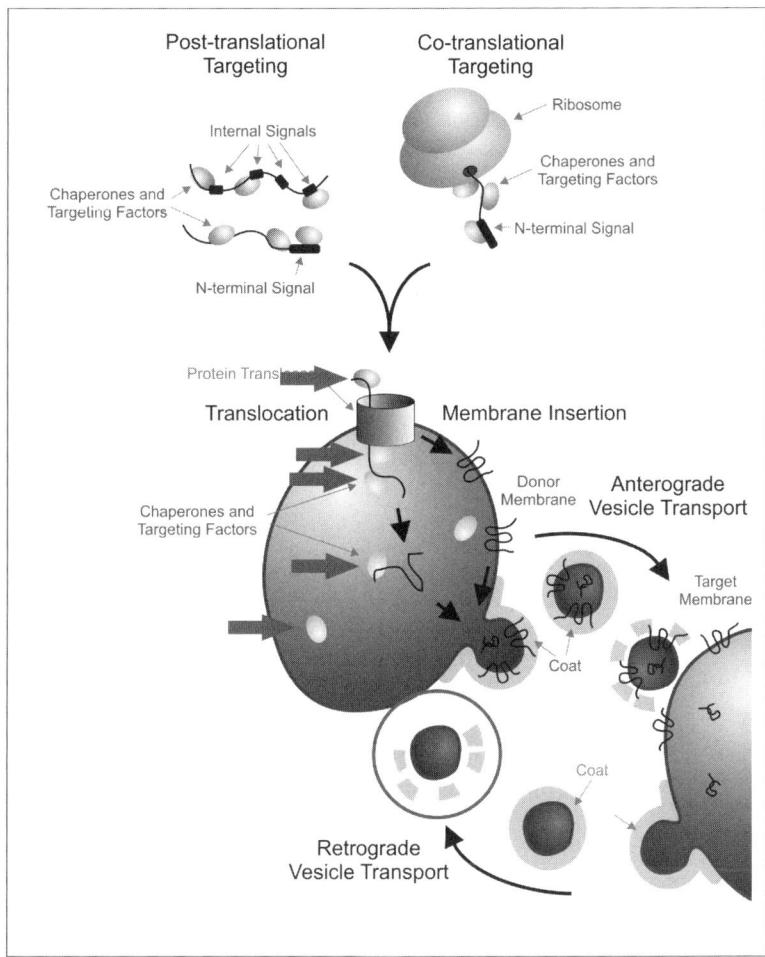

Fig. 1. Schematic representation of the processes of intracellular protein trafficking. In order to be translocated, proteins are recognized during or after their synthesis by membrane-localized complexes. These translocases mediate the protein transfer across cellular membranes in a process that typically is assisted by chaperones and targeting factors. Integral membrane proteins have to be recognized by the translocases and inserted into the lipid bilayer of the membrane. To be transported along the secretory pathway, proteins are taken up into transport vesicles and passed by budding and fusion events from one compartment to the next. The principles of these processes are largely conserved between the different translocation and transport systems of the cell, but each system owes its specific set of factors and requirements.

membrane. The proteins can be recognized by additional translocases in subsequent membranes (e.g., in mitochondria or chloroplasts), or they are folded into the correct three-dimensional structure with the help of chaperones. In the case of the ER, the proteins accumulate at dedicated budding sites to embark on the secretory pathway. Coat proteins on the surface of the ER mediate the budding of vesicles and are released before the vesicles fuse with target compartments. An individual protein might be forwarded via different sorting stations before it reaches its final destination.

There are two fundamentally different transport processes that decide about the final location of any given protein: (a) Translocation, which is the transport across or into the membranes of the ER or other membrane-bound cellular organelles, like mitochondria, peroxisomes, and the nucleus; or (b) secretion, which is the vesicle-mediated transport between cellular compartments. The information that sorts each polypeptide along its specific route is encoded in its sequence. Many of the signals that address proteins to specific compartments of the secretory pathway are complex and still not well understood.

2.1. Protein Translocation

Only nonfolded polypeptides can be translocated into the ER, mitochondria, and chloroplasts. Therefore, translocation and protein synthesis often occur in a mechanistically or kinetically coupled process *(7–9)*. To allow the initiation of translocation before translation is completed, the targeting signals on these proteins are typically positioned on their very N termini *(10)*. Translocation is then driven across the translocases in a process that often relies on the activity of chaperones, namely of members of the Hsp70 family *(11–13)*.

Nuclear and peroxisomal proteins are folded already in the cytosol and, hence, cross the membrane in a folded conformation. In the case of the nucleus, the nuclear pore complex is large enough to allow translocation of folded proteins and even of fully assembled ribosomes. The translocation across peroxisomal membranes, however, is still one of the big mysteries in cell biology.

2.2. Vesicular Transport

Along the anterograde secretory pathway, proteins are transported from the ER via the Golgi to lysosomes, endosomes, peroxisomes, the plasma membrane, or the extracellular space. Along the opposite direction, the retrograde pathway mediates the uptake of peptides, proteins, fluids, and nutrients from the extracellular space by mechanisms collectively termed *endocytosis*. Moreover, the retrograde transport is critical for the recycling of components of the transport machinery.

All these transport steps occur in dedicated vesicles that serve as specialized transport carriers that shield their content form the cytoplasm. Transport vesicles are formed on a donor membrane by the concerted action of small GTPases of the Arf/Sar family and cytosolic coat proteins *(14–17)*. The coat proteins can interact selectively with potential cargo proteins, either directly (in case of membrane proteins) or via specialized membrane receptors. These interactions in turn stabilize the soluble coat components on the membrane, so that accumulations of cargo on the luminal site of the membrane can induce the polymerization of the coat on the cytosolic site. The formation of such coat complexes lead to a deformation of the membrane either by their mechanical properties or by the sequestration of certain lipids which promote the curvature of the membrane *(18)*. The membrane deforming capacity of coat proteins was first demonstrated in the laboratory of Randy Schekman by the formation of COPII-coated vesicles from synthetic, chemically defined liposomes *(19)*. Components of the COPII coat form 40- to 70-nm vesicles at the ER membrane, which transport proteins to the Golgi apparatus. Similar activities were reported for COPI and clathrin coats, the two other classes of coat complexes *(20–22)*. The vesicles are finally pinched off from donor membranes in a process which in some cases is facilitated by dynamin proteins, specific membrane-binding GTPases *(23,24)*.

After their release from the donor membrane, vesicles can interact with motor proteins that hook them to the cytoskeleton. The transport on tracks of the cytoskeleton is critical for long-range transport as in neurons and other specialized mammalian cells. In fungi, vesicles from the Golgi are directed to the plasma membrane on a motor-driven transport route. The interaction of vesicles with motors often is dependent on coat proteins, which adds to the debate about the time when the coat comes off the transport vesicle. Three scenarios for the coordination of uncoating were suggested, which are not mutually exclusive. First, the coat could start to peel off while the vesicles form and the small GTPases would only initiate the polymerisation of such meta-stable coats *(25,26)*. Second, the GTPase cycle may act as timer that releases the coat after a certain time period *(27–29)*. Finally, the coat may be removed directly prior to the fusion reaction with the target membrane *(30,31)*. One point is generally accepted, however: The coat has to be released from the vesicle before fusion can take place.

Albeit critical for the formation of vesicles, cargo does not directly determine the target compartment of a given transport vesicle. The selectivity is at least in part provided by cognate pairs of receptor proteins on the vesicle (v-SNARE) and on the target membrane (t-SNARE) *(32–35)*. The achievement of a reconstituted fusion assay using proteoliposomes of defined composition clearly showed that *in vitro* SNARE proteins are sufficient to initiate membrane fusion *(36)*. *In vivo* the system is much more complex and many additional layers of selectivity regulate the fusion processes. It appears unlikely that *in vivo* vesicles try to fuse with all the compartments they encounter until they finally find the compartment that exposes the matching t-SNAREs. Presumably, additional, yet so far unidentified, components confer a vectorial transport of vesicles. In most *in vitro* systems used to study intracellular transport, such components may not be essential, and therefore missed. It also remains unclear how SNARE proteins are included into vesicles and how t-SNAREs get to their final compartment without causing fusion along the way. More sophisticated methods will need to be employed to shed light on these important issues.

3. Future Directions

Most of what is known about protein translocation is deduced from the study of a small number of model proteins that were analyzed in a few model organisms. In order to generalize these findings, it is very important to base our knowledge on a broader ground. The advent of the postgenomic era now allows the analysis of processes on genome- and proteome-wide scales. For example, the use of yeast knockout collections offers a straight-forward and cheap strategy to identify the full complement of components required for a given process. A wonderful example for the influence of such a screen is the study by the laboratory of Bill Wickner, which, in one study, identified 137 genes that are needed for normal vacuole size and copy number in growing cells *(37)*. The use of such yeast knockout arrays is described in Chapters 2 and 3 in this book.

A second novel spike for translocation research came from the identification of entire proteomes of individual cellular compartments. For example, the analysis of the mitochondrial proteome *(38)* initiated a boom in the search for novel components which significantly improved our understanding of the composition and function of mitochondrial protein translocases. Comprehensive proteomic studies were recently published for most compartments of animal, yeast, or plant cells (for review *see* **refs. *39–43***).

The impact of systems biology on the current understanding of how the cell works is invaluable. Nevertheless, without the detailed analysis of the molecular structure and function of individual components, all genomic and proteomic approaches (collectively referred to as "omic" approaches) would only provide inventory lists and interaction networks. Therefore, a thorough characterization of specific factors is more important than ever. In addition to classical biochemical and cell biological strategies, which are outlined in many chapters of this book, insights into the structure of proteins has shown to be very helpful. A good example is the recent solution of the structure of the Sec translocase by the group of Tom Rapoport *(44)*, which significantly inspired the entire field of protein translocation. In addition to crystallization studies, the use of cryoelectron microscopy and modern spectroscopic strategies allowed stimulating insights into the structure of translocation complexes (for examples, *see* **refs**. *45–48*). Also the vesicle budding field has greatly benefited from recently published crystal structures like those of the coat components *(49–51)*. These studies provided the first evidence for a direct interaction of SNARE proteins with a subunit of the COPII coat.

Two major findings revolutionized the way cell biologists conduct their experiments today: the discovery of green fluorescence protein as a localization tool and the applicability of siRNA in different experimental systems. These technologies combined with high-throughput automated microscopy will increase the knowledge on the communication between and the dynamics of intracellular organelles tremendously. All the puzzle pieces contributed by the different strategies of systems biology, biochemistry, cell biology, genetics, and structural biology promise—when combined into one big picture—an exciting full view on the processes by which eukaryotic cells achieve the intracellular trafficking of their constituents.

References

1. Palade, G. (1975) Intracellular aspects of the process of protein synthesis. *Science* **189**, 347–358.
2. Novikoff, P. M., Novikoff, A. B., Quintana, N., and Hauw, J. J. (1971) Golgi apparatus, GERL, and lysosomes of neurons in rat dorsal root ganglia, studied by thick section and thin section cytochemistry. *J. Cell Biol.* **50**, 859–886.
3. Ladinsky, M. S., Mastronarde, D. N., McIntosh, J. R., Howell, K. E., and Staehelin, L. A. (1999) Golgi structure in three dimensions: functional insights from the normal rat kidney cell. *J. Cell Biol.* **144**, 1135–449.
4. Griffiths, G., Fuller, S. D., Back, R., Hollinshead, M., Pfeiffer, S., and Simons, K. (1989) The dynamic nature of the Golgi complex. *J. Cell Biol.* **108**, 277–297.
5. Wickner, W., and Schekman, R. (2005) Protein translocation across biological membranes. *Science* **310**, 1452–1456.
6. Schatz, G., and Dobberstein, B. (1996) Common principles of protein translocation across membranes. *Science* **271**, 1519–1526.
7. Walter, P., Gilmore, R., and Blobel, G. (1984) Protein translocation across the endoplasmic reticulum. *Cell* **38**, 5–8.
8. Wild, K., Halic, M., Sinning, I., and Beckmann, R. (2004) SRP meets the ribosome. *Nat. Struct. Mol. Biol.* **11**, 1049–1053.
9. Johnson, A. E., and van Waes, M. A. (1999) The translocon: a dynamic gateway at the ER membrane. *Annu. Rev. Cell Dev. Biol.* **15**, 799–842.
10. von Heijne, G. (1990) Protein targeting signals. *Curr. Opin. Cell Biol.* **2**, 604–608.
11. Sheffield, W. P., Shore, G. C., and Randall, S. K. (1990) Mitochondrial precursor protein. Effects of 70-kD heat shock protein on polypeptide folding, aggregation, and import competence. *J. Biol. Chem.* **265**, 11069–11076.

12. Brodsky, J. L., Hamamoto, S., Feldheim, D., and Schekman, R. (1993) Reconstitution of protein translocation from solubilized yeast membranes reveals topologically distinct roles for BiP and cytosolic Hsc70. *J. Cell Biol.* **120**, 95–102.
13. Young, J. C., Hoogenraad, N. J., and Hartl, F. U. (2003) Molecular chaperones Hsp90 and Hsp70 deliver preproteins to the mitochondrial import receptor Tom70. *Cell* **112**, 41–50.
14. Bonifacino, J. S., and Glick, B. S. (2004) The mechanisms of vesicle budding and fusion. *Cell* **116**, 153–166.
15. Fromme, J. C., and Schekman, R. (2005) COPII-coated vesicles: flexible enough for large cargo? *Curr. Opin. Cell Biol.* **17**, 345–352.
16. Lippincott-Schwartz, J., and Liu, W. (2006) Insights into COPI coat assembly and function in living cells. *Trends Cell Biol.* **16**, e1–e4.
17. Spang, A. (2004) Vesicle transport: a close collaboration of Rabs and effectors. *Curr Biol.* **14**, R33–R34.
18. Farsad, K., Ringstad, N., Takei, K., Floyd, S. R., Rose, K., and De Camilli, P. (2001) Generation of high curvature membranes mediated by direct endophilin bilayer interactions. *J. Cell Biol.* **155**, 193–200.
19. Matsuoka, K., Orci, L., Amherdt, M., Bednarek, S. Y., Hamamoto, S., Schekman, R., and Yeung, T. (1998) COPII-coated vesicle formation reconstituted with purified coat proteins and chemically defined liposomes. *Cell* **93**, 263–75.
20. Spang, A., Matsuoka, K., Hamamoto, S., Schekman, R., and Orci, L. (1998) Coatomer, Arf1p, and nucleotide are required to bud coat protein complex I-coated vesicles from large synthetic liposomes. *Proc. Natl. Acad. Sci. USA* **95**, 11199–11204.
21. Bremser, M., Nickel, W., Schweikert, M., et al. (1999) Coupling of coat assembly and vesicle budding to packaging of putative cargo receptors. *Cell* **96**, 495–506.
22. Takei, K., Haucke, V., Slepnev, V., et al. (1998) Generation of coated intermediates of clathrin-mediated endocytosis on protein-free liposomes. *Cell* **94**, 131–141.
23. Hinshaw, J. E. (2000) Dynamin and its role in membrane fission. *Annu. Rev. Cell Dev. Biol.* **16**, 483–519.
24. Roux, A., Uyhazi, K., Frost, A., and De Camilli, P. (2006) GTP-dependent twisting of dynamin implicates constriction and tension in membrane fission. *Nature* **441**, 528–531.
25. Antonny, B., Bigay, J., Casella, J. F., Drin, G., Mesmin, B., and Gounon, P. (2005) Membrane curvature and the control of GTP hydrolysis in Arf1 during COPI vesicle formation. *Biochem. Soc. Trans.* **33**, 619–622.
26. Spang, A. (2002) ARF1 regulatory factors and COPI vesicle formation. *Curr. Opin. Cell Biol.* **14**, 423–427.
27. Tanigawa, G., Orci, L., Amherdt, M., Ravazzola, M., Helms, J. B., and Rothman, J. E. (1993) Hydrolysis of bound GTP by ARF protein triggers uncoating of Golgi-derived COP-coated vesicles. *J. Cell Biol.* **123**, 1365–1371.
28. Kirchhausen, T. (2000) Three ways to make a vesicle. *Nat. Rev. Mol. Cell Biol.* **1**, 187–198.
29. Salama, N. R., and Schekman, R. W. (1995) The role of coat proteins in the biosynthesis of secretory proteins. *Curr. Opin. Cell Biol.* **7**, 536–543.

30. Andag, U., Neumann, T., and Schmitt, H. D. (2001) The coatomer-interacting protein Dsl1p is required for Golgi-to-endoplasmic reticulum retrieval in yeast. *J. Biol. Chem.* **276**, 39150–39160.
31. Behnia, R., Barr, F. A., Flanagan, J. J., Barlowe, C., and Munro, S. (2007) The yeast orthologue of GRASP65 forms a complex with a coiled-coil protein that contributes to ER to Golgi traffic. *J. Cell Biol.* **176**, 255–261.
32. Söllner, T., Whiteheart, S. W., Brunner, M., et al. (1993) SNAP receptors implicated in vesicle targeting and fusion. *Nature* **362**, 318–324.
33. Sollner, T., Bennett, M. K., Whiteheart, S. W., Scheller, R. H., and Rothman, J. E. (1993) A protein assembly–disassembly pathway in vitro that may correspond to sequential steps of synaptic vesicle docking, activation, and fusion. *Cell* **75**, 409–418.
34. Jahn, R., and Scheller, R. H. (2006) SNAREs—engines for membrane fusion.*Nat. Rev. Mol. Cell Biol.* **7**, 631–643.
35. Sudhof, T. C. (2004) The synaptic vesicle cycle. *Annu. Rev. Neurosci.* **27**, 509–547.
36. Weber, T., Zemelman, B. V., McNew, J. A., et al. (1998) SNAREpins: minimal machinery for membrane fusion. *Cell* **92**, 759–772.
37. Seeley, E. S., Kato, M., Margolis, N., Wickner, W., and Eitzen, G. (2002) Genomic analysis of homotypic vacuole fusion. *Mol Biol Cell* **13**, 782–794.
38. Sickmann, A., Reinders, J., Wagner, Y., et al. (2003) The proteome of Saccharomyces cerevisiae mitochondria. *Proc. Natl. Acad. Sci. USA* **100**, 13207–13212.
39. Andersen, J. S., and Mann, M. (2006) Organellar proteomics: turning inventories into insights. *EMBO Rep.* **7**, 874–879.
40. Eisenstein, M. (2006) Exploring how the organelles are organized. *Nat. Methods* **3**, 420–421.
41. Simpson, J. C., and Pepperkok, R. (2006) The subcellular localization of the mammalian proteome comes a fraction closer. *Genome Biol.* **7**, 222.
42. Yates, J. R., 3rd, Gilchrist, A., Howell, K. E., and Bergeron, J. J. (2005) Proteomics of organelles and large cellular structures. *Nat. Rev. Mol. Cell Biol.* **6**, 702–714.
43. Millar, A. H., Whelan, J., and Small, I. (2006) Recent surprises in protein targeting to mitochondria and plastids. *Curr. Opin. Plant Biol.* **9**, 610–615.
44. Van den Berg, B., Clemons, W. M., Jr., et al. (2004) X-ray structure of a protein-conducting channel. *Nature* **427**, 36–44.
45. Beckmann, R., Spahn, C. M., Eswar, N., et al. (2001) Architecture of the protein-conducting channel associated with the translating 80S ribosome. *Cell* **107**, 361–372.
46. Ménétret, J. F., Neuhof, A., Morgan, D. G., Plath et al. (2000) The structure of ribosome-channel complexes engaged in protein translocation. *Mol. Cell* **6**, 1219–1232.
47. Giepmans, B. N., Adams, S. R., Ellisman, M. H., and Tsien, R. Y. (2006) The fluorescent toolbox for assessing protein location and function. *Science* **312**, 217–224.
48. Presley, J. F. (2005) Imaging the secretory pathway: the past and future impact of live cell optical techniques. *Biochim. Biophys. Acta* **1744**, 259–272.

49. Stagg, S. M., Gurkan, C., Fowler, D. M., et al. (2006) Structure of the Sec13/31 COPII coat cage. *Nature* **439**, 234–238.
50. Miller, E. A., Beilharz, T. H., Malkus, P. N., et al. (2003) Multiple cargo binding sites on the COPII subunit Sec24p ensure capture of diverse membrane proteins into transport vesicles. *Cell* **114**, 497–509.
51. Bi, X., Corpina, R. A., and Goldberg, J. (2002) Structure of the Sec23/24-Sar1 pre-budding complex of the COPII vesicle coat. *Nature* **419**, 271–277.

2

High-Throughput Protein Extraction and Immunoblotting Analysis in *Saccharomyces cerevisiae*

Todd C. Lorenz, Vikram C. Anand, and Gregory S. Payne

Summary

A variety of *Saccharomyces cerevisiae* strain libraries allow for systematic analysis of strains bearing gene deletions, repressible genes, overexpressed genes, or modified genes on a genome-wide scale. Here we introduce a method for culturing yeast strains in 96-well format to achieve log-phase growth and a high-throughput technique for generating whole-cell protein extracts from these cultures using sodium dodecyl sulfate and heat lysis. We subsequently describe a procedure to analyze these whole-cell extracts by immunoblotting for alkaline phosphatase and carboxypeptidase yscS to identify strains with defects in protein transport pathways or protein glycosylation. These methods should be readily adaptable to many different areas of interest.

Key words: Genome deletion library; glycosylation; high-throughput proteomic analysis; immunoblot; protein transport; *Saccharomyces cerevisae*.

1. Introduction

Complete sequences of the genomes of many organisms have allowed for the systematic analysis and manipulation of genes within those organisms on a genome-wide scale. A worldwide consortium of laboratories used the annotated genome of the budding yeast *Saccharomyces cerevisiae* to construct deletions of each putative open reading frame not essential for viability and assembled the resulting single-gene deletion strains into commercially available libraries *(1)*.

T.C.L. and V.C.A. contributed equally to the experiments and to the writing of this manuscript.

From: *Methods in Molecular Biology, vol. 457: Membrane Trafficking,*
Edited by: A. Vancura, DOI: 10.1007/978-1-59745-261-8_2, © Humana Press, Totowa, NJ

Numerous other *S. cerevisiae* genome-wide libraries have been constructed and are commercially available including a single-gene green fluorescent protein (GFP)-tagged library, tandem-affinity purification (TAP) tagged single-gene library, and libraries of strains with modified genes (*see* **Note 1**).

Here we describe a method that utilizes the single-gene deletion library to obtain whole-cell protein extracts from rapidly dividing cultures in a high-throughput manner. We subsequently describe a method for systematic immunoblotting to detect defects in protein transport and protein glycosylation in each deletion strain extract. These methods should be readily adaptable to obtain extracts from a variety of strain libraries grown in a range of environmental, nutrient, and stress-inducing conditions. These extracts may then be immunologically probed for specific proteins and protein modifications.

2. Materials

2.1. Cell Culture and Lysis

1. *MAT*a single-gene deletion library (Open Biosystems).
2. "Standard" 96-well plates. Corning 360 μL/well 96-well clear flat-bottom polystyrene untreated microplates, sterile, with Lid (Corning Life Sciences).
3. "Deep-well" 96-well plates. ScreenMates 2 mL per well 96-well storage block, round-bottom, polypropylene, sterile. (Matrix Technologies).
4. 96-well polymerase chain reaction (PCR) plates. Twin.tec 250 μL per well 96-well PCR plate, skirted (Eppendorf).
5. 96-pin replicator. Flat square pin design, $1.5 \times 1.5 \times 25.4$ mm per pin dimension (Boekel).
6. Bleach Solution: commercial bleach (e.g. Chlorax) diluted to 10% in water.
7. 95% ethanol.
8. Stainless-steel micro-stir bars $4.5 \times 1.5 \times 1$ mm (VP-Scientific).
9. Neodymium magnet, $2.5 \times 1.9 \times 5.1$ cm (United Nuclear).
10. Motorized magnetic tumble stirrer. Alligator 710C1 (V&P Scientific).
11. Yeast extract/peptone/dextrose (YPD): 1% bacto-yeast extract, 2% bacto-peptone, 2% Dextrose (BD Diagnostic Systems).
12. G418 sulfate (CALBIOCHEM).
13. Lysis buffer: 5% sodium dodecyl sulfate (SDS), 2% β-mercaptoethanol (β-ME), 50 mM Tris-HCl, pH 6.8, 10% glycerol, 0.03% bromophenol blue. β-ME is both toxic and caustic and should be handled with gloves at all times (*see* **Note 2**).
14. 96-well PCR machine or suitable boiling apparatus.
15. Sorvall RT-6000B with 96-well plate adaptors (Sorvall).

2.2. SDS–Polyacrylamide Gel Electrophoresis (SDS-PAGE)

1. Resolving gel buffer: 1.5 M Tris-HCl pH 8.8.
2. Stacking buffer: 1 M Tris-HCl pH 6.8.

3. 30% (w/v) acrylamide/*bis*-acrylamide solution (37.5/1) (National Diagnostics). Acrylamide and *bis*-acrylamide are neurotoxins and should be handled with gloves at all times.
4. N,N,N',N'–Tetramethyl-ethylenediamine (TEMED). TEMED is a caustic reagent that should be handled with gloves.
5. 10% (w/v) ammonium persulfate freshly prepared each time of use.
6. 10% SDS.
7. 0.1% SDS.
8. Running buffer: 25 mM Tris (untitrated), 0.192 M glycine, 0.1% SDS.
9. Prestained protein molecular-weight markers (New England Biolabs).

2.3. Immunoblotting

1. 100% methanol.
2. Transfer buffer: 20 mM Tris-base, untitrated, 150 mM glycine, 20% methanol.
3. 0.45 μm polyvinylidene fluoride membrane (PVDF; Millipore Billerica).
4. 3MM-Chr chromatography paper (Whatman Inc).
5. Tris-buffered saline with Tween-20 (TBS-T): 20 mM Tris-HCl pH 7.6, 137 mM sodium chloride, 0.1% Tween-20.
6. Blocking buffer: 5% (w/v) nonfat dry milk in TBS-T.
7. Primary antibody.
8. Secondary antibody: goat anti-rabbit immunoglobulin G-horseradish peroxidase conjugate (Biorad). This procedure assumes that the primary antibody is derived from a rabbit source (*see* **Note 3**).
9. ECL plus detection reagents (Amersham Biosciences).

3. Methods

This method involves growing yeast strains in a 96-well format that matches the organization of most commercially available yeast strain libraries. The procedure can be used to generate whole-cell extracts from strains during any phase of yeast growth: stationary, logarithmic-phase (log-phase) growth, and post-diauxic shift growth.

Yeast cultures may exist in one of these three distinct growth phases depending on nutrient availability in the growth medium (*see* **Fig. 1**; reviewed in **ref.** *2*). In the stationary phase, cells in a nutrient-deficient environment remain dormant and expend little energy. During fermentative growth, cells exponentially divide in a sugar-rich environment, rapidly generating energy by fermenting sugars into ethanol. After the complete consumption of sugars, yeast cells begin to undergo respiratory growth where ethanol is consumed. This is known as the "diauxic shift" and yeast cultures in post-diauxic shift growth divide significantly more slowly than cells fermenting sugars.

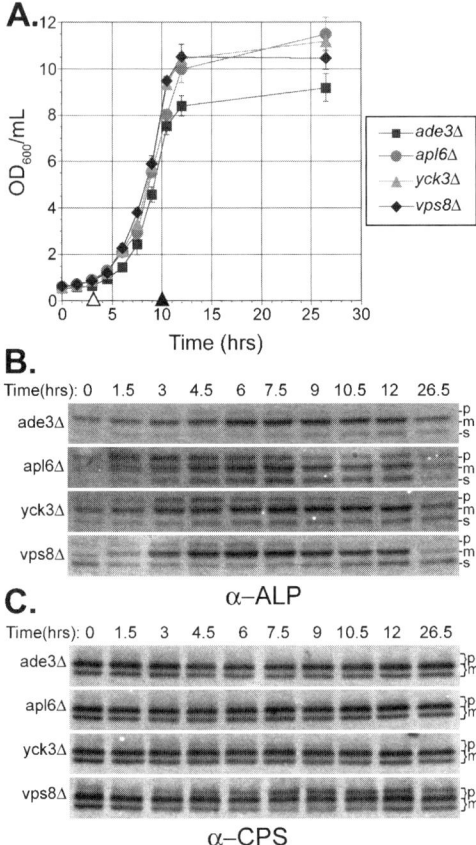

Fig. 1. Effects of growth phase on zymogen precursor accumulation. (A) Growth curve of strains cultured in deep-well 96-well format. Semi-log analysis (not shown) indicates fermentative growth between 3 and 10 h after inoculation indicated by the interval between the open arrowhead (△) and closed arrowhead (▲). Cells enter the post-diauxic phase after 10 h of growth indicated by the interval after the closed arrowhead (▲). Stationary phase cultures are analyzed at T = 0. Stationary cultures of congenic *ade3*△, *apl6*△, *yck3*△, and *vps8*△ cells were inoculated to a final concentration of 0.5 OD_{600}/mL in 2 mL of YPD+200 μg/mL G418 in a 96-deep-well plate. Concentration of cells (OD_{600}/mL) was determined for three independent cultures of each strain at T = 0 and at 90 min intervals. At each time point 1 OD_{600} from each of the three independent cultures was removed and combined together. The combined samples were processed using glass-bead mechanical lysis (*see* **Note 16**) with a final volume of 300 μL and immunoblotted as described in **Subheading 3.3** (*see* panels B and C). (B) Accumulation of precursor ALP fluctuates in strains deficient in ALP transport at various stages of growth. Extracts were analyzed by SDS-PAGE and immunoblotted for ALP. ALP is synthesized as a precursor (p) that is proteolytically processed into mature membrane (m) and soluble (s) forms upon delivery to the vacuole. Note that the ALP signal is highest in all strains

Cells within each growth phase may display significant differences in gene expression, protein levels, and protein modifications *(3)*. For example, some single-gene deletion strains only exhibit protein-transport deficiencies during fermentative growth (*see* **Fig. 1B,C**). Thus, it is important in genome-wide assays to analyze all strains at the same growth phase as much as possible (*see* **Note 4**). To maintain active growth, yeast cultures are continuously mixed in a volume of media sufficient to prevent nutrient deprivation and large enough to supply the necessary quantity of cells for immunoblotting analysis. To achieve mixing, a micro stir-bar magnetic agitation system is used in each well of a 96-"deep-well" plate. These deep-well plates have similar well diameters and well spacing as standard 96-well plates, yet are able to hold more liquid as a result of the larger depth of the well. Another issue for high-throughput analysis is how to generate whole-cell protein extracts. The yeast cell wall poses a lysis barrier that can be removed enzymatically or shredded by agitation with glass beads. For large-scale analysis, we demonstrate that SDS and heat lysis alone are sufficient to extract most proteins from yeast cells, particularly if the cells are in fermentative growth (*see* **Fig. 2** and **Note 5**).

Extracts from strains are subsequently subjected to immunoblotting analysis. To investigate protein transport we follow the proteolytic maturation of two zymogens, the protease carboxypeptidase yscS (CPS), and the phosphate hydrolase alkaline phosphatase (ALP). These proteins are synthesized as higher molecular-weight precursors into the endoplasmic reticulum (ER) and are then transported to the Golgi apparatus. There, CPS and ALP are sorted into two distinct routes to the vacuole *(4)*, a lytic organelle analogous to the mammalian lysosome *(5,6)*. At the vacuole, CPS and ALP are processed to lower molecular-weight enzymatically active forms *(7)*.

Disruption of intracellular transport can affect delivery of ALP and CPS to the vacuole resulting in accumulation of the precursor forms. The relative ratio

---

Fig. 1. (Continued) during fermentative growth. In the *yck3Δ* strain precursor accumulation is most apparent from 1.5 to 9 h. The *ade3Δ* and *vps8Δ* strains contain only trace amounts of precursor ALP. (C) Accumulation of precursor CPS at various growth phases. Extracts were analyzed by SDS-PAGE and immunoblotted for CPS. CPS is synthesized as a precursor that is differentially glycosylated to yield two forms (p). Delivery to the vacuole results in proteolytic maturation that produces two mature glycosylated forms (m), one of which comigrates by SDS-PAGE with the lower molecular-weight precursor form (*see* Fig. 4). Thus, precursor accumulation can be detected by the appearance of the higher molecular weight precursor form. Only *vps8Δ* exhibits a defect in CPS maturation. The highest levels of precursor CPS accumulation occur during fermentative growth from 4.5 to 12 h.

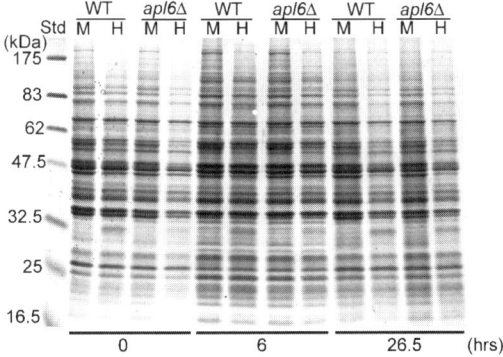

Fig. 2. Effectiveness of heat lysis during fermentative growth. 3 OD$_{600}$ of a mutant strain (apl6Δ) and a wild-type strain (WT) (see **Note 17**) were harvested at T = 0, 6 h, and 26.5 h then lysed either using mechanical (M) glass-bead lysis (see **Note 16**) or using heat (H) lysis. Both samples were brought to a final volume of 300 μL in lysis buffer and 5 μL of each was analyzed by SDS-PAGE and stained with Coomassie blue. The resulting gel was digitally scanned. Both strains were adequately lysed by the heat method although mechanical lysis was more effective at extracting higher molecular weight proteins, particularly during stationary phase (see **Note 5**).

of precursor to mature form can be determined by electrophoretic separation of the two forms by SDS-PAGE and immunoblotting (8).

In addition to proteolytic processing, both CPS and ALP undergo glycosylation in the ER and Golgi by glycosyltransferase enzymes (5). Defects in protein glycosylation may result in lower molecular-weight, aberrantly glycosylated species of these proteins (see **Fig. 3**).

These methods may be readily adapted to investigating other protein modifications using antibodies or electrophoretic mobility changes specific to that modification. Extracts may be derived from any library of choice and during any desired phase of growth.

3.1. Cell Culture and Extraction

1. Add 200 μL of YPD containing 200 U/mL G418 to standard 96-well plate. Media dispensing is facilitated by use of a multichannel pipette.
2. Place the replicator for 30 sec into a container with bleach solution that has been filled to a level such that the entire length of the pins is covered. Remove the replicator from the bleach solution container and shake off residual bleach solution from the pins. Then place the replicator into a container filled with sterile water to the same depth. Gently stir the replicator in the water container for 30 sec. Remove the replicator from the water container and shake off any residual liquid. Place the replicator into a second container with water filled to the same

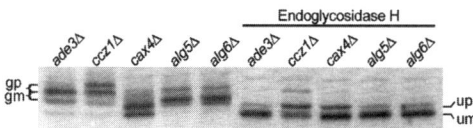

Fig. 3. High-throughput immunoblotting can reveal defects in glycosylation. Several strains with potential glycosylation defects were identified in a high-throughput screen including *cax4Δ*, *alg5Δ*, and *alg6Δ*. This figure illustrates the mobility differences detected in the high-throughput screen and demonstrates that the differences are caused by glycosylation. Extracts from *ade3Δ*, *ccz1Δ*, *cax4Δ*, *alg5Δ*, and *alg6Δ*? cells were generated by glass-bead mechanical lysis. Cells were grown fermentatively (1–5 OD_{600}) and 5 OD_{600} of cells were sedimented at approx 2000 × g for 3 min in a glass 13 × 100-mm tube. The supernatant was aspirated off and acid-washed glass beads were added to each tube. Cells were resuspended in 50 μL 2% SDS, vortexed, and incubated at 100°C for 5 min. To each sample, we added 190 μL water and 60 μL of 250 mM sodium citrate buffer pH 5.5. The samples were divided and one sample of each strain extract was incubated with endoglycosidase H (endo H) for 1 h at 37°C. *Samples were analyzed by SDS-PAGE and immunoblotted for CPS as described in* **Subheadings** **3.2** *and* **3.3**. *The glycosylated precursor (gp) and glycosylated mature (gm) as well as the endo H treated glycosylated precursor (up) and unglycosylated mature (um) forms are indicated. The two WT mature glycosylated CPS forms are present in ade3Δ cells. Two normally glycosylated precursor forms are present in ccz1Δ strains. The other strains exhibit changes in the pattern of CPS bands that are mostly the result of glycosylation as opposed to alternate protein modification as indicated by the conversion to a common pattern in each strain by endo H mediated removal of N-linked oligosaccharides.*

depth as noted before. Gently stir the replicator in water for 30 sec. Remove the replicator from water container and gently shake off residual liquid (*see* **Note 6**).
3. Place pins into a container filled with 95% ethanol of sufficient depth as described in **Subheading 3.1.2**. Stir the replicator briefly in the ethanol container and remove. Do not shake.
4. Hold the replicator such that the pins are oriented horizontally and your hand is not above the pins. Ignite residual ethanol on the pins by passing the replicator over a lit Bunsen burner gently to avoid any dropping or splashing of flaming liquid. Ensure that all pins have been ignited. Once all remaining ethanol on the pins of the replicator has burned off, place the replicator on a sterile surface to cool. Cooling may take up to 5 min (*see* **Note 7**).
5. Remove a deletion library plate from –80°C frozen storage. If a seal is present over the wells of the plate, remove the seal prior to thawing (*see* **Note 8**).
6. Place the plate at room temperature until the cultures in all the wells have completely thawed.
7. Remove the lid of the plate and place the replicator into the wells. Stir gently and avoid splashing that may cause cross contamination.

8. Carefully remove the replicator in a vertical manner to avoid contact with the sides of the well as much as possible. This will reduce the likelihood of cross-contamination.
9. Place the replicator into a standard 96-well plate containing 200 μL/well of YPD with 200 μg/mL of G418 from **step 1**.
10. Stir the replicator gently and avoid splashing. Remove the replicator while being mindful of the sides (*see* **Note 9**).
11. Cover both plates with the respective lid of each.
12. Return the original plate to −80°C frozen storage. Plates may be thawed and frozen up to four times.
13. Place the recently inoculated 96-well plate into a plastic bag with a moist paper towel to keep the edge wells from drying out. The plate is placed into the incubator with the open bag end folded underneath the plate.
14. Incubate the plate for 2 days at 30°C or until the wells are saturated (*see* **Note 4**).
15. After two days prepare a 2 mL deep-well plate by adding a micro stir-bar to each well (*see* **Note 10**)
16. Add 2 mL of YPD-G418 into each well. We use a repeat pipetor to expedite this step.
17. Transfer 100 μL of stationary cell culture from each well of the standard 96-well plate to the corresponding well of the filled, 2 mL deep-well plate. A repeat pipetor expedites this step. Most stationary-phase cultures grown in standard 96-well plates have an absorbance at 600 nm of 10, so the 100 μL aliquot contains 1 OD_{600} cells (*see* **Notes 11** and **12**).
18. Place the inoculated deep-well plate onto the motorized tumble stirrer. Set the tumble stirrer to spin at approx 300 cycles per minute. Incubate the stirring plate at 30°C for 5 h (*see* **Note 10**).
19. Switch the tumble stirrer off and remove the plate.
20. Remove stir-bars by passing a strong neodymium magnet over the wells (*see* **Note 13**).
21. Place the plate in a centrifuge using the 96-well plate adaptor as a platform. Pellet the cells at 2000 × g for 5 min and decant media.
22. Resuspend cells by adding 100 μL of ice-cold lysis buffer and transfer to a 96-well PCR plate set on ice.
23. Incubate the 96-well PCR plate at 99°C for 5 min.
24. Analyze samples by SDS-PAGE and immunoblotting (*see* **Sections 3.2** and **3.3**). Sample-containing plates may be stored for up to 1 mo at −80°C and thawed completely before use.

3.2. SDS-PAGE

1. This procedure assumes the use of BioRad mini-PROTEAN II gel-casting apparatus, although it may be easily adaptable to other such systems.
2. Clean eight long (8 × 10 cm) and eight short (7 × 10 cm) glass plates with a laboratory detergent (e.g., Dri-clean, Decon Labs) and water while scrubbing with

a sponge. Rinse thoroughly with water and dry. Ensure that no acrylamide gel from previous uses remains on glass surface or gels may tear upon removal.
3. Assemble eight cassettes each with a long and short glass plate separated by two 0.75 mm spacers on either edge. Place cassettes on casting trays that seal the bottom edge with a rubber gasket.
4. Insert a 15-well comb into each assembled cassette and mark 0.5 cm below the comb on the glass plate using a permanent marker. This mark indicates the target level of resolving gel mixture. Remove the comb.
5. Prepare 40 mL of a 10% resolving gel mixture: 15.9 mL of H_2O, 13.3 mL of 30% acrylamide/*bis*-acrylamide solution, 10 mL of resolving gel buffer, 400 µL of 10% SDS, 400 µL of 10% APS, and 20 µL of TEMED).
6. Pour the resolving gel mixture between the long and short plates of each cassette to the height indicated by the mark.
7. Overlay each gel with 500 µL 0.1% SDS.
8. Allow the gels to polymerize at room temperature for 30 min.
9. Remove overlays by removing gel cassettes from the trays, inverting, and aspirating thoroughly using a vacuum aspirator (*see* **Note 14**).
10. Return the gel cassettes to casting trays.
11. Prepare 20 mL of a 4% stacking gel mixture: 11.9 mL of H_2O, 5 mL of stacking gel buffer, 2.7 mL of 30% acrylamide, 200 µL of 10% SDS, 200 µL of APS, and 20 µL TEMED.
12. Pour the stacking gel mixture onto the resolving gel.
13. Working quickly, insert a 15-well comb into each cassette.
14. Allow the stacking gels to polymerize at room temperature for 15 min.
15. Remove the combs and aspirate each well of each cassette to remove unpolymerized liquid.
16. Fill each well of each cassette with water and aspirate the wells again.
17. Place the cassettes into BioRad mini-PROTEAN II electrode assemblies and place the assemblies into buffer tanks. Fill both internal and external chambers with running buffer.
18. Load 5 µL of a protein molecular-weight standard and 5 µL of samples onto the gels using a 10 µL Hamilton syringe or disposable 1 mm diameter gel-loading pipet tips (e.g. Fisher Scientific). To distinguish each gel, it is helpful to use a distinct marking system by varying the lane(s) used for the molecular-weight standard in each gel (*see* **Note 15**).
19. Cover the tanks with lids and attach the electrodes to a suitable power supply. Apply 200 V (constant voltage) at room temperature to each tank until the bromophenol blue dye band migrates beyond the bottom of the gel.
20. Turn off the power supply, remove the lids and electrode assemblies from the tanks.
21. Disassemble the cassettes and remove the glass plates containing the gels.
22. Separate the long and short plate of each gel by gently twisting one of the spacers. The gels will remain adhered to one of the two plates.
23. Remove stacking gel by cutting with a razor blade and discard.

24. Remove the gels from the remaining glass plate and place the gels into a container with transfer buffer. Incubate the gels for 20 min at room temperature.

3.3. Immunoblotting for CPS and ALP

1. This procedure assumes the use of a Biorad Trans-blot cell.
2. Cut two 20 × 15-cm pieces of PVDF membrane and four pieces of 20 × 15-cm chromatography paper. Membrane must be handled with clean gloves at all times to prevent transfer of proteins and lipids to the hydrophobic membrane.
3. Wet the PVDF membranes for 15 s in 100% methanol.
4. Rinse the membranes three times with water then place them into a separate container with transfer buffer and equilibrate for 10 min.
5. Briefly submerge two fiber pads and two pieces of pre-cut chromatography paper in transfer buffer.
6. Place one soaked fiber pad on one side of the transfer cassette. Then, place one piece of pre-cut, soaked chromatography paper on top the fiber pad in the transfer cassette.
7. Place one piece of pre-cut equilibrated PVDF membrane on top the chromatography paper in the transfer cassette.
8. Arrange four of the equilibrated mini-gels onto the PVDF membrane in the transfer cassette. Gently press the gels against the membrane with a moistened glove to remove any air bubbles trapped between the gels and the membrane to prevent disruption of protein transfer.
9. Place a second piece of soaked, pre-cut chromatography paper over the arranged gels and then place a soaked fiber pad over the second piece of chromatography paper in the cassette.
10. Close the transfer cassette and place it into the Trans-blot cell tank with the side of the cassette nearest to the membrane at the anode (positive pole) and the side of the cassette nearest to the gels at the cathode (negative pole).
11. Repeat **steps 6 to 12** to assemble the additional four gels onto the second piece of PVDF membrane.
12. Fill the Trans-blot cell tank with transfer buffer and place at 4°C.
13. Close the tank with a lid, attach the electrodes to a suitable power supply, and apply 100 V (constant voltage) for 1 h.
14. Turn off the power supply and remove the lid and electrodes once transferring is complete.
15. Pour off the transfer buffer, remove the gel-holder cassettes, and disassemble the cassettes by first removing fiber pad and chromatography paper on the membrane side of the transfer cassette.
16. Carefully cut the PVDF membrane around each gel with a razor blade and peel the membrane piece away from the gel.
17. Place each cut membrane in a separate container with a sufficient amount of blocking buffer to submerge the membrane. Ensure that the side of the membrane previously in contact with the gel is facing up.

High-Throughput Protein Extraction and Immunoblotting Analysis 23

18. Incubate for 1 hour at room temperature on a gently rocking platform.
19. Decant blocking buffer and add primary antibody diluted in blocking buffer. Incubate for at least 1 h at room temperature with rocking (or overnight at 4°C).
20. Wash each membrane with TBS-T six times, incubating at least 5 min at room temperature for each wash.
21. Add a secondary antibody diluted in blocking buffer to each membrane for at least 1 h at room temperature.
22. Wash each membrane with TBS-T six times, incubating at least 5 min at room temperature for each wash.
23. Remove the membranes from their containers and shake off residual liquid. Place the membranes onto a clear plastic sheet ensuring that the side of the membrane formerly in contact with the gel is facing up.
24. Mix 16 mL of ECL plus reagent A with 400 µL of reagent B. Using a 1 mL pipetor, dispense 2 mL of this mixture onto each membrane and incubate for 5 min at room temperature.
25. Shake residual liquid from the membranes and place the side of the membranes facing up, down onto the glass surface of the Typhoon 9410 imager. Place a clear plastic sheet on top of the membranes and press the plastic with a finger or card to remove any air bubbles.
26. Obtain a digital image by scanning on the Typhoon with 100 µm pixel size using 457 nm laser excitation through the 520 band pass (BP) 40 Cy2 filter and detect with 500 V setting on photomultiplier tube (PMT). Alternatively, images can be captured by exposing the membranes to x-ray film.
27. Examples of strains analyzed by these growth, lysis, and immunoblotting methods are shown in **Fig. 4**.

4. Notes

1. Examples of additional commercially available *S. cerevisiae*, genomic libraries
    a. Yeast *MAT*α Haploid Single Deletion Collection (Open Biosystems and Invitrogen)
    b. Yeast *MAT*a Haploid Single-Collection Deletion (Invitrogen)
    c. Yeast Heterozygous Knockout Strain Collection (Open Biosystems or Invitrogen Carlsbad, Ca)
    d. Yeast Tandem Affinity Purification (TAP)-Tagged Collection for profiling the yeast proteome (Open Biosystems)
    e. Yeast Insertional Mini Transposon Mutant Strains (Open Biosystems Catalog)
    f. Yeast HA-Tagged Strains (Open Biosystems)
    g. Yeast Tet-Promoter Strains for control of transcriptional activity with a tetracycline rheostat (Open Biosystems Catalog)
    h. Yeast GFP Collection (Invitrogen)

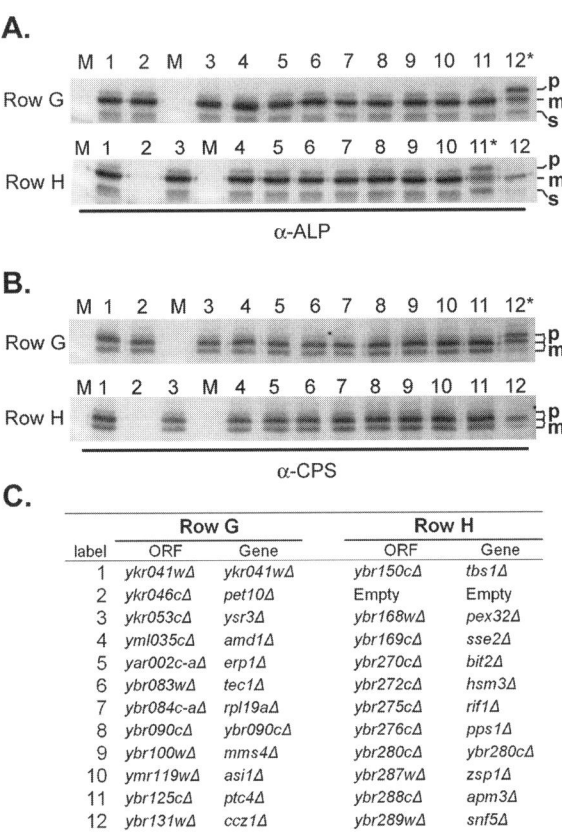

Fig. 4. Example of high-throughput extraction and immunoblotting reveals genes involved in protein trafficking. Plate 44 of the yeast *MAT*a single-gene deletion library was grown and lysed as described in **Subheadings 3.1** to **3.3**. Rows G and H were analyzed by SDS-PAGE and immunoblotting. (A) Immunoblots were probed for ALP. The precursor (p) mature (m) and soluble (s) forms are indicated. (M) designates protein molecular-weight standards. **Note 15** details a systematic marker method for labeling each gel to avoid confusion. Lanes with precursor accumulation are marked with *. (B) CPS precursor (p) and mature (m) forms have been designated. Lanes with precursor accumulation are marked with *. (C) Table indicates both the systematic open reading frame names as well as the corresponding gene names for the strains analyzed in each well in panels A and B.

2. All water used to make solutions in this protocol should be purified so that it has a resistance of 18 MΩ/cm and has a total organic content less than six parts per billion.
3. Occasionally, direct binding of commercially available secondary antibody to yeast proteins can be significant. If the secondary antibodies have not been affinity purified against the Fc region of the target antibody then additional host anti-

bodies are present and may include antibodies that cross-react with yeast proteins perhaps due to a prior yeast/fungal infection and immune response. Addition of concentrated wild-type cell extracts to the secondary antibody dilution before adding to the membrane may reduce background band signals because the soluble extract competes for binding of these background antibodies.

4. Certain gene disruptions cause yeast to grow slowly. The yeast single-deletion libraries have two plates that contain slow-growing strains. Many of the strains in these plates may be cultured as above and attain sufficient growth to analyze by immunoblotting. However, some strains grow too poorly to obtain a sufficient quantity of cells for immunoblotting analysis and must be individually cultured over a greater length of time to attain sufficient density.

5. Most proteins are adequately extracted by heat lysis, although a number of proteins are more abundant in extracts obtained by glass-bead mechanical disruption, especially if cells are in the stationary phase (*see* **Fig. 2**). It is recommended when using heat lysis to determine the cell density that yields an optimal signal for the protein of interest by immunoblotting as done in **Fig. 1**.

6. If replicating more than one plate, after removing replicator from the bleach solution, be sure to rinse the replicator first in the same container of water that was used first from the previous replication before placing into the second container. It may be helpful to label them. This ensures that no bleach remains on the replicator after the second rinse. Replace water in both reservoirs after four uses.

7. It may be helpful before beginning to determine the exact length of time required for replicator pins to return to room temperature after sterilizing procedure. Use a thermometer or even hand (if only very briefly touched to avoid serious burn) after 5 min to ensure pins are cool to the touch. Then sterilze the replicator again and wait the appropriate amount of time. A sterile surface to place the replicator on may be achieved by autoclaving a metal pan, or cotton cloths or velvets that may be folded several times and spread over any desired surface.

8. Many libraries are packaged with adhesive aluminum foil seals that cover the 96-well plates. These seals are useful to reduce sublimation of frozen strain cultures in the wells. However, even mild accidental jarring may cause droplets of cultures to adhere to the surface of the seal and freeze. This adherence presents a problem if the droplets melt to liquid before the seal is removed because they may drop into neighboring wells. To avoid this situation, immediately after the plate is removed from the –80°C freezer and still frozen, invert the plate so that the seal side is facing down. Peel the seal in a downward motion so any adhering drops fall away from the plate.

9. The *MAT***a** deletion library contains a blank orientation well for each plate in row H column 2. The *MAT*α deletion library contains a blank orientation well for each plate in row H column 3. A strain used for control purposes for the phenotype of interest may be inoculated into this well and grown within the plate alongside the other strains to ensure growth conditions are properly maintained. Care must then be given to maintain proper orientation of the plate.

10. If strains are to be cultured for less than 12 h, strict sterility is not necessary. Both the presence of the antibiotic G418 and the large innoculum of yeast will prevent culture contamination within this time frame. We recommend, soaking the resuable stir bars between uses in bleach solution for 5 min, and then rinsing thoroughly with water and drying. If sterility is required, stir bars may be autoclaved and removed carefully with sterile forceps and dropped into wells. Additionally, sterile lids may be rested on the tops of the deep-well plates. This will inhibit air exchange and may alter strain growth. Therefore, the authors recommend that the growth time to observe the phenotype of interest be evaluated using control strains with a lid present.
11. Cell densities are measured by dilution into water and determination of absorbance at 600 nm (in a linear range) in a spectophotometer using a 1 cm cuvette. With the Ultrospec 3000 spectophotometer (Pharmacia Biotech) 1 OD_{600} is equivalent to approx 10^7 cells.
12. To make a viable replica of a library plate for long-term storage, 100 µL 30% glycerol can be added to each well of the saturated strains after 100 µL of culture from each well is removed for culture in the deep-well 96-well plate. The plate should then be sealed with an adhesive foil (available from Corning Life Sciences) and placed at –80°C.
13. Stir bars must first be removed before centrifugation to avoid disruption of the pellet by the stir bar when supernatant is decanted.
14. A suitable aspiration device may be constructed using a micro-pipet tip, side-arm flask, Pasteur pipet, rubber stopper of sufficient diameter to seal top of side-arm flask and with hole in the center to accommodate the Pasteur pipet, rubber tubing, and a vacuum line or pump. The vacuum line is connected with rubber tubing to the side-arm of the side-arm flask. The Pasteur pipet is placed through the hole in the rubber stopper such that when the rubber stopper is placed on the top of the side-arm flask, the flask-end does not touch the bottom of the flask, and the external end is a minimum length of 3 cm. An additional rubber tube is connected with the micropipet tip on one side and the external side of the Pasteur pipet on the other.
15. A gel-marking system is useful to avoid confusion when working with many gels at once. Because each 96-well plate requires eight 15-well polyacrylamide gels, each plate row is analyzed on a separate gel. To distinguish between the gels, the protein molecular-weight standard is applied to different lanes in different gels. The authors used the following system. The protein standard for gel 1 (row A) is applied to lane 1. The standard for gel 2 (row B) is placed in lane 2. The standard for gel 3 (row C) is placed in lane 3. The standard for gel 4 (row D) is placed in lane 4. To designate row E, a standard is placed in lanes 1 and 2. To designate row F a standard is placed in lanes 1 and 3. To designate row G, a standard is placed in lanes 1 and 4. To designate row H, a standard is placed in lanes 1 and 5 of the eighth gel.
16. Mechanical disruption by glass beads is performed as follows. Sediment 3 OD_{600} cells at 2000 × g in 13 × 100 mm glass tubes and decant the supernatant. Briefly

pellet the cells again to collect residual liquid and aspirate the supernatant. Add 250 μL acid-washed 0.5 mm glass beads and 50 μL lysis buffer and agitate at full speed with a vortex mixer for 90 sec. Minimal liquid is needed during the lysis step to ensure a thick suspension of glass beads with cells. This maximizes the time and number of cells that are pressed between glass beads. After 90 sec, add 250 μL lysis buffer so that the overall volume matches that used for heat lysis. Incubate samples at 100°C for 5 min.

17. Wild-type strain is BY4741 (MATa, *his31△, leu20△, ura30△, met150△*) and is the background strain used to generate the *MAT*a single-gene deletion library.

Acknowledgments

This work was supported by NIH RO1 GM39040 and GM67911 to G.P. and NIH-MSTP T32-GM08042 to V.A.

References

1. Winzeler, E. A., Shoemaker, D. D., Astromoff, A., et al. (1999) Functional characterization of the S. cerevisiae genome by gene deletion and parallel analysis. *Science* **285**, 901–906.
2. Rose, A. H., and Harrison, J. S. (1987) *The yeasts*. Academic Press, London; San Diego.
3. Broach, J. R., Pringle, J. R., and Jones, E. W. (1991) *The molecular and cellular biology of the yeast* Saccharomyces. Cole Spring Harbor Laboratory Press, Cold Spring Harbor, N.Y.
4. Cowles, C. R., Snyder, W. B., Burd, C. G., and Emr, S. D. (1997) Novel Golgi to vacuole delivery pathway in yeast: identification of a sorting determinant and required transport component. *Embo. J.* **16**, 2769–2782.
5. Spormann, D. O., Heim, J., and Wolf, D. H. (1992) Biogenesis of the yeast vacuole (lysosome). The precursor forms of the soluble hydrolase carboxypeptidase yscS are associated with the vacuolar membrane. *J. Biol. Chem.* **267**, 8021–8029.
6. Klionsky, D. J., and Emr, S. D. (1989) Membrane protein sorting: biosynthesis, transport and processing of yeast vacuolar alkaline phosphatase. *Embo. J.* **8**, 2241–2250.
7. Jones, E. W. (1984) The synthesis and function of proteases in Saccharomyces: genetic approaches. *Annu. Rev. Genet.* **18**, 233–270.
8. Burnette, W. N. (1981) "Western blotting": electrophoretic transfer of proteins from sodium dodecyl sulfate–polyacrylamide gels to unmodified nitrocellulose and radiographic detection with antibody and radioiodinated protein A. *Anal. Biochem.* **112**, 195–203.

3

Genome-Wide Analysis of Membrane Transport Using Yeast Knockout Arrays

Helen E. Burston, Michael Davey, and Elizabeth Conibear

Summary

The transport of membrane-bound proteins through post-Golgi compartments depends on the coordinated function of multiple genes that direct the recognition and routing of protein cargoes to their final cellular destination. As many of these sorting components are nonessential for viability, genome-wide screening of the yeast gene-deletion mutant collection provides a useful strategy for their identification. The potential of this approach is limited only by the availability of transport assays suitable for the high-throughput screening of yeast colony arrays. Two large-scale phenotypic screens to identify novel transport genes are described here. The fluorescence-based Calcofluor white assay identifies mutants with altered plasma membrane localization of the chitin synthase Chs3, which recycles between the cell surface, endosomes, and the late Golgi. The carboxypeptidase Y (CPY) assay allows mutants of a distinct Golgi-to-vacuole transport pathway to be identified, due to the missorting and secretion of the vacuolar hydrolase CPY from the cell.

Key words: Chs3; CPY; deletion collection; endosome; Golgi; phenotypic screening; *Saccharomyces cerevisiae*.

1. Introduction

Following synthesis, proteins that are translocated into the endoplasmic reticulum (ER) are transported through the various membrane-bound organelles within the secretory and endosomal systems. Sorting at each of these compartments is tightly regulated and depends on the concerted action of many molecular components *(1–3)*. Although defects in trafficking between the late Golgi, endosome, and vacuole lead to protein mislocalization, they generally do not cause cell death, making the genome-wide deletion collection of yeast mutants a valuable resource for identifying genes involved in these cellular transport pathways.

We have developed an assay for the systematic screening of yeast colony arrays that identifies mutants with altered plasma membrane levels of the chitin synthase Chs3 *(4)* based on the chitin-binding fluorescent compound Calcofluor white (CW). Chs3 is responsible for the majority of chitin production at the plasma membrane, and its transport is tightly regulated at different stages of the secretory and endocytic pathways *(4,5)*. Chs3 export from the ER requires palmitoylation and interaction with a dedicated chaperone, Chs7. At steady state, most Chs3 is contained within specialized endosomes called chitosomes, but it is rapidly mobilized to the cell surface in response to plasma membrane stress (**Fig. 1**). Delivery to the plasma membrane involves a newly discovered putative Golgi vesicle coat complex, the exomer *(6)*. Inactivation of Chs3-mediated chitin production also occurs rapidly, by a process of endocytosis. Chs3 is thus a useful reporter for identifying genes required for multiple trafficking events that underlie regulated plasma membrane delivery.

The fluorescent chitin-binding antifungal drug CW is traditionally used to visualize bud scars by fluorescent microscopy. The authors found that mutants with altered chitin production can readily be identified based on the fluorescence of colonies grown on CW plates and imaged under ultraviolet (UV) light.

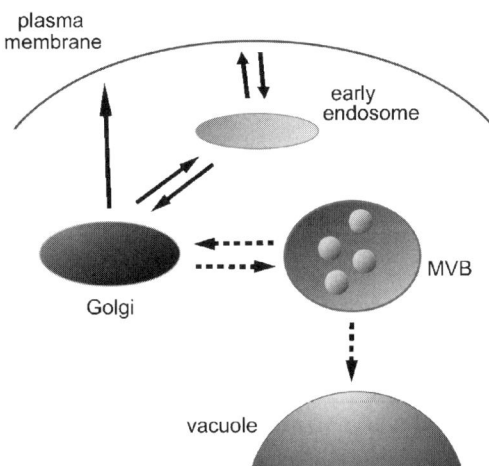

Fig. 1. Chs3 and CPY are sorted on different intracellular pathways. Chs3 recycles between the Golgi, early endosome, and plasma membrane (solid arrows), whereas CPY is transported from the Golgi to the vacuole by a receptor that recycles between the Golgi and the multivesicular body (MVB) (dashed arrows).

Genome-wide application of this assay has led to the identification of new genes that regulate Chs3 transport *(4)*.

The soluble vacuolar hydrolase carboxypeptidase Y (CPY) is recognized by its receptor at the *trans*-Golgi-network (TGN) and transported to the vacuole via an endosomal intermediate referred to as the multivesicular body (MVB; *see* **Fig. 1**). When the CPY sorting pathway is impaired, a portion of newly synthesized CPY enters the secretory pathway and is secreted from the cell *(7)*. Classical genetic screens based on CPY missorting led to the identification of the *vps* mutants that are defective in multiple steps in transport between the Golgi, MVB, and vacuole. The transit of proteins through these organelles is easily monitored on a genome-wide scale, using an array-based CPY immunoblotting assay *(8,9)*.

Here, a detailed methodology is provided for handling the mutant strain collections, followed by a description of two phenotypic assays—the CW assay and the CPY secretion assay—that can be applied to study distinct trafficking pathways.

2. Materials

2.1. Constructing Phenotypic Screening Arrays: Manual Pinning Method

1. Yeast Deletion Collection Arrays: collections of homozygous diploid, *MATa* haploid or *MATα* haploid deletion strains are available from commercial sources including Open Biosystems (in North America) or Euroscarf (in Europe), as colony arrays on agar plates (in 96-array format), or as frozen stocks in 96-well plates (*see* **Note 1**). Store agar-based arrays at 4°C, and frozen stocks at −80°C.
2. Plates: Omnitrays (Nunc, 242811).
3. Manual replicators: the following items can be purchased from V&P Scientific, Inc.

    a. 96-floating pin E-clip manual replicator (VP408FH).
    b. 384-floating pin E-clip manual replicator (VP384FH).
    c. Replicator templates: Library copier™ (VP381), Colony copier™ (VP380).
    d. Pin-cleaning supplies: pin-cleaning brush (VP425), water reservoirs (VP421), alcohol reservoir with lid (VP420) (*see* **Note 2**).

4. G418 stock solution: Dissolve in water to a final concentration of 200 mg/mL, filter sterilize and store aliquots at −20°C (*see* **Note 3**).
5. Yeast extract/peptone/dextrose (YPD): Add 10 g of yeast extract, 20 g of peptone, 20 g of agar, and 335 mg tryptophan to 900 mL of water in a 2L flask. Autoclave and add 100 mL sterile 20% dextrose; mix thoroughly and pour plates by pipeting 35 mL of media per Omnitray (*see* **Note 4**).
6. YPD+G418: Add 10 g of yeast extract, 20 g of peptone, 20 g of agar, and 335 mg tryptophan to 900 mL water in a 2 L flask. Autoclave and add 100 mL sterile 20% dextrose. Cool to 60°C, then add 0.5 mL G418 stock solution to a final concentra-

tion of 100 μg/mL G418 (*see* **Note 5**). Mix thoroughly and pour plates by pipeting 35 mL of media per omnitray.

2.2. Genome-Wide Calcofluor White Screen

1. CW stock solution: Dissolve calcofluor white (Sigma, F3543) in 0.5 M Tris-HCl pH 9.6 with gentle heat to a final concentration of 10 mg/mLmL. Store at room temperature in the dark for up to 1 wk (*see* **Note 6**).
2. YPD+CW: Add 10 g of yeast extract, 20 g of peptone, 20 g of agar, and 335 mg tryptophan to 900 mL water in a 2L flask. Autoclave, and add 100 mL sterile 20% dextrose. Cool to 60°C, and add 5 mL of CW stock solution to a final concentration of 50 μg/mL (*see* **Note 7**). Mix thoroughly, and pour into omnitrays by pipeting 35 mL of media into each plate. Store plates at room temperature in the dark.
3. Flat-bed scanner for acquisition of white-light images.
4. Fluor-S Max™ multi-imager (Bio-Rad Laboratories) and the Quantity One™ software package for acquisition of fluorescent-light images.

2.3. Genome-Wide CPY Secretion Screen

1. Phosphate-buffered saline (1X PBS) per 1 L: 1 mM KH_2PO_4 (0.136 g/L), 10 mM $Na_2HPO_4 \cdot 7H_2O$ (2.681 g/L), 137 mM NaCl (8.006 g/L), 2.7 mM KCl (0.201 g/L).
2. Washing buffer (TTBS): 1X PBS+1% Tween-20.
3. Blocking buffer: 5% (w/v) nonfat dry milk in TTBS.
4. Nitrocellulose blotting membrane, cut to 7.3 × 11.3 cm to fit inside Omnitray. Store at 4°C.
5. Anti-CPY monoclonal antibody 10A5 (Molecular Probes, cat. no. A-6428). Store at –20°C in small aliquots; once thawed, keep at 4°C.
6. Anti-mouse immunoglobulin IgG conjugated to horseradish peroxidase (HRP; e.g., goat anti- mouse IgG-HRP, Biorad cat. no 170-6516, used at a 1:5000 dilution). Store small aliquots at –20°C, and thawed aliquots at 4°C.
7. Enhanced chemiluminescent (ECL) reagents. Store solutions at 4°C.

3. Methods

Yeast gene-deletion collections are provided commercially either as 96-density colony arrays on YPD+G418 agar plates, or as frozen stocks in 96-well plates containing YPD+15% glycerol. For ease of maintenance and handling, the original 96-array plates are first condensed fourfold, producing a set of 384-density source arrays. This considerably reduces the number of plates required for subsequent pinning and screening steps.

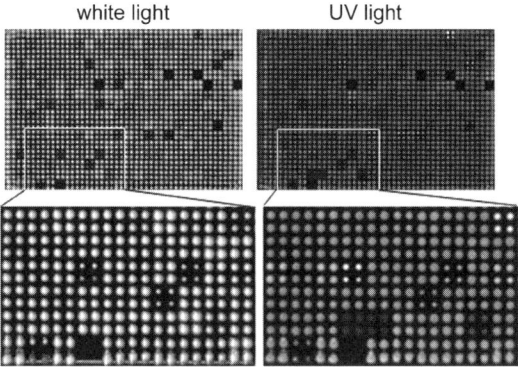

Fig. 2. Calcofluor white assay identifies mutants defective in trafficking of the chitin synthase Chs3 to the plasma membrane. Yeast deletion mutants pinned as a 1,536-array to a plate containing the fluorescent chitin-binding compound calcofluor white are imaged under white or ultraviolet light (top panels). Mutants with higher or lower levels of cell-surface chitin exhibit different fluorescent intensities compared to wild-type strains (lower panels show detail from 1,536 array). Each mutant is represented by a group of four colonies.

For phenotypic screens, the 384 source array is replicated fourfold on suitable media to produce 1,536-density arrays. The higher density format results in more even growth of colonies and provides four replicates for each mutant, which improves subsequent data analysis. Although all pinning steps can be carried out manually, a pinning robot is preferred for high-throughput screening in 1,536-array format. For the Chs3 activity assay, fresh deletion arrays are pinned to plates containing CW. As the colonies grow, they produce cell-wall chitin that is bound by CW, resulting in fluorescence under UV light (**Fig. 2**). For the CPY secretion assay, arrays are grown on a nitrocellulose filter that covers the agar surface; any CPY that is missorted to the outside of the cell is captured on the filter and detected using an immunoblot assay (**Fig. 3**). This assay can easily be adapted to the detection of other secreted molecules for which highly specific antibodies are available.

3.1. Condensing the Gene-Deletion Collection Into 384-Density Arrays by Manual Pinning

1. Prepare a sufficient number of YPD+G418 plates and ensure they are clearly labeled.
2. Prepare deletion arrays for condensing, and arrange in numerical order. If the frozen collection is being used, *see* **Note 8**.

3. The 96-pin manual replicator must be sterilized immediately prior to each pinning step to avoid contamination of the arrays. Set up five reservoirs of increasing depths (volumes of reservoirs may vary): three reservoirs of sterile water (suggested depths of 13 mm, 15 mm, 16 mm), one reservoir of 10% bleach (14 mm) and one reservoir of 100% ethanol (17 mm; *see* **Note 9**).
4. Place the replicator in each reservoir as follows: 13 mm water reservoir (1 min), 10% bleach (20 sec), 15 mm water reservoir (5 sec), 16 mm water reservoir (5 sec). Finally, place the replicator in the 100% ethanol reservoir (5 sec). Allow the excess ethanol to drip from the pins, and then ignite with a flame while holding the replicator in a horizontal position. Allow the replicator to cool before contacting the colonies (*see* **Note 10**).
5. Condense every four plates with 96-density arrays to make a single 384-density array with a sterile 96 pin replicator, using alignment holes A through D on the Library Copier™ for successive pinnings (*see* **Note 11** if pinning from frozen stocks).
6. If desired, add appropriate control strains to empty positions of each 384-array plate by hand (*see* **Note 12**).
7. Incubate 384-density array plates for 2 d at 30°C and store at 4°C for up to 2 mo.
8. Source arrays must be replicated to fresh media and grown for 2 d before use in phenotypic screens (*see* **Note 13**). Replicate the 384-density array onto fresh YPD+G418 plates using a sterile 384-pin replicator. Place the Library Copier™ over an open omnitray with the pair of nine-alignment holes on the front of the frame. Use the middle hole (E) to align the replicator with the yeast colonies.
9. Incubate the plates for 2 d at 30°C. These 384-density "source" arrays are now ready for use in phenotypic screens.

3.2. Genome-Wide Phenotypic Screening Using a Pinning Robot

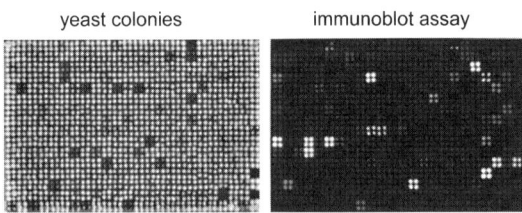

Fig. 3. Vacuolar protein sorting mutants secrete the vacuolar hydrolase CPY. In the left panel, yeast colonies were arrayed on an agar plate whereas in the right panel, the same yeast strains were grown on a nitrocellulose filter, and secreted CPY was detected by immunoblotting and digital imaging.

3.2.1. Robotic Pinning System

Large-scale replication of source arrays in 1,536 format can be laborious using the manual pinning method. Robotic systems are preferred for high-throughput screens, as all steps of pin sterilization and colony replication are fully automated once the appropriate reagents and programs are set up. Procedures for use of pinning robots vary with the manufacturer; however, it is critical to establish proper sterilization routines to prevent contamination of source and screening arrays (*see* **Note 14**). We program the BioRad colony arrayer to sterilize the robotic pinning tool prior to each pinning step as follows:

1. Place the pinning tool in a reservoir containing high-purity water (140 mL), raising and lowering the pinning tool four times for duration of 1 min to remove adherent yeast.
2. Repeat step 2 with a second water reservoir (150 mL).
3. Sterilize the pinning tool by sonication in 70% ethanol (320 mL) for 1 min (*see* **Note 15**).
4. Place the pinning tool in the 100% ethanol reservoir (160 mL) for 30 sec.
5. Dry the pinning tool over the fan for 30 sec.

3.2.2. Calcofluor White Assay

1. Prepare YPD+CW omnitrays.
2. Prepare a fresh 384-density source array (grown for 2 d on YPD+G418).
3. Replicate each colony on the 384-density source array four times to the YPD+CW plate in 1,536 condensed format (*see* **Note 16**).
4. Wrap the YPD+CW plates in foil and incubate for 3 to 5 d at 30°C.
5. Acquire white light images by scanning YPD+CW plates on a standard flatbed scanner (*see* **Note 17**).
6. Acquire fluorescent images using a Biorad Fluor-S Max™ multi imager using UV epifluorescence and the 530 DF60 filter. Place each 1,536-array plate face up without the lid in the center of the imaging platform, and adjust the focus using an aperture setting of F4. Acquire the image with an aperture of F11 and an exposure time of 2 s (*see* **Note 18**).
7. Densitometry of digital images (e.g., using the array tool in Quantity One software) is more effective at identifying mutants with subtle differences in fluorescence intensity effective compared to visual inspection (*see* **Note 19**). Slow-growing or absent strains are identified by densitometry of white light images and removed from the analysis. Average growth and fluorescence values can be compared for each strain to identify mutants with higher or lower fluorescence intensities compared to wild-type controls.

3.2.3. CPY Secretion Assay

1. Prepare YPD+G418 plates. Cut nitrocellulose to match the inside dimensions of the omnitray (113 mm × 73 mm with notched corners), hold at the edges with forceps, and place carefully over top of the agar, ensuring that there are no bubbles.
2. Pin a fresh 384-density source array in 1,536 format first onto the YPD plates overlaid with nitrocellulose, and then to a second YPD+G418 plate. Pinning to the second plate (without first washing the pinning tool) allows the growth of each strain in the collection to be measured (*see* **Note 20**).
3. Incubate the YPD+G418+nitrocellulose plates for 12 to 15 g at 30°C, and the YPD+G418 plates for 2 d at 30°C.
4. Scan YPD plates on a standard flatbed scanner. These images can later be used to normalize each colony with respect to growth.
5. Carefully remove the filters from the YPD+G418+nitrocellulose plates. Rinse off adherent yeast cells under a stream of distilled water.
6. Incubate each filter in blocking buffer (TTBS+5% milk powder) for 15 to 30 min while shaking on an orbital platform to block nonspecific binding sites.
7. Place each filter in anti-CPY primary antibody, at a dilution of 1:500 in TTBS+5% milk. Incubate the filters at room temperature for at least 2 h while shaking.
8. Remove the primary antibody solution, and wash the filters three times (for a total of 15 min per wash) in 1X PBS while shaking.
9. Place each filter in a solution of HRP-labeled anti-mouse secondary antibody diluted in TTBS+5% milk. Incubate filters at room temperature for 1 h while shaking.
10. Remove the secondary antibody solution, and wash the filters three times (for a total of 15 min per wash) in 1X PBS while shaking.
11. Place the filter (protein-side up) on Saran wrap, ensuring that there are no air bubbles. Pour the ECL solution (~5 mL) onto the filter, and swirl around for approx 1 min. Detect the signal immediately using x-ray film or a digital imaging system such as the Fluor-S Max™ multi-imager (*see* **Note 21**).

4. Notes

1. Colony arrays on solid agar are less expensive and are more convenient for immediate screening, but should be examined carefully for damage to the agar or cross-contamination of colonies during shipping. Frozen stocks should be made for long-term storage of yeast arrays. Note that as many as 5% of the knockout strains may have the incorrect genotype. Homozygous diploid deletion collections contain fewer second-site suppressor mutations relative to haploid strains and therefore give more reliable knockout phenotypes. Screening two or more independent deletion sets reduces error. Genotypes can be confirmed by polymerase chain reaction, and mutant phenotypes should be tested for linkage to the *Kan^R* marker and/or for complementation by a wild-type copy of the gene.

2. Empty pipette box lids can also be used as reservoirs for pin cleaning.
3. G418 is sold by Invitrogen under the name Geneticin™. Each batch of G418 varies in its activity. The amount of G418 powder used to make the stock should be adjusted so that the final solution contains 200 mg/mL of active G418. Thawed G418 aliquots can be stored at 4°C for long periods without noticeable loss of activity.
4. 35 mL agar is appropriate for Omnitrays. A thinner layer of agar is more likely to crack during incubation. Ensure plates are poured on a level surface and are completely dry prior to pinning (usually allow 1 to 3 d depending on ambient temperature and humidity). If agar is too moist, the pressure from the pinning (especially using a robot) causes moisture to bead on the agar surface, disturbing colony growth and allowing cross-contamination of colonies.
5. Cooling YPD to 60°C before adding G418 minimizes the possibility of thermal inactivation. 100 μg/mL G418 is preferred for maintaining Kan^R yeast deletion arrays because many Kan^R trafficking mutants show impaired growth at 200 μg/mL G418, the concentration typically used to for Kan^R selection.
6. CW is light-sensitive and will precipitate out of solution if stored at less than 20°C.
7. 50 μg/mL CW gives good chitin staining without drastically reducing the growth of sensitive strains.
8. Yeast arrays frozen in 96-well plates are typically covered with cryogenic sealing tape that should be removed while plates are still frozen to prevent cross-contamination. Cover plates with lid and allow them to thaw at room temperature (approx 15 min) before pinning. After pinning, re-seal plates with fresh sealing tape and freeze. Yeast can survive several (at least three) freeze-thaw cycles without significant loss of viability.
9. The wash reservoirs are set up to cover the pins in incremental depths in each step. For example, the first wash should submerge the pins to the depth of a standard microtiter plate. The final ethanol wash should submerge the lower half of the pins. This prevents the dirtier initial wash liquid from dripping down the pins during later sterilization steps.
10. Careful cleaning and sterilization routines prevent the buildup of residue on the replicator pins. It is a good idea to ensure the pins move freely before each pinning, to ensure good contact with the source colonies and with the agar surface. Using more than one replicator can reduce wait times during the sterilization and pinning procedures. Process the replicators in rotation.
11. Liquid cultures frozen in 96-well plates should be pinned to agar in 96-density format before condensing to make 384 stock arrays. Because most robotic systems cannot resuspend liquid yeast cultures that have settled to the bottom of 96-well plates, this is best carried out by hand. Dip the sterile replicator into the wells and swirl to resuspend the yeast before pinning.
12. The *MATa* and *MATα* deletion arrays have strains missing from the H2 and H3 coordinates, respectively. Control strains placed at these positions are useful for quality control and plate normalization. We use one wild-type strain (with Kan^R

marker), and two control strains for every experiment (*chs3* and *vps10* mutants are appropriate controls for the CW and CPY screens, respectively), leaving the remaining position blank. Note that each 96-density plate has a unique blank position that is useful for tracking plates and assessing the effectiveness of sterilization procedures.

13. Using fresh source arrays (2 d old) results in more even patch size when constructing 1,536 arrays.
14. Alternative robotic systems suitable for pinning yeast arrays include the Singer Rotor HAD bench top robot (Singer Instruments) and the QBot, QpixXT, MegaPix (Genetix). All condensing and replicating steps required to carry out the phenotypic screens described here can be accomplished by manual pinning, using a 384-pin replicator, the sterilization procedure described in **Subheading 3.1.**, and the Library Copier™, with the pair of four-alignment holes on the front of the frame. Use alignment holes A through D for each pinning.
15. If a sonicator is not available, pins can be sterilized in 10% bleach and rinsed with distilled water and ethanol; however, bleach may corrode pins over time. Sterilization procedures should be tested thoroughly to make sure they are effective.
16. This procedure can also be used to test the sensitivity of mutant strains to a chemical inhibitor. Replicate each 384-density source array first to YPD, and then to the YPD+drug plate in 1,536 condensed format. Pinning first to a control plate removes some of the yeast from the pins such that fewer cells are transferred to the drug plate, thus increasing the sensitivity of the assay. Furthermore, the amount of yeast pinned to the control plate will be proportional to the amount pinned to the experimental plate, thus allowing more accurate assessment of growth differences.
17. Inverting the plates on the flatbed scanner (i.e., imaging the tops of the colonies) gives the best data. Because this can produce fogging, plates should be cooled to room temperature (approx 2 h) before scanning.
18. Images should be acquired with "highlight-saturated pixels" selected. If pixel saturation is a problem, re-scan all the plates at a much lower exposure time. Alternatively, fluorescent images can be captured using a handheld long-wave UV lamp and a digital camera. Apparent gradients in fluorescence intensity across the plate scans can result from a UV bulb failure or exposure of the CW plates to light during drying, storage, or incubation. Bacterial or fungal contamination can also affect CW fluorescence.
19. Many commercial and open source image-analysis programs designed for the analysis of microarray images can be used for densitometry of digital images (e.g., CellProfiler, GenePix).
20. The amount of yeast pinned to the control plate will be proportional to the amount pinned to the experimental plate, thus allowing more accurate assessment of growth differences.
21. Because x-ray film doesn't give a linear response and is easily saturated, digital imaging systems yield higher quality data.

References

1. Keller, P., and Simons, K. (1997) Post-Golgi biosynthetic trafficking. *J. Cell. Sci.* **110**, 3001–3009.
2. Lemmon, S. K., and Traub, L. M. (2000) Sorting in the endosomal system in yeast and animal cells. *Curr. Opin. Cell Biol.* **12**, 457–466.
3. van der Goot, F. G., and Gruenberg, J. (2006) Intra-endosomal membrane traffic. *Trends Cel Biol.* **16**, 514–521.
4. Lam, K. K. Y., Davey, M., Sun, B., Roth, A. F., Davis, N. G., and Conibear, E. (2006) Palmitoylation by the DHHC protein Pfa4 regulates the ER exit of Chs3. *J. Cell Biol.* **174**, 19–25.
5. Cesar, R. (2002) The genetic complexity of chitin synthesis in fungi. *Curr. Genet.* **41**, 367–378.
6. Wang, C.-W., Hamamoto, S., Orci, L., and Schekman, R. (2006) Exomer: a coat complex for transport of select membrane proteins from the trans-Golgi network to the plasma membrane in yeast. *J. Cell. Biol.* **174**, 973–983.
7. Bowers, K., and Stevens, T. H. (2005) Protein transport from the late Golgi to the vacuole in the yeast Saccharomyces cerevisiae. *Biochim. Biophys. Acta.* **1744**, 438–454.
8. Bonangelino, C. J., Chavez, E. M., and Bonifacino, J. S. (2002) Genomic Screen for Vacuolar Protein Sorting Genes in Saccharomyces cerevisiae. *Mol. Biol. Cell* **13**, 2486–2501.
9. Conibear, E., and Stevens, T. H. (2002) Studying yeast vacuoles. In Christine Guthrie and Gerald R. Fink (eds.) Guide to Yeast Genetics and Molecular and Cell Biology, Pt C, Methods in Enzymology, Vol. 351. Academic Press, CA. pp. 408–432.

4

A Cell-Free System for Reconstitution of Transport Between Prevacuolar Compartments and Vacuoles in *Saccharomyces cerevisiae*

Thomas A. Vida

Summary

Genetic approaches have revealed more than 50 genes involved in the delivery of soluble zymogens like carboxypeptidase Y (CPY) to the lysosome-like vacuole in *Saccharomyces cerevisiae*. At least 20 of these genes function in transport between the prevacuolar endosome-like compartment (PVC) and the vacuole. To gain biochemical access to these functions, the authors developed a cell-free assay that measures transport-coupled proteolytic maturation of soluble zymogens *in vitro*. A polycarbonate filter with a defined pore size is used to lyse yeast spheroplasts after pulse-chase radiolabeling. Differential centrifugation enriches for PVCs containing proCPY (p2CPY) in a 125,000g membrane pellet and is used as *donor* membranes. Nonradiolabeled spheroplasts are also lysed with a polycarbonate filter but a 15,000g membrane pellet enriched for vacuoles is collected and used as *acceptor* membranes. When these two crude membrane pellets are incubated together with adenosine triphosphate and cytosolic protein extracts, nearly 50% of the radiolabeled p2CPY can be processed to the mature vacuolar form, mCPY. This cell-free system allows reconstitution of intercompartmental transport coupled to the function of *VPS* gene products.

Key words: Carboxypeptidase Y; cell-free assay; lysosome; membrane fusion; *Saccharomyces cerevisiae*; vacuole.

1. Introduction

The secretory pathway and endocytosis in eukaryotic cells involve intercompartmental transport events. Carrier vesicles mediate the traffic of cargo between subcellular compartments for most steps in the secretory and endocytic pathways. Genetic and biochemical approaches have elucidated many molecular details of vesicle-mediated transport in the secretory and endocytic pathways, often in a complementary fashion *(2)*. Mutant isolation screens and selections in the yeast, *Saccharomyces cerevisiae*, aimed at uncovering defects in protein traffic have uncovered many genes and helped reveal common mechanisms of secretion and endocytosis among eukaryotic organisms. The *SEC* (SECretory), *VPS* (Vacuolar Protein Sorting), and *END* (ENDocytosis) are among the most common genes from these genetic efforts *(3–10)*. Biochemical efforts have uncovered many proteins involved in the secretory and endocytic pathways using cell-free reconstitution assays. These assays measure either heterotypic fusion or vesicle budding from almost every known step in intercompartmental transport *(11)*.

This protocol describes methods for an intercompartmental protein transport assay using partially purified organelles from yeast *(1)*. This cell-free system measures proteolytic maturation of soluble vacuolar proenzymes such as carboxypeptidase Y (CPY) and proteinase A (PrA) after transfer from the prevacuolar endosome-like compartment PVC to the vacuole. Membranes enriched for PVCs with radiolabeled proenzymes are mixed with membranes enriched for nonradiolabeled vacuoles. After incubation with adenosine triphosphate (ATP) and cytosolic extracts, the membranes are solubilized, immunoprecipitated with antisera against the radiolabeled vacuolar proteins, subjected to sodium dodecyl sulfate-polyacrylamide gel electrophoresis (SDS-PAGE), and autoradiography. An essential technique is described that uses extrusion of yeast spheroplasts through pores of defined size in a polycarbonate filter. This lysis method should be applicable to many studies of organelle function in yeast and mammalian cells.

2. Materials

2.1. Yeast Strains, Growth Media, Spheroplast Production, and Radiolabeling

2.1.1. Yeast Strains

1. SEY6210 *MATα leu2-3,112 ura3-52 his3-Δ200 trpl-Δ901 lys2-801 suc2-Δ9 GAL* (available from the ATCC); TVY614 *MATα ura3-52 leu2-3,112 his3-Δ200 trp 1-Δ901 lys2-801 suc2-Δ9 prc1Δ::HIS3 prb1Δ::hisG pep4Δ::LEU2* (available from the author).

2.1.2. Growth Media

1. Rich yeast growth media: 1.0% yeast extract, 2.0% bacto-Peptone (YP). Leave room (in terms of volume) for the addition of glucose. Sterilize in autoclave and store at room temperature.

A Cell-Free System for Reconstitution of Transport 43

2. 40% glucose; sterilize in an autoclave and store at room temperature. When needed, add to YP for a final concentration of 2% to make yeast extract/peptone/dextrose (YPD; *see* **Note 1**).
3. Wickerham's minimal proline media (WIMP) base: 1% L-proline (Sigma-Aldrich), 2% glucose, and 1X salts (*see* **Subheading 2.1.2.;step 4**). Sterilize in autoclave and store at room temperature.
4. WIMP salts (100X): 100 g KH_2PO_4, 10 g NaCl, 83 g, $MgCl_2$•$6H_2O$ dissolve in 1 L. Sterilize in autoclave and store at room temperature.
5. Trace elements (1,000X): 500 mg boric acid, 40 mg copper chloride, 100 mg potassium iodide, 200 mg ferric chloride, 400 mg manganese chloride, 200 mg sodium molybdate, 400 mg zinc chloride (all from Sigma-Aldrich) dissolve in 1 L. Sterilize in autoclave. Store at room temperature. Over time, undissolved material may become apparent but is not a problem (*see* **Note 2**).
6. Vitamins (200X): 0.4 mg biotin, 80 mg calcium pantothenate, 0.4 mg folic acid, 400 mg inositol, 80 mg niacin, 40 mg *p*-aminobenzoic acid, 80 mg pyridoxine HCl, 40 mg riboflavin, 80 mg thiamine-HCl, (all from Sigma-Aldrich) dissolve in 1 L and sterilize with filtration. Store at store at 4°C in 5 mL aliquots.
7. 20% yeast extract: 20 g yeast extract per 100 mL total volume. Sterilize in autoclave and store at room temperature.
8. Auxotrophic supplements (all 100X): a) Uracil: 2 g/L; b) Amino acids: 2 g/L L-tryptophan, 2 g/L L-histidine-HCl, 6 g/L L-leucine, 3 g/L L-lysine-HCl. Sterilize in autoclave and store both solutions at room temperature.
9. WIMP complete: WIMP base (**Subheading 2.1.2.**; **item 3**), 1X vitamins (**Subheading 2.1.2.**; **item 6**), 0.5% yeast extract, 1X uracil, 1X amino acid supplements (*see* **Note 3**).

2.1.3. Spheroplast Production

1. Pretreatment buffer (10X): 1.0 M Tris-HCl pH 9.4 (Trizma base from Sigma-Aldrich adjusted to pH with 6 M HCl). Sterilize in an autoclave and store at room temperature.
2. 1 M DTT in water, sterilize with filtration and store in 500 μL aliquots at –20°C.
3. 10% L-proline, sterilize in autoclave and store at room temperature.
4. 2.0 M sorbitol, high purity: 97% (Acros Organic available from Fisher Scientific) sterilize in an autoclave and store at room temperature.
3. Spheroplast conversion buffers: for radiolabeling purposes use 1% proline, 1X WIMP salts, 1X vitamins, 1.0 M sorbitol, 25 mM Tris-HCl pH 7.4, 0.2% glucose, 1X auxotrophic supplements. For rich media use 1.0 M sorbitol, 25 mM Tris-HCl pH 7.4, 0.5X YP, 0.2% glucose. Reduced glucose (0.2%) is used because 2% glucose inhibits enzymes that digest the cell wall.
4. Zymolyase 100T (Seikagaku America, Inc. available through Associates of Cape Cod Inc.): 25 mg/mL in 1.0 M sorbitol, 20 mM Hepes-KOH pH 7.0, 150 mM potassium acetate, and 5 mM magnesium acetate (i.e., freezing buffer, *see* **Subheading 2.1.4.**; **item 5**).
5. Glusulase: β-glucuronidase (90,000 U/mL) and sulfatase (19,000 U/mL), 10 mL (Perkin-Elmer).

2.1.4. Radiolabeling Yeast Cells

1. Tran^{35}S-label (ICN, Inc.) or EXPRE^{35}S^{35}S Protein Labeling Mix (Perkin-Elmer).
2. Labeling media: WIMP complete (**Subheading 2.1.2.**; **item 9**) with 1.0 M sorbitol (*see* **Note 6**).
3. Chase solution (50X): 250 mM methionine, 50 mM cysteine in water. Sterilize with filtration and store at room temperature.
4. Cell freezing solution: 1.0 M sorbitol, 20 mM Hepes-KOH pH 7.2, 150 mM potassium acetate, and 5 mM magnesium acetate.
5. Trichloroacetic acid (TCA): 100% (wt/vol); store at 4°C.

2.2. Cell-Free Assay

1. Buffer stocks: 200 mM Hepes-KOH, pH 7.0; 1.5 M potassium acetate pH 7.0; and 50 mM magnesium acetate. Sterilize with filtration and store at room temperature.
2. Transport buffer (TB): 250 mM sorbitol, 20 mM Hepes-KOH, pH 7.0, 150 mM potassium acetate pH 7.0 (adjust the pH to 7.0 with dilute acetic acid), 5.0 mM magnesium acetate. Sterilize with filtration and store at 4°C.
3. Polycarbonate filter lysis buffer: 0.6 M sorbitol, 5 mM Hepes-KOH pH 7.0.
4. Energy solutions (all 100X): 100 mM ATP; 4 M creatine phosphate; and 20 mg/mL creatine kinase all made in transport buffer. Store in small aliquots at –70°C.
5. Phenylmethylsulfonylfluoride (PMSF): 100 mM in isopropanol. Store at room temperature; stable for at least 9 mo.
6. Stop solution: 8 M urea, 5% SDS, 5% nonidet P-40 (NP-40).

2.3. Immunoprecipitation

1. Immunoprecipitation (IP) buffer: 50 mM Tris-HCl pH 7.5, 150 mM NaCl, and 0.1 mM EDTA. Store at 4°C.
2. IP buffer with Tween-20: IP buffer with 0.5% Tween-20. Store at 4°C.
3. Protein A-Sepharose suspension: swell 0.4 g protein A-Sepharose in 12.0 mL of 10 mM Tris, pH 7.5, 1 mg/mL bovine serum albumin (BSA), 1 mM NaN$_3$ (Protein A buffer) for at least 30 min with continuous rocking at room temperature. Harvest beads with centrifugation at 500*g* for 2 min. Aspirate the supernatant and add fresh buffer to 11 mL. Store at 4°C.
4. Rabbit antibodies against CPY (Abcam, Inc.) (*see* **Note 4**).

2.4. SDS-PAGE and Autoradiography

2.4.1. SDS-PAGE

1. Acrylamide: 30% acrylamide, 0.8% methylene bisacrylamide (acrylamide-*bis* [30:0.8]). Dissolve 75 g acrylamide and 2 g methylene bisacrylamide in 250 mL

of H_2O. Stir in a beaker covered with aluminum foil until the acrylamide is in solution, sterilize with filtration using a 0.2 μm filter and store at 4°C in a brown glass bottle.
2. Lower Tris buffer: 1.5 M Tris-HCl pH 8.8. Sterilize with filtration using a 0.2 μm filter and store at room temperature (see **Note 5**).
3. Upper Tris buffer: 0.5 M Tris-HCl pH 6.8. Sterilize with filtration using a 0.2 μm filter and store at room temperature (see **Note 5**).
4. 10% SDS: sterilize by filtration using a 0.2 μm filter and store at room temperature.
5. 10% Ammonium persulfate. Divide into 0.25 mL aliquots and store at –20°C. The frozen aliquots will last indefinitely.
6. 75% Glycerol (wt/vol).
7. 1% Bromophenol blue. Sterilize with filtration using a 0.2 μm filter and store at room temperature.
8. Tetramethyl-ethylenediamine (TEMED). Store at 4°C.
9. Sample buffer: 2.5 mL Upper Tris buffer, 1.0 mL β-mercaptoethanol, 1.5 mL 1% bromophenol blue, 5.0 mL 10% SDS, 1.5 mL 75% glycerol.
10. Overlay solution: 0.1% SDS.
11. Gel fix solution: 10% acetic acid, 40% methanol.
12. SDS-PAGE buffer: 192 mM glycine, 25 mM Tris-base, 0.1% SDS.

2.4.2. Autoradiography

1. Autoradiography-enhance solution: 1.0 M sodium salicylate, 10% glycerol.

3. Methods

The following steps can be accomplished in 4 to 5 d. Cytosol should be prepared first (see **Subheading 3.4**). Freezing cells facilitates flexibility in planning experiments.

3.1. Radiolabeling Yeast Spheroplasts for Donor Membrane Preparation

All quantitation of cell number or growth uses measurements of optical density (OD; turbidity) at 600 nm (OD_{600}). One OD_{600} unit corresponds to approx 1 to 2×10^7 yeast cells. The following represents the radiolabeling of 25 OD_{600} units of cells and can be scaled up proportionately.

1. Grow a liquid culture of SEY6210 to an OD_{600} between 0.8 and 1.2 in WIMP-complete media with 0.5% yeast extract at 30°C (generation time is approx 1.75 h). The best way to do this is dilute a fresh (< 24-h old) culture to an OD_{600} of 0.0017 and grow for 16 h (5 p.m. to 9 a.m.). Then dilute this culture

to an OD_{600} of 0.25 and grow until the desired concentration of cells; usually about 3 h.
2. Harvest the cells are by centrifugation at 1,500g for 5 min and wash once with 25 mL of sterile distilled water.
3. Resuspend the cell pellet in 2.5 mL of 0.1 M Tris-HCl pH 9.4 plus 10 mM dithiothreitol (DTT) and incubate with shaking at 30°C for 15 to 30 min. The high pH disrupts protein–protein interactions and the DTT reduces disulphide bonds in the cell wall proteins, which facilitate enzymatic digestion.
4. Centrifuge the cells (1,500g for 5 min) and resuspend the cell pellet in Wickerham's minimal proline media containing 0.2% glucose, 1.0 M sorbitol and 25 mM Tris-HCl pH 7.5 to a *total* volume of 1 mL. Use a 15 mL round-bottom tube (i.e., Falcon 2059), loosely capped (*see* **Note 6**).
5. Remove 10 µL of the cells, dilute to 1 mL of water, and determine the OD_{600}. The undiluted OD_{600} should be approx 25 or greater.
6. Add zymolyase 100T (5 µg/1 OD_{600} unit of cells) and glusulase to 0.5% (v/v) (*see* **Note 7**).
7. Gently agitate the cell suspension for 20 to 30 min at 30°C. A New Brunswick G76 shaker is preferred at a setting between 4 and 6.
8. Repeat **step 5**, but wait approx 1 min for the cells to lyse before determining the OD_{600}, which will result in clearing of the turbidity. The undiluted OD_{600} should be less than 0.025 (*see* **Note 8**).
8. Harvest the cells (1,500g for 5 min) and resuspend in 2.5 mL of Wickerham's minimal proline media containing 2% glucose and 1.0 M sorbitol.
9. Incubate the cells are with gentle shaking at 30°C for 15 min. A New Brunswick G76 shaker is preferred at a setting between 4 and 6.
10 Add 50 µL of 40% glucose and then Tran^{35}S-label (or EXPRE^{35}S^{35}S Protein Labeling Mix) to a final concentration of 200 µCi/mL and incubate for 5 min with gentle agitation.
11. After this pulse, add 50X chase solution to 1X (methionine, 5 mM final; cysteine 1 mM final), and yeast extract (0.5% final) and incubate for 2 min with gentle agitation. The goal is to achieve the labeled CPY precursors shown in **Fig. 1**, lane 3.
12. After the chase, transfer the cells to 10 mL of ice-cold 1.0 M sorbitol, 20 mM Hepes-KOH, 150 mM potassium acetate, and 5 mM magnesium acetate (freezing buffer) and incubate on ice for 5 min. The cold temperature rapidly halts all protein transport processes (*see* **Note 17**).
13. Harvest the cells with centrifugation at 1,500g for 5 min at 4°C. Perform all remaining steps on ice or at 4°C.
14. Resuspend the cells in 1 mL freezing buffer and transfer to a 1.7 mL polypropylene tube.
15. Harvest in a microcentrifuge (1 min; 16,000g) at 4°C and aspirate the supernatant.
16. Add 45 µL of freezing buffer to the washed cell pellet, gently resuspend, and place in a Nalgene 1°C cryofreezer jacketed with isopropanol (prechilled to 4°C). The

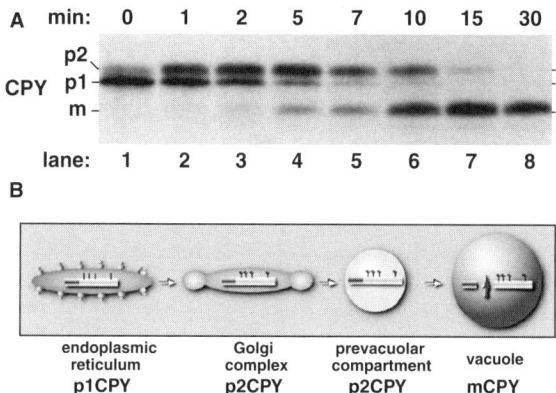

Fig. 1. Biosynthesis and processing of carboxypeptidase Y (CPY) in yeast spheroplasts. (**A**) Wild-type yeast spheroplasts (SEY6210) were radiolabeled with Tran^{35}S-label for 5 min and chased with methionine and cysteine for the indicated times at 30C. The cells were lysed and immunoprecipitated with antisera against CPY. Each immunoprecipitate was subjected to SDS-PAGE and autoradiography. The positions of the 67 kDa ER form of CPY (p1); the 69 kDa Golgi and PVC form (p2); and the 61 kDa vacuolar form (m) are indicated. (**B**) Schematic diagram depicting the organelle compartments where the different CPY forms reside.

cryofreezer allows very slow freezing of the cells preventing rupture from rapid ice crystal formation (*see* **Note 9**).

17. Place the cryofreezer at –70°C. The cells need at least 45 min for complete freezing and can then be kept at –70°C for several weeks before thawing to prepare radiolabeled donor membranes.

3.2. Preparation of Acceptor Membranes From Yeast Spheroplasts

All steps to prepare cells for acceptor membranes are identical to the above steps for donor membranes except for the following changes:

- Use rich media, YPD, for cell growth instead of WIMP complete media.
- Use rich spheroplast conversion media for making spheroplasts (*see* **Subheading 2.1.3.; step 3**).
- Do *NOT* radiolabel the cells.
- After cell wall removal (**Subheading 3.1; step 8**), incubate the spheroplasts at 30°C (10 OD$_{600}$ U/mL) in YPD plus 1.0 M sorbitol for 60 min *WITHOUT* constant shaking. However, the cells will settle so every approx 15 min very gently agitate them to ensure complete usage of the media. This step allows for regeneration of all cellular processes, which becomes very slow during the enzymatic removal to the cell wall.

3.3. Cell Lysis Via Extrusion Through Polycarbonate Filters

1. Frozen cells: remove tube(s) of cells from the –70°C freezer and thaw in a circulating water bath for 1 min, then place on ice.
2. Collect cells with centrifugation at 16,000g for 1 min at 4°C. If cells are not prefrozen, then start at this step.
3. Add 600 μL of ice-cold polycarbonate filter lysis buffer to the cells and resuspend thoroughly. Do this by pipeting up and down; *NOT* with a vortex mixer.
4. Collect cells with centrifugation at 16,000g for 1 min at 4°C.
5. Resuspend the cell pellet to approx 8 OD_{600} units/mL in lysis buffer (approx 3 mL for 25 OD_{600} units of cells).
6. Transfer to the resuspended cells to a 3-mL syringe with a preassembled 13-mm polycarbonate filter (Nucleopore™; Corning; available through Fisher Scientific) with 3-μm pores (*see* **Fig. 2**; *Important, see* **Note 10** for assembly tips).
7. With slow and steady pressure, push the cells through the filter and collect the effluent into a 15 mL polypropylene tube (conical or round bottom) on ice (*see* **Note 11**).
8. Add 0.5 to 0.75 mL of PFL buffer to the syringe and push through as in **step 7**, collecting into the same 15-mL tube.
9. Pipete up and down with a 1-mL Rainin Pipetman (or similar device) to gently mix the contents of the 15-mL tube. This is now the cell lysate.
10. Subject the cell lysate to centrifugation at 440g for 5 min to generate a P1 pellet with unbroken cells plus cell wall debris and a S1 supernatant fraction (**Fig. 2**).
11. Carefully remove the S1 supernatant and subject it to centrifugation at 15,000g for 10 min to generate a P2 pellet and a S2 supernatant fraction. The P2 pellet contains vacuoles to be used as acceptor membranes in cell-free assays (**Fig. 2**).

Fig. 2. Lysis of yeast spheroplasts via extrusion through a polycarbonate filter. (**A**) Schematic diagram for polycarbonate filter lysis and subsequent differential centrifugation. After pushing a yeast spheroplast suspension in a syringe through a polycarbonate filter with 3-μm pores, the crude cell lysate is subjected to differential centrifugation using the indicated g forces and times. The P1, P2, and P3 pellets are enriched in the indicated organelles. (**B**) Fractionation of vacuolar marker proteins. Wild-type yeast spheroplasts (SEY6210) were radiolabeled with Tran[35]S-label for 5 min and chased with methionine and cysteine for 2 min at 30°C. The cells were subjected to lysis through a polycarbonate filter with subsequent differential centrifugation. Each supernatant and pellet from the 440 (S1 and P1), 15,000 (S2 and P2), and 125,000g (S3 and P3) centrifugation steps was sequentially immunoprecipitated with antisera against carboxypeptidase Y (CPY), proteinase A (PrA), and alkaline phosphatase (ALP). Each immunoprecipitate was subjected to SDS-PAGE and autoradiography. The mid and late Golgi complex-modified precursor zymogen is designated p2, the endoplasmic reticulum and early Golgi complex precursor is designated p1 (CPY only), and the mature hydrolase is designated m for each protein. "Reproduced from **The Journal of Cell Biology, 1999, 146: 85–98**. Copyright 1999 The Rockefeller University Press."

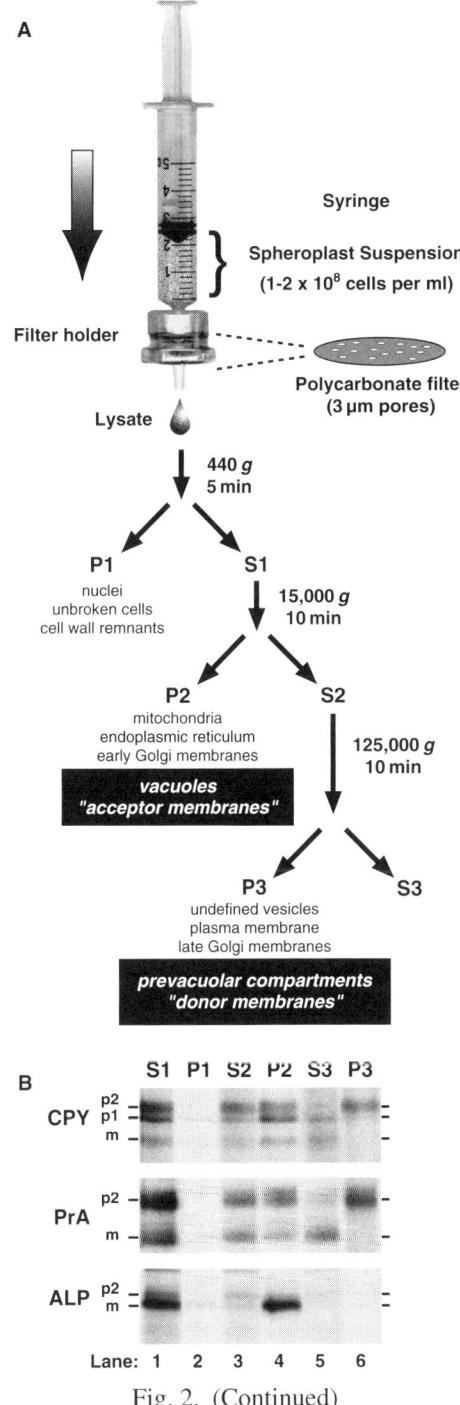

Fig. 2. (Continued)

12. Carefully remove the S2 supernatant and subject it to centrifugation at 125,000g for 10 min to generate a P3 pellet and a S3 supernatant fraction (Fig. 2). A tabletop ultracentrifuge (Beckman TL-100) is preferred using 2 or 3-mL capacity polycarbonate (clear) tubes. The P3 pellet contains prevacuolar compartments to be used as donor membranes in cell-free assays.
13. Resuspend the P2 pellet in transport buffer (200 µL total for 25 OD_{600} units of cells). Keep on ice until the donor membranes are prepared.
14. Resuspend the P3 pellet in transport buffer (200 µL total for 25 OD_{600} units of cells). Keep on ice (*see* **Note 12**).

3.4. Preparation of Cytosol

1. Yeast strain TVY614 is grown at 30°C in YPD (with 5% glucose) to an OD_{600} of 4 to 6; usually use 800 to 1,600 total OD_{600} units of cells (200–400 mL of YPD culture media).
2. Harvest the cells with centrifugation at 4,000g for 15 min.
3. Wash the cells once with sterile distilled water using 50% of the original volume of media and harvest again as above.
4. Resuspend the washed cell pellet in approx 35 mL of ice-cold 0.25 M sorbitol, 20 mM Hepes-KOH, 150 mM potassium acetate, and 5 mM magnesium acetate pH 7.0 (standard transport buffer[TB]).
5. Transfer the resuspended cells to a 50 mL conical tube and harvest via centrifugation at approx 1,750g for 5' at 4°C.
6. Resuspend the cells in TB to approx 200 OD_{600}/mL and transfer to 2 mL polypropylene tubes (with screw caps and O-rings) in 1 mL aliquots.
7. Add approx 1 g of acid-washed glass beads (0.425-0.610 mm; Sigma-Aldrich) and then agitate for three 30 s intervals in a Mini Bead-Beater set on high (BioSpec Products, Inc., Bartlesville, OK) at 4°C*see* **Note 13**).
8. Subject all tubes to centrifugation at 15,000g for 2 min.
9. Remove the supernatant from each tube and add 1 mL of TB to the pellet. Agitate briefly (15–30 s) on a vortex mixer, and subject to centrifugation at 15,000g for 2 min.
10. Pool the second supernatant with the first supernatant and subject to centrifugation at approx 103,000g (average) for 10 min.
11. Remove the supernatant, dispense into small aliquots (200 µL), and snap-freeze in liquid nitrogen.
12. Save a very small amount from freezing (25-50 µL) and dilute 1/100 for determination of protein concentration, which should be in the range of 25 to 50 mg/mL. The authors use a modified Bradford Coomassie blue binding assay available from BioRad with BSA as a standard. This is done in a microtiter plate. The assay uses 160 µL sample or BSA standard (1, 2, 4, 8, and 10 µg) plus 40 µL of concentrated Coomassie blue dye (BioRad). Mix well, incubate 5 min at room temperature, and read the absorbance at 595 nm.

3.5. Cell-Free Assays

1. Assemble all reactions on ice to a total volume of 50 µL.
2. Make a 10X ATP regeneration mix from thawed aliquots of the solutions in **Subheading 2.2.**, **item 5**. Mix 1/10 volume 100 mM ATP, 1/10 volume 4 M creatine phosphate, and 1/10 volume 20 mg/mL creatine kinase 7/10 volume TB to make a solution with 10 mM ATP, 400 mM creatine phosphate, and 2 mg/mL creatine kinase. Keep this on ice.
3. Use 1.7 mL capped polypropylene tubes. For a complete transport reaction add the following in order:

   - TB to bring final volume to 50 µL (volume added is dependent on cytosol concentration)
   - 5 µL 10X ATP regeneration mix
   - 13 µL Nonradiolabeled acceptor membranes (P2 pellet from **Subheading 3.2.**)
   - Cytosol at 5 mg/mL final (volume dependent on concentration; **Subheading 3.4.**)
   - 13 µL Radiolabeled donor membranes (P3 pellet from **Subheading 3.1.**).

   For controls, leave out acceptor membranes, cytosol, and ATP (*see* **Fig. 3**).
4. Place the tubes in a circulating water bath for 60 min at 25°C.
5. To stop the reactions, add 3 µL of 100 mM PMSF and mix with vortex mixer. The PMSF inhibits any processing of pro-CPY that may occur once the membranes are solubilized. This can happen even in the presence of SDS.
6. Add 25 µL of 8.0 M urea, 5% SDS, 5% NP-40 and boil for 5 min. This solubilizes the membranes and completely denatures all protein activity.
7. Subject the tubes to centrifugation at 16,000g for 1 min.

3.6. Immunopreciptiation

1. Add 1 mL of IP buffer *with* Tween-20 to the solubilized cell-free reactions and mix thoroughly.
2. Subject the tubes to centrifugation at 16,000g for 5 min.
3. Remove 1 mL to a new tube that contains 75 µL of the Protein A-Sepharose suspension and the appropriate antibody. Mix the Protein A-Sepharose frequently when adding it to the tubes to make sure a constant amount of the beads is added, which settle very fast.
4. Rock the tubes for 3 h to overnight at 4°C.
5. Subject the tubes to centrifugation at 16,000g for 1 min to collect the Protein A-Sepharose beads.
6. Carefully aspirate the supernatant so as to not disturb the beads.
7. Add 1 mL of IP buffer *with* Tween-20 and mix thoroughly to wash away unbound radioactive proteins.
8. Repeat **steps 5** and **6**.
9. Add 1 mL of IP buffer *without* Tween-20 and mix thoroughly to wash away unbound radioactive proteins and remove the Tween-20 detergent.

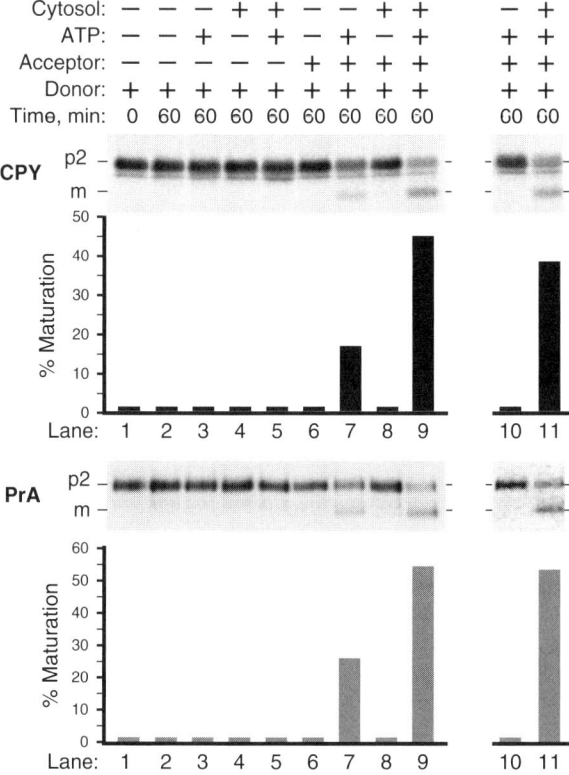

Fig. 3. Vacuolar precursor proteins undergo maturation after incubating donor and acceptor membranes in a cell-free system. Wild-type yeast spheroplasts were radiolabeled (as in Fig. 1A). The cells were subjected to lysis through a polycarbonate filter with subsequent differential centrifugation to generate a 125,000g, P3 donor membrane pellet. The same yeast strain was used to make a 15,000g, P2 acceptor membrane pellet from nonradiolabeled spheroplasts. The radiolabeled donor membranes (from approx 5×10^7 spheroplasts per reaction) were incubated at 25°C for 60 min with various combinations of buffer, ATP (plus a regeneration system), cytosol (5 mg/mL), and acceptor membranes (approx 100 μg) in a total volume of 50 μL, as indicated. All reactions were sequentially immunoprecipitated for CPY and proteinase A (PrA), subjected to SDS-PAGE, and autoradiography. For the reactions in lanes 10 and 11, both donor and acceptor membranes were washed once with lysis buffer and re-harvested before incubation with ATP and cytosol. "Reproduced from **The Journal of Cell Biology, 1999, 146: 85–98**. Copyright 1999 The Rockefeller University Press."

10. Repeat **steps 5** and **6**.
11. Dry the wet Protein A-Sepharose beads under vacuum in a Speed-Vac with warming for 5 to 10 min.
12. Add 50 μL of SDS-PAGE sample buffer and heat for 5 min at 90 to 100°C.

13. Subject the tubes to centrifugation at 16,000g for 1 min. The samples are now ready for SDS-PAGE or can be stored at −20°C for several weeks.

3.7. SDS-PAGE

1. Prepare a resolving gel solution of 9% acrylamide, 0.375 M lower gel Tris-HCl solution, 0.1% SDS to the volume needed for gel apparatus (*see* **Note 14**).
2. Add 10% ammonium persulfate to a final concentration of 0.067% and 100% TEMED to 0.067%.
3. Pour mixture into a gel-casting apparatus and overlay with 0.1% SDS such that it is 1 to 2 cm above the gel mixture.
4. Allow the gel mixture solution to polymerize for 30 to 45 min.
5. Prepare a "stacking" gel solution of 5% acrylamide, 0.125 M upper gel Tris-HCl solution, 0.1% SDS, and 0.1% ammonium persulfate. Make twice as much needed for the stacking gel volume.
6. Drain the 0.1% SDS from the polymerized resolving gel and rinse with half of the stacking gel solution.
7. Add 100% TEMED to 0.1% final concentration to the remaining stacking gel solution and pour on top of the polymerized resolving gel.
8. Insert comb of choice for sample lanes and allow the solution to polymerize for 30 to 45 min.
9. Assemble the gel into the apparatus of choice and add SDS electrophoresis buffer to both upper and lower reservoirs.
10. Remove the comb for sample lane formation and thoroughly rinse the empty lanes with SDS-PAGE buffer from the upper reservoir.
11. Load half of the IP samples onto each lane. This is best done by drawing up the entire amount of solubilized Protein A-Sepharose beads into a 50 µL Hamilton syringe with a blunt-end needle to measure the total volume (it is often not 50 µL from re-swelling of the dried Sepharose) and applying half into the lane. Save the remaining half in the same tube just in case another gel separation needs to be performed.
12. Apply a constant current (approx 1.5 mA per cm of total distance gel length) and allow the bromophenol blue tracking dye to exit the gel. Continue electrophoresis for 20 to 30 min (*see* **Note 15**).
13. Disassemble the gel apparatus and remove the gel from the glass plates.
14. Rinse the gel with deionized water in a suitably sized tray for 30 s and discard the water.
15. Add enough gel fix solution to completely cover the gel, allowing it to float and rock for 10 to 15 min.
16. Discard the gel fix solution and rinse the gel with deionized water three times for 2 to 5 min on each rinse. Rock the gel during the rinses.
17. Add enough autoradiography-enhance solution to cover the gel well and rock for 1 to 15 min (*see* **Note 16**).
18. Discard the solution and place gel on a piece of Whatman 3 mm paper.

19. Cover with plastic wrap and dry the gel for 45 to 90 min under vacuum and with a heated gel dryer.
20. Put a piece of x-ray film and intensifying screen on top of the dried (in a darkroom) and place at –70°C for 4 to 16 h. Alternatively, use a phosphorimage plate and expose at room temperature.
21. Develop the x-ray film with a commercial processor.
22. Scan the x-ray film with a high quality photo scanner. Use at least 300 dpi for publication purposes.
23. Perform densitometry on the bands with NIH Image or other image quantitation software.

4. Notes

1. Do not add glucose to YP before sterilization in an autoclave, the sugar will burn due to the peptides in bacto-peptone. When making cytosol, increased glucose to 5% for increased cell yields per mL of media.
2. The use of trace elements is *optional* and only used when cell growth or adequate incorporation of radioactivity is a problem.
3. As an alternative to minimal proline media, yeast nitrogen base (YNB) without amino acids can be used. It is a simpler formulation composed of 0.67% YNB without amino acids, 2% glucose, and the appropriate autotrophic supplements. Yeast extract at 0.2 to 0.5% should still be included, which facilitates faster growth and subsequent removal of the cell wall.
4. This antisera is untested by the author but is supposed to work in immunoprecipitation.
5. The pH of the Tris solutions is critical for fine resolution of proteins with small differences in molecular weight. It should be accurate to ±0.05 pH units. It is best to adjust the pH with HCl to the proper values and then let the solutions set for 30 to 60 min and check it again.
6. A *conical* shaped tube does not provide adequate cell suspension during the duration of cell-wall removal. However, when scaling up the procedure to more than 25 OD_{600} units of cells, a 50–mL conical tube works well because the volume geometry is not overly vertical. Do *NOT* include yeast extract, as this will prevent incorporation of ^{35}S-amino acids into protein.
7. Small amounts of glusulase (0.5–3%) help remove surface agglutinating moieties and aids in dispersing cells, which frequently clump tightly with centrifugal harvesting after using just Zymolyase 100T. With Zymolyase 100T alone, cell homogenates are often aggregated, impeding organelle resolution.
8. *This is a critical step*. The goal is to convert at least 90% of the yeast cells to spheroplasts. This is measured with cell lysis in water. If this is not achieved during the first 20 to 30 min of enzymatic digestion, then add more zymolyase 100T and incubate for another 10 min. However, after this point if at least 90% conversion is not obtained, then the procedure should be stopped. Going further will present major problems when attempting to lyse the spheroplasts via extrusion through polycarbonate filters. Furthermore, spheroplast conversion efficiency is yeast strain-

dependent and highly tied to the growth phase of cultures. Growing cells beyond an OD_{600} of 1.2 to 1.5 in WIMP-complete media with 0.5% yeast extract will hamper efficient removal of the cell wall. It may be necessary to perform a pilot experiment on a smaller scale of cells and titrate zymolyase 100T to determine the optimal amount for efficient spheroplast conversion.
9. It is best to mix the cell pellet with the pipette tip before drawing them up and down. This will be very thick and slurry.
10. The polycarbonate filters have a glossy side and a matte side. The glossy side is to be used facing up. Remove a filter from the pack with flat forceps and place shiny side down on the surface of approx 200 mL of water in a 250-mL beaker. Let it set a few minutes to hydrate. Take the bottom of the filter holder and wet the surface of it with water. Remove the filter from the beaker with flat forceps, turn it over (shiny side up), and position on the middle of the bottom filer holder. Assemble the filter holder; for small-scale preparations (approx 25 OD_{600} units of cells) use a snap-lock holder and snap the top into place. Slide the retaining ring over the outside until it tightens. Check for leaks by attaching a 3-mL syringe and filling it with water. If water rapidly drips out, then dismantle the assembly and repeat, making sure that water does not drip. A very, very slow drip is normal ({LT}2 drops per minute). Once the assembly is correct, rinse the filter with 3 mL of lysis buffer and place at 4°C before use.
11. The maximum cell concentration when pushing through the 13-mm filters is 20 OD_{600} U/mL . This is used on larger preparations of donor or acceptor membranes with larger diameter filters. About 3 to 5 min should be used for the 3 mL to travel through the filter. Some backpressure may occur put not enough to require strong force for pushing. If this does happen, then the increased force may blow the seal on the filter and all contents will leak out fast and spill. This is especially important to avoid when dealing with radiolabeled cells.
12. This pellet is tightly impacted and somewhat translucent. The best way to resuspend it completely is to seal the end of a Pasteur pipete with a flame. After cooling, use this to scrape the pellet off the tube bottom in approx 75 µL of buffer. Allow it to sit on ice for a few minutes and scrape again. Continue doing this until "globs" of the pellet are no longer visible. This may take 10 to 15 min. Transfer the resuspended pellet to a 1.7-mL tube and then rinse the centrifuge tube twice with buffer in portions that will not exceed the 200 µL total volume.
13. For success in this reconstitution the cytosolic protein extract must be highly concentrated (at least 30 mg/mL) *directly* from the cell lysis. The Bead Beater can reciprocate tubes at nearly 3,000 oscillations per minute and achieves very efficient cell lysis. Other disruption methods like agitation on a vortex mixer or a French press usually do not result in such concentrated extracts.
14. Many systems exist for SDS-PAGE. We use a system from R. Shadel, Inc. that makes 0.75-mm thick gels. Furthermore, the resolving gel is about 10 to 11 cm, which is much longer than most "mini"-type gels that are most popular today. Smaller systems provide about half the resolving length than the larger format system that we use. These may be adequate but experience has demonstrated that they

do not provide adequate resolution between the 67 and 69 kDa ER and Golgi forms of CPY, respectively.
15. Using a constant current for electrophoresis of SDS polyacrylamide gels with this buffer system results in a voltage gradient that starts low and ends high. This can have benefits in resolving power.
16. The sodium salicylate solution is a water-based flour that is very inexpensive compared to organic solvent-based commercial solutions. It results in about a fivefold enhancement of radioactivity detection at $-70°C$. If exposure at $-70°C$ is beyond the linear range of the film, then expose at room temperature.
17. The incorporation of ^{35}S-amino acids into protein should yield *at least* 1×10^7 cpm per OD_{600} unit of cells. To determine this follow the steps listed here:

- Remove 1 OD_{600} unit of cells just after **step 11** or before **step 12**(**Subheading 3.1.**) and add 100% TCA to 10% final concentration.
- Hold on ice for 5 to 10 min, then collect the cell pellet in a microcentrifuge on the highest setting for 2 to 5 min.
- Aspirate the supernatant and wash the pellet twice with ice-cold acetone to remove residual TCA, collecting the cells each time in a microcentrifuge on the highest setting for 2 to 5 min.
- Dry the cell pellet thoroughly (a Speed Vac is preferred) then add 50 μL of 50 mM Tris-HCl pH 7.8, 6 M urea, 2% SDS, 2% NP-40 and solubilize the cell pellet (a water bath sonicator may help).
- Heat the solubilized pellet to 90–100°C for 5 min then spin in a microcentrifuge on the highest setting for 30 s.
- Add 1 mL IP buffer plus Tween-20 then spin in a microcentrifuge on the highest setting for 2 min.
- Remove 10 μL (approx 0.01 OD_{600} unit of cells) and determine the ^{35}S amount in a scintillation counter. The reading should be at least 1×10^5 cpm (1×10^7 per OD_{600} unit of cells).

Acknowledgments

The author is funded through a NIGMS grant; GM052092 and is grateful to Brenda Gerhardt for technical assistance.

References

1. Vida, T., and B. Gerhardt, (1999) A cell-free assay allows reconstitution of Vps33p-dependent transport to the yeast vacuole/lysosome. *J. Cel.l Biol.* **146**, 85–98.
2. Wilson, D.W., C.A. Wilcox, G.C. Flynn, et al. (1989) A fusion protein required for vesicle-mediated transport in both mammalian cells and yeast. *Nature.* **339**, 355–359.
3. Bowers, K., and T.H. Stevens (2005) Protein transport from the late Golgi to the vacuole in the yeast Saccharomyces cerevisiae. *Biochim. Biophys. Acta.* **1744**, 438–454.

4. Novick, P., S. Ferro, and R. Schekman. (1981) Order of events in the yeast secretory pathway. *Cell.* **25**, 461–469.
5. Novick, P., C. Field, and R. Schekman (1980) Identification of 23 complementation groups required for post-translational events in the yeast secretory pathway. *Cell.* **21**, 205–215.
6. Robinson, J.S., D.J. Klionsky, L.M. Banta, and S.D. Emr, (1988) Protein sorting in Saccharomyces cerevisiae: isolation of mutants defective in the delivery and processing of multiple vacuolar hydrolases. *Mol. Cell. Biol.* **8,** 4936–4948.
7. Rothman, J.H. and T.H. Stevens, (1986) Protein sorting in yeast: mutants defective in vacuole biogenesis mislocalize vacuolar proteins into the late secretory pathway. *Cell.* **47**, 1041–1051.
8. Schekman, R. and P. Novick, (2004) 23 genes, 23 years later. *Cell.* **116**, S13–S15, 1 p following S19.
9. Shaw, J.D., K.B. Cummings, G. Huyer, S. Michaelis, and B. Wendland, (2001) Yeast as a model system for studying endocytosis. *Exp. Cell. Res.* **271**, 1–9.
10. Wendland, B., S.D. Emr, and H. Riezman, (1998) Protein traffic in the yeast endocytic and vacuolar protein sorting pathways. *Curr. Opin. Cell. Biol.* **10**, 513–522.
11. Cook, N.R. and H.W. Davidson, (2001) In vitro assays of vesicular transport. *Traffic.* **2**, 19–25.

5

In vitro Analysis of the Mitochondrial Preprotein Import Machinery Using Recombinant Precursor Polypeptides

Dorothea Becker, Martin Krayl, and Wolfgang Voos

Summary

The import of precursor proteins into mitochondria represents a cell biological process that is absolutely required for the survival of an eukaryotic cell. A complex chain of reactions needs to be followed to achieve a successful transport of mitochondrial proteins from the cytosol through the double membrane system to their final destination. In order to elucidate the details of the translocation process, *in vitro* import assays have been developed that are based on the incubation of isolated active mitochondria with natural or artificial precursor proteins containing the appropriate targeting information. Although most of the protein components of the import machinery have been identified and functionally characterized using this basic system, the definition of the molecular mechanisms requires more specialized assay techniques. Here we describe modifications of the standard *in vitro* import assay technique that are based on the utilization of recombinant preprotein constructs. The application of saturating amounts of substrate preproteins is a prerequisite for the determination of translocation kinetics and energy requirements of the import process. Accumulation of preproteins as membrane-spanning translocation intermediates further provides a basis for the functional and structural characterization of the active translocation machinery.

Key words: Import kinetics; mitochondria; protein import; precursor proteins; *Saccharomyces cerevisiae*; translocation intermediates

1. Introduction

According to proteomic studies, yeast mitochondria contain roughly 1,000 different proteins. Only about 1% of these proteins are encoded on the mitochondrial genome. Consequently, the vast majority of proteins have to be imported

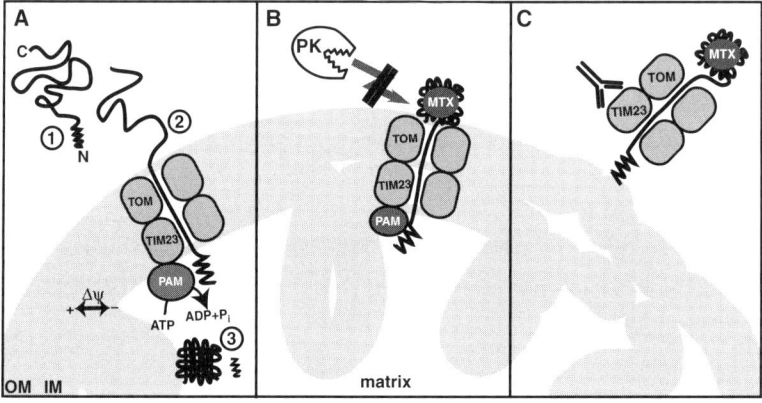

Fig. 1. Principles of the analysis of mitochondrial preprotein import. (**A**) Import of a presequence-containing protein into the mitochondrial matrix. Matrix-targeted preproteins contain N-terminal signal sequences (1) that are recognized by receptors on the outer face of the outer mitochondrial membrane. The preprotein is inserted into the outer membrane by the translocase of the outer membrane (TOM complex), and subsequently handed over to the translocase of the inner membrane (TIM23 complex), where the presequence is inserted in a reaction dependent on the electric potential $\triangle\psi$ (2). Completion of matrix-import requires the mtHsp70-containing import motor (PAM) that couples protein import with ATP hydrolysis. Cleavage of the presequence and folding of the protein occurs in the matrix, resulting in the mature size protein in its native state (3). (**B**) Generation of protease-resistant translocation intermediates using preproteins with a C-terminal dihydrofolate reductase (DHFR) domain. Addition of the specific ligand methotrexate (MTX) stabilizes the folding state of the DHFR-domain, thereby preventing its translocation. The N-terminal b_2-part crosses both membranes and contacts the import motor. Stable folding of the DHFR domain inhibits proteolysis by externally added proteinase K (PK) when the translocation machinery is active. (**C**) Isolation of the MTX-generated TOM–TIM supercomplex by use of detergent. Addition of digitonin solubilizes the mitochondrial membranes, whereas the protein complex remains stable and can be extracted intact. Detection of the complex can be done by use of anti-Tim23 antibodies. OM, outer membrane; IM, inner membrane.

into the organelle after their synthesis on cytosolic ribosomes (*see* **Fig. 1A**). To ensure their proper destination, mitochondrial proteins are generated as precursors carrying specific targeting signals. Most matrix-destined precursors, as well as monotopic inner membrane proteins possess amino-terminal, typically cleavable presequences. Polytopic outer and inner membrane proteins, by contrast, carry internal signals that are usually distributed throughout the mature protein. The first step of the preprotein import process is the specific recognition of the targeting information by receptor proteins at the mitochondrial surface. This reaction governs the insertion of preproteins into the translocase of the outer mitochondrial membrane (TOM complex), which represents the general entry pore for essentially all nuclear-encoded proteins. Depending on the targeting signal, insertion into or translocation across the inner membrane is mediated by either of the two translocases of the inner membrane (TIM) complexes *(1)*. Proteins destined for the inner membrane or the matrix, require a membrane potential across the inner membrane as initial energy source. Completion of matrix import furthermore requires adenosine triphosphate (ATP) hydrolysis, catalysed by the matrix heat-shock protein (mtHsp) 70 kDa (mtHsp70, Ssc1 in yeast), providing the driving force for the unfolding of preprotein domains and the completion of membrane translocation *(2)*. MtHsp70 represents the core component of the presequence-translocase-associated motor (PAM). It is recruited to the import sites by the peripheral membrane protein Tim44. Further essential PAM components are the nucleotide exchange factor Mge1 (mitochondrial GrpE homolog), as well as the ATPase-stimulating J-domain protein Pam18, and the degenerate J-protein Pam16. In the matrix, the presequence is cleaved off (often by the matrix-processing peptidase), resulting in the mature size protein. Folding of imported proteins is initially promoted by mtHsp70, which stabilizes, and thereby protects unfolded proteins until they acquire their native conformation (assisted by the Hsp60 chaperonin system) or (in the case of multimeric proteins) assemble into the respective protein complexes *(3)*. Because of its convenience, especially regarding genetic manipulation, the yeast *Saccharomyces cerevisiae* represents an ideal model organism for the study of the mitochondrial biogenesis processes typical for eukaryotic cells. An experimental analysis of the mitochondrial protein import first requires the isolation of functionally intact mitochondria, as well as an assessment of their structural integrity. Second, an appropriate precursor protein, consisting of a mitochondrial signal sequence and a reporter component has to be generated. The preprotein can be either synthesized as a radiolabeled precursor by translation in a cell-free system, or isolated from *Escherichia coli* cells as a recombinant precursor. The protein import reaction *in vitro* is performed by combining intact mitochondria with precursor polypeptides under appropriate buffer conditions. A successful translocation reaction is typically monitored by polyacrylamide gel electrophoresis (PAGE) analyzing the gain of protease resistance of the preprotein and the removal of the targeting sequence *(4)*. The isolation of yeast mitochondria includes an enzymatic digestion of the cell wall, followed by a mechanical disruption of the resulting spheroplast. Enrichment of the mitochondrial fraction is achieved by subsequent differential centrifugation of the crude cell lysate. The structural integrity of the mitochondrial preparation is assessed by checking the membrane potential across the inner mitochondrial membrane, which is of particular importance for the comparative analysis of mitochondria derived from different yeast strains.

The assay described here is based on the reversible interaction of a fluorescent dye with membranes in a potential-sensitive manner *(5,6)*. For the import reaction, mitochondria are additionally energized by ATP, as well as nicotinamide adenine dinucleotide (NADH), which represents the main substrate of the respiratory chain in yeast.

Precursors synthesized *in vitro* or generated by heterologous expression are functionally equivalent but their different properties have major implications for the import reaction. Generally speaking, the *in vitro* synthesis of a radiolabeled precursor in a cell-free system is more convenient and less time-consuming. It is ideally performed in commercially available rabbit reticulocyte lysate, supplemented with [^{35}S]methionine and/or [^{35}S]cysteine, after *in vitro* transcription using the SP6 or T7 RNA polymerase. However, the *in vitro* import of radioactive precursor proteins, which are usually produced in substoichiometric amounts, only permits a qualitative analysis of the import reaction. In order to apply more stringent conditions, or to obtain reliable kinetical data, saturating amounts of precursor polypeptides have to be used. Recombinant preproteins obtained by expression and purification from *E. coli* cells have been shown to be import competent even in the absence of cytosolic cofactors and are virtually unlimited in their amounts. Using saturating amounts of precursor proteins, conditions can be achieved that represent the maximal rate of the translocation reaction (V_{max}). These conditions are a prerequisite for the quantitative characterization of the preprotein translocation kinetics *(7)*. Besides, applying excess preprotein accentuates potential import phenotypes in the characterization of mutants of individual components of the import machinery.

This chapter describes expression and isolation, as well as *in vitro* import of a fusion protein, consisting of a modified mitochondrial presequence (yeast pre-cytochrome b_2), and a heterologous passenger protein (mouse dihydrofolate reductase [DHFR]). These b_2-DHFR fusion proteins are standardized for the *in vitro* import into mitochondria *(8,9)*. The use of a heterologous protein moiety allows to take advantage of the highly specific antigen-antibody recognition for analysis of the import reaction, as interference from endogenous proteins is bypassed. As an alternative to the protein detection by immunodecoration, a direct assessment of imported material based on fluorescence labeling has been recently reported *(10)*. The fluorescence labeling of the fusion protein is performed using the SNAP-tag technology (i.e., the b_2-DHFR construct is fused at its C-terminus to a modified version of the human DNA repair enzyme O6-alkylguanine-DNA-alkyltransferase). Attachment of a single fluorescein molecule is achieved by use of the fluorescent dye benzyl-guanine-fluorescein (BG-FL). Because only a single fluorophore is attached to the protein, the labeling reaction is highly reproducible and well defined. The direct detection of the fluorescent preprotein allows a more reliable quantification of imported protein amounts and significantly reduces the overall experimental time.

Based on the import of precursor proteins carrying a DHFR-domain, the mitochondrial protein import can be analyzed in greater detail by the generation of specific translocation intermediates (*see* **Fig. 1B**). To that end, the DHFR-ligand methotrexate (MTX) is employed that strongly stabilizes the native folding state of DHFR. Thereby, the full translocation of preproteins containing a C-terminal DHFR moiety is prevented while the insertion of N-terminal segments into the translocation channel is not affected. Thus, the precursor accumulates in an intermediate state in which the N-terminal b_2-part contacts the TIM23 complex (also termed presequence translocase), whereas the C-terminal DHFR domain remains in the cytosol *(9,11)*. Because these intermediates represent preproteins engaged in active translocation, they can be used for an assessment of the inward-directed translocation force exerted by the import motor. Provided that the mtHsp70-containing PAM is functional and active, the preprotein is stably held in the intermediate complex, whereas it is lost in case of a motor defect *(12)*. Because the preprotein spans both translocation complexes (TOM and TIM23), a physical connection between the machinery of the outer and inner membrane is generated that allows the extraction and analysis of the whole translocation machinery in its active state, termed the TOM–TIM supercomplex (*see* **Fig. 1C**). Formation of this supercomplex is typically analysed by Blue Native PAGE (BN-PAGE) that allows for the electrophoretic separation of multimeric proteins in their native state *(13)*.

2. Materials

Prepare all buffers in distilled water.

2.1. Growth of Yeast Cells and Isolation of Mitochondria From Yeast Cells

1. Culture medium (autoclaved): Yeast extract/peptone/glycerol (YPG; for growth on nonfermentable carbon source): 1% (w/v) yeast extract (Difco), 2% (w/v) bacto-peptone (Difco), 3% glycerol, pH 5.0 (HCl). Yeast extract/peptone/dextrose (for growth on fermentable carbon source): 1% (w/v) yeast extract (Difco), 2% (w/v) bacto-peptone (Difco), 2% glucose, pH 5.0 (HCl).
2. Dithiotreitol (DTT) buffer: 100 mM Tris-H_2SO_4, pH 9.4, 10 mM DTT, prewarmed to 30°C. Add DTT just prior to use.
3. Zymolase buffer: 1.2 M sorbitol, 20 mM potassium phosphate buffer, pH 7.4. Prewarm to 30°C.
4. Zymolase 20T (Seikagaku Kogyo Co., Japan).
5. Homogenization buffer: 0.6 M sorbitol, 10 mM Tris-HCl, pH 7.4, 1 mM ethylene diaminetetraacetic acid (EDTA), 1 mM phenylmethylsulfonyl fluoride (PMSF),

0.2% (w/v) bovine serum albumine (BSA, essentially fatty acid-free, Sigma-Aldrich). Add PMSF from a freshly prepared stock in ethanol just prior to use (*see* **Note 2**). Pre-cool homogenization buffer to 4°C.

6. SEM buffer: 250 mM sucrose, 1 mM EDTA, 10 mM morpholinepropanesulfonic acid (MOPS-KOH), pH 7.2. Pre-cool SEM to 4°C.

2.2. Measurement of the Mitochondrial Membrane Potential by Fluorescence Spectrometry

1. Isolated mitochondria in SEM buffer (*see* **Subheading 2.1.**): 10 mg/mL protein, freshly prepared or thawed on ice.
2. 1 M potassium phosphate buffer, pH 7.2: Prepare 50 mL each of 1 M KH$_2$PO$_4$ and of K$_2$HPO$_4$, mix solutions by adding the appropriate volume of KH$_2$PO$_4$ (low pH) to 50 mL of K$_2$HPO$_4$ (high pH), until the required pH 7.2 is reached.
3. Potential buffer: 0.6 M sorbitol, 0.1% (w/v) BSA (for molecular biology, e.g., Roche), 10 mM MgCl$_2$, 20 mM potassium phosphate buffer, pH 7.2.
4. 3,3'-dipropylthiadicarbocyanine iodide (diSC$_3$[5], Molecular Probes, Invitrogen), prepare fresh as 1 mM stock in ethanol, wrap tube in aluminium foil as the dye is light sensitive.
5. 100x inhibitor mix: 800 μM antimycin A (blocks electron transfer within complex III of the respiratory chain), store as 8 mM stock in ethanol at –20°C, 50 μM valinomycin (dissipates membrane potential by acting as K$^+$-ionophore), store as 1 mM stock in ethanol at –20°C, 2 mM oligomycin (inhibits F$_1$ subunit of F$_1$F$_o$-ATP synthase), store as 10 mM stock in ethanol at –20°C; for preparation of 1 mL 100x working solution, mix 100 μL of antimycin A stock with 50 μL of valinomycin stock and 200 μL of oligomycin stock in 650 μL ethanol, store at –20°C, stable for years.
6. 100x valinomycin: dilute 50 μL of valinomycin stock (*see* **Heading 4.**) in 950 μL ethanol, store at –20°C, stable for years.
7. Substrates: 1 M succinate in H$_2$O, pH 8.5 (L-succinic acid, di-sodium salt), 1 M malate in H$_2$O, pH 8.0 (L-malic acid, di-sodium salt), prepare 1 mL each, store at –20°C, stable for years.
8. Fluorescence spectrometer (e.g., Aminco Bowman), cuvette (1-cm pathlength) for fluorescence measurements (e.g., Hellma 101-OS).
9. Aminco Bowman II software.

2.3. Preparation of Precursor Proteins From Escherichia coli

2.3.1. Cytochrome b$_2$-Fusion Proteins Containing a C-Terminal DHFR Domain

1. Luria-Bertani (LB)-medium: 1% tryptone, 0.5% yeast extract, 0.5% NaCl.

In vitro *Analysis of the Mitochondrial Preprotein Import Machinery*   65

2. Isopropyl-β-D-thiogalactopyranoside (IPTG): prepare a 1 M stock and store at −20°C, stable for years.
3. Pre-lysis buffer: 30% sucrose, 20 mM potassium phosphate buffer, pH 8.0, 1 mM EDTA, 10 mM DTT (*see* **Note 1**), 1 mM PMSF (added from 0.1 M stock in ethanol, add just prior to use, se*e* **Note 2**); pre-cool to 4°C before use.
4. Buffer A: 20 mM Tris-HCl, pH 7.0, 1 mM EDTA, 10 mM DTT (*see* **Note 1**), 1 mM PMSF (*see* **Note 2**), protease-inhibitors (1.25 μg/mL leupeptin, 2 μg/mL antipain, 0.25 μg/mL chymostatin, 0.25 μg/mL elastinal, 5 μg/mL pepstatin, e.g., "Complete EDTA-free," Roche); add 1 mg/mL lysozyme, 0.1% Triton X-100 (keep as 10% [v/v] solution at room temperature), and 0.1 mg/mL DNAse, as well as 50% glycerol, as indicated; pre-cool to 4°C before use.
5. Buffer B: Buffer A plus 1 M NaCl.
6. Sonifier (e.g., Branson)
7. Cellulose acetate membranes, 0.22 μm pore size (e.g., Millipore).
8. MonoS HR 10/10 cation-exchange column for fast protein liquid chromatography (FPLC), 6 mL (e.g., Pharmacia).
9. Low-molecular-weight calibration kit for gel filtration (e.g., GE Healthcare) consisting of marker proteins for determination of absolute protein amounts.

2.3.2. Cytochrome-b_2-DHFR-Fusion Proteins Containig a C-Terminal SNAP-Tag

1. LB medium (**Subheading 2.3.1.**) supplemented with 1 mM ZnCl$_2$ (*see* **Note 3**).
2. Isopropyl β-D-thiogalactoside (IPTG; **Subheading 2.3.1.**).
3. Buffer A: 50 mM Tris-HCl, pH 7.0, 50 mM NaCl, 1 mM ZnCl$_2$, 0.02% β-mercaptoethanol, 1 mM PMSF (*see* **Note 2**); where indicated add 0.5% and 2% of Triton X-100, respectively, 0.1 mg/mL DNAse, and 8 M urea.
4. Buffer B: Buffer A plus 1 M NaCl.
5. Sonifier (e.g., Branson).
6. MonoS HR 10/10 cation-exchange column for FPLC, 6 mL (e.g., Pharmacia).
7. Low-molecular-weight calibration kit for gel filtration (*see* **Subheading 2.3.1.**).
8. Labeling buffer: 50 mM Tris-HCl, pH 7.0, 1 mM ZnCl$_2$, 50 mM NaCl, 2 mM DTT (*see* **Note 1**), 1 μM BG-FL, SNAP substrate(Covalys Biosciences).
9. Fluorescence scanner (e.g., GE Healthcare), ImageQuant Software (GE Healthcare).

2.4. In Vitro *Import of Recombinant Precursor Proteins Into Isolated Mitochondria*

2.4.1. Import of DHFR-Fused Cytochrome b_2 Precursors

1. Isolated mitochondria in SEM buffer (*see* **Subheading 2.1.**): 10 mg/mL protein content, freshly prepared or thawed on ice.

2. Preprotein to be imported, freshly prepared, or thawed on ice.
3. Import buffer: 250 mM sucrose, 80 mM KCl (*see* **Note 4**), 5 mM MgCl$_2$, 2 mM KH$_2$PO$_4$, 10 mM MOPS-KOH, pH 7.2, 3% (w/v) fatty-acid free BSA, store in ten 20 mL aliquots at –20°C.
4. 0.2 M NADH, prepare fresh, keep solution on ice.
5. 0.2 M ATP, store as stock solution after pH adjustment to 7.0 with 5 M KOH in working aliquots at –20°C, stable for years, do not refreeze.
6. 1 M potassium phosphate buffer, pH 7.2 (*see* **Subheading 2.2., item 2**).
7. 100x inhibitor mix (*see* **Subheading 2.2.,item 5**).
8. 100x valinomycin (*see* **Subheading 2.2.,item 6**).
9. 0.1 M PMSF in ethanol (*see* **Note 2**).
10. 2.5 mg/mL Proteinase K (PK, specificity: hydrophobic and aromatic residues), dissolved in SEM buffer, store in working aliquots at –20°C, stable for years.
11. Import samples need to be incubated on ice for various experimental purposes; we use metal blocks with holes corresponding to the diameter of Eppendorf tubes, placed on ice.

2.4.2. Import of Cytochrome-b$_2$-DHFR-Fusion Proteins Containing a SNAP-Tag

Materials are the same as listed in **Subheading 2.4.1, items 1–11**, except for one variation: Import buffer is supplemented with 1 mM ZnCl for the import reaction (*see* **Note 3**).

2.4.3. Import of Urea-Denatured Precursor Proteins

1. Cold-saturated ammonium sulphate, pH 7.2; store at 4°C, stable for years.
2. Urea buffer: 7 M urea, 30 mM MOPS-KOH, pH 7.2, 1 mM DTT (*see* **Note 1**), prepare fresh, and keep on ice.

2.5. Analysis of the Import Reaction

2.5.1. Quantification of Western blot Signals for Import of b$_2$(△)-DHFR

TotalLab 1D Software (GE Healthcare)

2.5.2. Quantification of Fluorescence Signals for Import of b$_2$(△)-DHFR-SNAP

1. Fluorescence scanner (*see* **Subheading 2.3.2.**).
2. ImageQuant Software (*see* **Subheading 2.3.2.**).

2.6. Dissection of a Functional Import Intermediate on BN-PAGE

2.6.1. Gel Preparation and Running Settings for BN-PAGE

1. ATP-regenerating system: creatine kinase, 10 mg/mL, prepare fresh; creatine phosphate, store as 1 M stock at –20°C, stable for years.
2. 5% Digitonine in water, prepare a 5% stock solution in water (*see* **Note 5**), keep at 4°C, stable for months.
3. Solubilization buffer: 1% digitonine (*see* **Note 5**), 20 mM Tris/HCl, pH 7.4, 0.1 mM EDTA, 50 mM NaCl, 10% glycerol, 1 mM PMSF. Prepare a 2x solution w/o digitonin, and add detergent prior to use.
4. Acrylamide solution: 49.5% Acrylamide (e.g., Roth), 1.5% *bis*-acrylamide (e.g., Serva Electrophoresis GmbH), store at room temperature in the dark, stable for months.
5. 3x Gel buffer: 200 mM ϵ-amino-*n*-caproic acid, 150 mM Bis-Tris-HCl, pH 7.0, store at 4°C, stable for months.
6. 10x Loading dye: 5% Coomassie blue G250 (water soluble, whereas Coomassie R is not), 500 mM ϵ-amino-*n*-caproic acid, 100 mM Bis-Tris-HCl, pH 7.0. Prepare 10 mL and store at 4°C, stable for months.
7. 10x Anode buffer: 500 mM Bis-Tris-HCl, pH 7.0, store at 4°C, stable for months; 1x solution can be used up to three times.
8. 10x Cathode buffer: 500 mM tricine, 150 mM Bis-Tris-HCl, pH 7.0, prepare 1 L, divide into halves, and dissolve 0.2% Coomassie blue G250 in 500 mL.
9. Marker protein set (e.g., Sigma).
10. Gradient mixer (e.g., Roth), pump (e.g., Gilson), gel system (e.g., SE 600 Series, Amersham Pharmacia Biotech).

2.6.2. Generation of the TOM–TIM-Supercomplex

Materials are the same as listed in **Subheading 2.4.1.** with the following additions:

1. Import buffer as described in **Subheading 2.4.1.** w/o BSA.
2. MTX: keep as 10 mM solution in 0.1 M MOPS-KOH, pH 7.2, store at –20°C, stable for years; dilute in import buffer w/o BSA as indicated.

2.7. Determination of Matrix-Directed Import Forces Exerted by mtHsp70

1. Isolated mitochondria in SEM buffer (*see* **Subheading 2.1.**): 10 mg/mL protein content, freshly prepared or thawed on ice.
2. Import buffer (*see* **Subheading 2.4.1.**, **item 3**).
3. 0.1 mM MTX in import buffer w/o BSA, store at –20°C, stable for years.
4. 0.2 M NADH, prepare fresh, keep on ice.

5. 0.2 M ATP (see **Subheading 2.4.1.**, item 5).
6. 1 U/μl apyrase (e.g., Sigma-Aldrich), store at –20°C, stable for years.
7. 10 mM oligomycin (see **Subheading 2.2.**, item 5).
8. ATP-regenerating system (see **Subheading 2.6.1.**, item 1).
9. Buffer R1 (-ATP): 20 μM oligomycin, 5 μM MTX, in P80 w/o BSA, prepare fresh.
10. Buffer R2 (+ATP): 2 mM ATP, 20 mM creatine phosphate, 200 μg/mL creatine kinase, 5 μM MTX, prepare fresh.
11. 100x inhibitor mix (see **Subheading 2.2.**, item 5).
12. 100x valinomycin (see **Subheading 2.2.**, item 5).
13. 0.1 M PMSF (see **Note 2**).
14. 2.5 mg/mL PK (see **Subheading 2.4.1.**, item 10).

Miscellaneous: Buffers, solutions, and equipment for SDS-PAGE, Western blot and immunodecoration.

3. Methods

All centrifugation steps are to be carried out at 4°C, unless stated otherwise.

3.1. Growth of Yeast Cells and Isolation of Mitochondria From Yeast Cells

For the isolation of mitochondria, the contents of yeast cells from mid-log phase are first made accessible by an enzymatic digestion of the cell walls. The resulting spheroplasts are then physically disrupted by use of a glass Teflon homogenizer. The isolation of an enriched mitochondrial fraction can be achieved by differential centrifugation as the different cellular components vary in their densities.

1. Prepare a preculture by inoculation of 100 mL YPG medium with yeast cells from a plate, and incubation overnight at 30°C with shaking.
2. Dilute cells from the preculture into 1.5 to 2 L of YPG medium in 5-L flasks to an OD_{600} of 0.05 to 0.1. Inoculate the culture at 30°C with shaking until it reaches an OD_{600} of 2 to 2.5.
3. Harvest the cells at room temperature by centrifugation at 3,000g for 5 min. Determine the weight of a fresh centrifugation bottle. Resuspend cells in distilled water, transfer them to the fresh centrifugation bottle, and pellet as above. Discard the water and determine the wet weight of the cells.
4. Resuspend the cell pellet in prewarmed DTT buffer (2 mL/g cell wet weight) and incubate with gentle agitation at 30°C for 30 min.
5. Centrifuge cells at 3000g for 5 min and resuspend the pellet in Zymolase buffer.
6. Spin down cells as in **step 5**, and resuspend the pellet thoroughly in Zymolase buffer (7 mL/g cell wet weight).
7. Dissolve 4 mg Zymolase per gram wet weight cells in a small aliquot of Zymolase buffer.

8. Add Zymolase to cell suspension and incubate at 30°C for 45 to 60 min. Incubate with gentle agitation.
9. Spin down spheroblasts at 2,000g for 10 min and wash pellet in Zymolase buffer (7 mL/g cell wet weight) by centrifugation at 2,000g for 10 min.

Pre-cool centrifugation bottles and rotors to be used from now to 4°C and perform all following steps on ice to avoid proteolysis.

10. Resuspend pellet in homogenization buffer (6.5 mL/g cell wet weight).
11. Homogenize spheroblasts with 15 up-and-down-passes using a glass Teflon homogenizer. Add previous volume of homogenizing buffer after homogenization, and keep on ice for 10 min with occasional shaking.
12. Centrifuge homogenate at 1,500g for 2 min, directly followed by centrifugation at 4,000g to pellet cell debris and nuclei.
13. Discard pellet and isolate mitochondria-containing fraction by centrifugation at 12,000g for 15 min.
14. Resuspend pellet with 1-mL-pipet tips in SEM buffer and centrifuge at 4,000g for 10 min.
15. Discard pellet and centrifuge supernatant at 12,000g for 10 min.
16. Resuspend pellet carefully in approx 0.2 to 0.5 mL of SEM buffer (depending on starting culture volume: 1 L of culture volume renders approx 10 to 20 mg of mitochondria. The yield depends on yeast strain and growth conditions; if using 20 to 200-μL pipet tips, cut off about 2 mm of the tip to avoid disruption of mitochondria).
17. Determine the protein concentration, then dilute mitochondria to a final concentration of 5 to 10 mg protein per milliliter SEM buffer. Aliquot the mitochondrial suspension in working volumes to avoid repeated freeze–thawing, flash-freeze samples in liquid N_2, and store them at –80°C.

3.2. Measurement of the mitochondrial membrane Potential by Fluorescence Spectrometry

The assessment of the membrane potential ($\triangle\psi$) by use of a cyanine dye is based on the changes in fluorescence emission in response to a membrane potential *(5,6)*. Upon addition of mitochondria to diSC$_3$(5)-containing buffer, the dye is potential-dependently partitioned between mitochondria and surrounding medium. The fluorescence emission from mitochondria-associated diSC$_3$(5) is decreased (quenched) with increasing $\triangle\psi$. This fluorescence quenching can be reverted by depolarization of the mitochondrial membranes. $\triangle\psi$ is reflected by the restoration in diSC$_3$(5) fluorescence emission at 670 nm, in response to the addition of uncoupling agents (*see* **Fig. 2**).

1. Add 3 mL of potential buffer to the cuvette.
2. At this step, you may need to add substrates to a final concentration of 5 m*M* each (*see* **Note 6**).

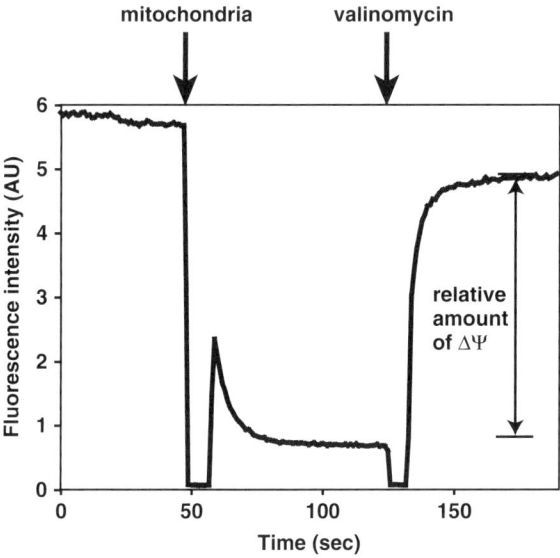

Fig. 2. Assessment of the membrane potential $\Delta\psi$ by use of the fluorescent dye diSC$_3$(5). Mitochondria are added to diSC$_3$(5)-containing potential buffer where indicated. A decrease of the fluorescent signal is observed due to uptake of the dye by the mitochondria and subsequent quenching of the fluorescence. After the signal has stabilized, uncoupling of the respiratory chain and dissipation of the membrane potential is achieved by addition of the K$^+$-ionophore valinomycin. The relative amount of the membrane potential ($\Delta\psi$) is reflected by the restoration of diSC$_3$(5)-fluorescence after addition of valinomycin.

3. Add 3 μL diSC$_3$(5), mix by closing the cuvette with a piece of laboratory film (e.g., Parafilm) and inverting once. Start your measurement using the following settings:
   - excitation wavelength 622 nm
   - emission wavelength 670 nm
   - measuring period of 200 to 250 s
   - response interval 1 s
   - Band pass filter 2 (for both excitation and emission)
4. When diSC$_3$(5) fluorescence has reached a stable value (~40–50 s), add 3 to 5 μL mitochondria, mix, and continue measurement until the decrease in fluorescence reaches a stable value.
5. Add 3 μL of 100x inhibitor mix to uncouple respiratory chain complexes, and trace fluorescence changes until the increase has reached its maximum again (*see* **Note 7**).

3.3. Preparation of Precursor Proteins From E.coli

Authentic cytochrome b_2 reaches the intermembrane space (IMS) of mitochondria via a stop-transfer mechanism that comprises the insertion of N-terminal segments into the matrix followed by a retranslocation and processing event in the IMS. The fusion proteins described here consist of the N-terminal 167 residues of the cytochrome b_2 precursor fused to the complete mouse DHFR *(8,9)*. The b_2-sequence encompasses the complete mitochondrial targeting signal but lacks residues 47 to 65 of the intermembrane spacing sorting signal (*see* **Fig. 3A**). Thus, the protein is targeted to the mitochondrial matrix instead of being exported to the IMS.

For isolation of $b_2(167)_\Delta$-DHFR, *E.coli* cell lysates are purified on an ion-exchange column, from which the protein is eluted by application of a salt gradient (0–0.5 M NaCl).

Because the solubility of $b_2(167)_\Delta$-DHFR-SNAP is significantly lower, the resulting inclusion bodies have to be solubilized and washed in urea-containing buffer. Refolding of the fusion protein to its native state is performed on an ion-exchange column by sequential washes with buffers containing decreasing urea concentrations. For subsequent labelling of the purified protein using the fluorescent dye BG-FL, a folded and enzymatically active SNAP-domain is required.

3.3.1. Purification of Cytochrome-b_2 Fusion Proteins Containing a C-Terminal DHFR Domain

1. Grow *E. coli* cells expressing the fusion protein $b_2(167)_\Delta$-DHFR at 37°C in LB medium to an OD_{600} of 0.6–0.8.
2. Add IPTG to a final concentration of 1 m*M*, and continue growth at 30°C for additional 2 to 3 hours.
3. Harvest cells by centrifugation at 3,500*g* for 8 min.
4. Wash cell pellet by resuspension in pre-lysis buffer and centrifugation as in **step 3**.
5. Resuspend cells in buffer A with 0.1 g cell pellet per milliliter lysis buffer. Destabilize cell walls by freezing the suspension in liquid N_2 and subsequent thawing at room temperature two times. Add lysozyme to a final concentration of 1 mg/mL, and 0.1 mg/mL DNAse, and incubate on ice for 10 min.
6. Add 0.1% (v/v) Triton X-100 and solubilize cells on ice for 10 min.
7. Disrupt cells by sonication on ice: three times 20 pulses (chill cells on ice briefly between pulse intervals), 40% duty cycle, micro-tip setting 7.
8. Pellet cell debris by centrifugation at 15,000*g* for 20 min. Take samples of both pellet and supernatant (lysate) to analyze lysis efficiency via SDS-PAGE.
9. Pour crude cell lysate over a cellulose acetate filter membrane. Take sample of filtered lysate for SDS-PAGE analysis.

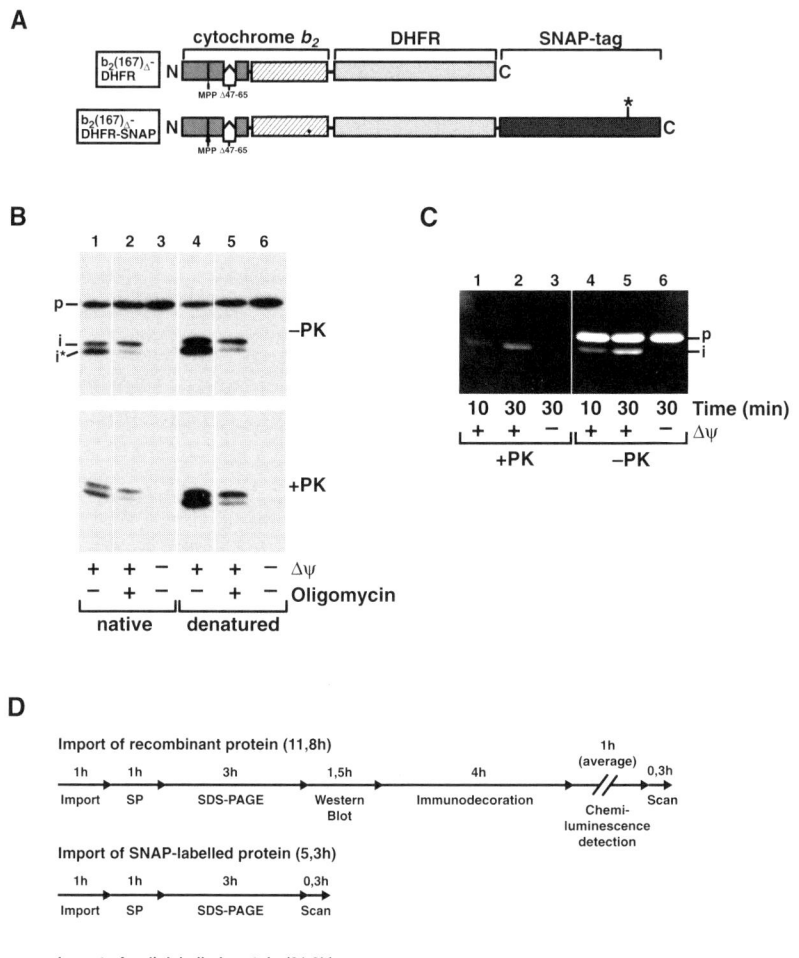

Fig. 3. Import of recombinant presequence-containing DHFR-fusion proteins. (**A**) Schematic drawing of reporter proteins $b_2(167)_\Delta$-DHFR and $b_2(167)_\Delta$-DHFR-SNAP representing fusion constructs of the N-terminal 167 residues of the cytochrome b_2 precursor and the complete mouse dihydrofolate reductase (DHFR). The b_2 precursor lacks residues 47-65 of the intermembrane space targeting signal. The cleavage site by the matrix-processing peptidase (MPP) is indicated. The lower construct additionally contains the C-terminal SNAP-tag. The site for fluorescence labeling is indicated by an asterisk. (**B**) *In vitro* import of native (lanes 1-3) and urea-denatured (lanes 4-6) pre-$b_2(167)_\Delta$-DHFR into the mitochondrial matrix in dependency of $\Delta\psi$. Lanes 1 and 4 represent a standard import reaction showing the appearance of processed, protease protected forms of the precurser protein. Lanes 2 and 5 show decreased protein import

10. Load-filtered lysate on a MonoS HR 10/10 cation exchange chromatography column. Take a sample of the flow through for SDS-PAGE.
11. Wash column with three column volumes of buffer A.
12. Elute column by addition of 10 column volumes applying a NaCl gradient from 0 to 50% (0.5 M NaCl). Collect eluate in 1-mL fractions, and take a sample each for SDS-PAGE.
13. Check the efficiency of your purification by SDS-PAGE and Western blot analysis using anti-mouse DHFR.
14. To the fractions containing the $b_2(\triangle)$-DHFR fusion protein, add glycerol to a final concentration of 50%, and snap-freeze them in small aliquots in liquid N_2 for storage at $-80°C$.

3.3.2. Purification of Cytochrome-b_2-DHFR Fusion Proteins Containing a C-Terminal SNAP-Tag

1. Grow *E. coli* cells harbouring the plasmid expressing the fusion protein $b_2(167)_\triangle$-DHFR-SNAP at 37°C in LB medium, supplemented with 1 mM ZnCl$_2$ to an OD$_{600}$ of 0.6–0.8.
2. Add IPTG to a final concentration of 1 mM, and continue growth at 30°C for additional 2 to 3 h.
3. Harvest cells by centrifugation at 3,500g for 8 min.
4. Wash cell pellet by resuspension in distilled water and centrifugation as in **step 3**.
5. Resuspend cells in buffer A with 0.1 g cell pellet per millilter lysis buffer. Destabilize cell walls by freezing the suspension in liquid N_2 and subsequent thawing at room temperature two times. Add lysozyme to a final concentration of 1 mg/mL, and 0.1 mg/mL DNAse, and incubate on ice for 10 min.

Fig. 3. (Continued) after inhibition of the F_1F_O-ATP-synthase by oligomycin, which reduces the ATP concentration in the matrix compartment. In lanes 3 and 6, the membrane potential was dissipated by addition of the inhibitors antimycin A, valinomycin and oligomycin, resulting in a complete block of the import reaction. Proteins were separated by SDS-PAGE, transferred to a PVDF-membrane, and immunodecorated with anti-mouse DHFR. The precursor (p) was processed twice (i, i*). In the lower panel, nonspecifically associated preproteins at the outer surface of mitochondria were degraded by treatment with proteinase K (PK). (**C**) Time-dependent *in vitro* import of pre-$b_2(167)_\triangle$-DHFR-SNAP into the mitochondrial matrix. The membrane potential in samples of lanes 3 and 6 has been dissipated by the inhibitor mix (*see* **A**). In lanes 1-3, nonspecifically associated precursor at the outer surface of mitochondria was degraded by treatment with PK. (**D**) Timelines for import and analysis of recombinant precursors, SNAP-tagged precursors, and radiolabeled precursors. Given are the time periods for each individual step, as well as the overall time of the assays presented. SP, sample processing for SDS-PAGE.

6. Pellet cell debris by centrifugation at 15,000g for 20 min.
7. Resuspend pellet containing the inclusion bodies in buffer A containing 0.5% (v/v) Triton X-100 and take a sample for SDS-PAGE. Centrifuge at 6,000g for 10 min.
8. Resuspend pellet in buffer A containing 2% (v/v) Triton X-100 and centrifuge as in **step 7**. Take a sample for SDS-PAGE.
9. Wash out Triton X-100 by resuspending pellet in buffer A w/o Triton X-100 and centrifugation as in **step 7**.
10 Resuspend the final pellet in buffer A containing 8 M urea. Take a sample for SDS-PAGE.
11. Load the protein solution on a MonoS HR 10/10 cation exchange column. Take a sample of the flow through.
12. Wash out urea over 15 column volumes of buffer A containing decreasing urea concentrations.
13. Elute column by addition of 10 column volumes applying a NaCl gradient from 0 ot 50% (0.5 M). Collect eluate in 1-mL-fractions, and take a sample each for SDS-PAGE.
14. Check the efficiency of your purification by SDS-PAGE.
15. Label purified protein by incubation in labelling buffer at 25°C in the dark for at least 1 h.

3.4. In Vitro *Import of Recombinant Precursor Proteins Into Isolated Mitochondria*

A standard import reaction contains 2 mM ATP, 4 mM NADH, as well as 10 mM potassium phosphate, and 25 µg of mitochondria in a total volume of 100 µL import buffer. Substrate-saturating conditions are reached by an addition of at least 10-fold excess of recombinant preprotein over available import sites. Based on the assumption of 0.025 pmol import sites per microgram of yeast mitochondria *(9)*, this corresponds to 6.25 pmol preprotein for a standard reaction (in the case of the described preprotein $b_2[167]_\triangle$-DHFR about 240 ng per reaction). After terminating the import reaction by dissipation of the membrane potential, nonspecifically bound and nonimported precursors are degraded by addition of PK. Further analysis requires Western transfer and immunodecoration, using anti-mouse DHFR antiserum. In the case of $b_2(\triangle)$-DHFR-SNAP, data analysis is done by use of a fluorescence scanner. **Figure 3B,C** shows the results from *in vitro* import of $b_2(167)_\triangle$-DHFR, and $b_2 (\triangle)$-DHFR-SNAP, respectively, gained by immunodecoration, and fluorescence scanning, respectively, under different import conditions. The direct detection of imported preproteins by fluorescence is highly sensitive and allows a reliable quantification. In addition, the overall experimental time is greatly reduced. The average time periods for the individual steps, as well as the overall times for the import assays described in this chapter, in comparison to the import of radiolabeled precursors, are given in **Fig. 3D**.

3.4.1. Import of DHFR-fused b_2 precursors

1. Thaw 17 μL (170 μg) mitochondria and 7 μL ATP on ice.
2. Prepare an import master mix for six time points on ice with a total volume of 660 μL, minus 6.6 times the amount of precursor added (*see* **Note 8**): 13.2 μL NADH, 6.6 μL 1 M potassium phosphate buffer, pH 7.2, ad 660 μL import buffer, minus the amount of mitochondria, and precursor added. Add 16.5 μL (165 μg) mitochondria, mix carefully by inverting the tube and spin down at low speed for 2 to 3 s (*see* **Note 9**). Take one 100 μL aliquot (smaple 6), and add 5.6 μL ATP to remaining master mix.
3. Divide your master mix into five portions of 100 μL, minus the amount of precursor added.
4. Dissipate the membrane potential in mitochondria of sample 6 by addition of 1 μL inhibitor mix from 100x stock; add 1 μL ethanol to samples 1-5 (mock). Mix carefully by flicking the tubes with your index finger.
5. Pre-incubate samples at 25°C for 3 min, add 240 ng precursor (corresponding to 6.25 pmol of b_2 (167)$_\Delta$-DHFR) to sample 6, mix, and leave at 25°C for the entire import incubation period.
6. Initiate import reaction by addition of precursor to sample 5 (longest incubation period) and subsequent precursor addition to samples 4-1 (with 1 being the shortest incubation period). Mix after addition by flicking the tubes.
7. Stop import reactions according to time scheme by addition of 1 μL valinomycin from 100x stock. Vortex sample shortly and move it on ice.
8. Prepare six fresh tubes (samples 7-12) with 1 μL PK (f.c.: 50 μg/mL) in 50 μL SEM buffer.
9. Add 50 μL each of samples 1-6 to PK-containing samples 7-12 and incubate for 15 min on ice; add 50 μL SEM buffer to samples 1-6 and likewise keep on ice.
10. Stop PK treatment by addition of 1 m*M* PMSF from 0.1 *M* stock (*see* **Note 2**); for equal treatment of all samples, add PMSF to non-PK-treated samples as well.
11. Keep samples on ice for other 10 min, before pelleting mitochondria by centrifugation at 12,000*g* for 10 min.
12. Remove supernatants carefully, wash mitochondrial pellet by addition of 150 μL SEM buffer containing 1 m*M* PMSF (*see* **Note 2**), and centrifugation at 12,000*g* for 8 min.
13. After removal of supernatants denature mitochondria in 1x Laemmli buffer.
14. Run samples on a polyacrylamide gel with a resolving gel concentration of 12%. Always run three control lanes with 1, 2, and 5% of the preprotein amount used in the assay for quantification.
15. Transfer proteins to a Western blot membrane (e.g., Millipore)
16. Immunodecorate with antibodies directed against mouse DHFR.
17. Evaluate raw data with the TotalLab 1D software (*see* **Subheading 2.3.2.**).

3.4.2. Import of Cytochrome-b_2-DHFR-Fusion Proteins Containing a SNAP-Tag

1. Import is performed as described in **Subheading 2.4.1,** with addition of 1 mM ZnCl to the import reaction (*see* **Note 3**).
2. For analysis of the gel using a fluorescence scanner, cover gel with a transparent plastic sheet. Analysis of raw data can be done using the ImageQuant Software (*see* **Subheadings 2.3.2.** and **3.5.2.**).

3.4.3. Import of Urea-Denatured Precursor Proteins

The import of precursor proteins in a denatured state can be useful for the analysis of mitochondrial import defects in more detail. Denaturation of the preprotein to be imported facilitates its import, as the import reaction is no longer dependent on the functionality of the mtHsp70 system for unfolding. ATP-driven precursor unfolding is of particular importance for the import of tightly folded proteins or protein domains. Thus, comparison of the import efficiencies of both natively folded and denatured preproteins in mitochondrial preparations from different strains can have implications both for the tertiary structure of the preprotein imported, and for the mtHsp70 protein (un)folding capacity.

3.4.3.1. DENATURATION OF PRECURSOR PROTEIN

1. Thaw the preprotein to be imported on ice.
2. Add three volumes of cold-saturated $NH_4(SO_4)_2$ and mix by vortexing at low speed.
3. Incubate on ice for 30 min with occasional vortexing.
4. Spin precipitates down at 20,000g for 15 min.
5. Take off supernatant carefully, and resuspend protein pellet in required volume of urea buffer by shaking at room temerature on a thermomixer at maximum speed for 15 min.
6. Spin down nondenatured protein material by centrifugation at 20,000g for 5 min
7. Transfer supernatant to a fresh tube and keep on ice until use.

3.4.3.2. IMPORT OF UREA-DENATURED PRECURSOR PROTEINS

Import is performed as described in **Subheadings 3.4.1.** and **3.4.2.**, respectively.

3.5. Analysis of the Import Reaction

3.5.1. Quantification of Western blot Signals for Import of $b_2(\triangle)$-DHFR

Evaluation of raw data can be done with the TotalLab 1D software provided that the signals have not saturated. Determination of the import efficiency is done from the protease-treated samples. Use the control lanes as reference.

3.5.2. Quantification of Fluorescence Signals for Import of $b_2(\triangle)$-DHFR-SNAP

1. Analysis of import and folding efficiency is done directly in the SDS gel using a fluorescence scanner. Keep gel wet during scanning procedure for subsequent check of protein levels by Coomassie staining. Fluorescence settings are as follows:
   - fluorescence mode
   - photomultiplier voltage of 600 V
   - normal sensitivity
   - excitation wavelength 532 nm

2. ImageQuant software (*see* **Subheading 2.3.2.**) can be used to evaluate obtained scanning data.

3.6. Dissection of a Functional Import Intermediate on BN-PAGE

The ligand MTX stabilizes the tertiary structure of the DHFR domain and thereby prevents translocation of fusion proteins containing a C-terminal DHFR domain across the mitochondrial membranes *(14,15)* When $b_2(\triangle)$-DHFR is imported in the presence of MTX, it accumulates in a translocation intermediate that spans both mitochondrial membranes. Determination of the constituents of this complex by BN-PAGE revealed the presence of TOM components, as well as the core components of the TIM23 complex of the inner membrane. The translocase complexes are connected to each other by the preprotein in transit resulting in the formation of a large multiprotein complex comprising both outer and inner membrane components *(9,11)*. Formation of the TOM–TIM-supercomplex is dependent on the membrane potential across the inner membrane, and results in an almost quantitative shift of the TIM23 core subunits from a size of approx 90 kDa to a size of approx 600 kDa, on BN-PAGE (*see* **Fig. 4A**).

The accumulation of $b_2(167)_\triangle$-DHFR in import sites is achieved by preincubation of the preprotein in the presence of 5 µM MTX, and addition of MTX in the import reaction, which results in the stable arrest of the preprotein inside the import channels. Formation of the supercomplex can be monitored by Western blot using anti-Tim23 antiserum. BN-PAGE is specifically suited for the analysis of membrane bound protein complexes *(13)*. The BN technique includes the solubilization of mitochondria by the non-ionic detergent digitonine (*see* **Note 10**), and subsequent substitution of the detergent for water-soluble Coomassie blue. Thereby, proteins become negatively charged, and are thus separated mainly by molecular size in the electrophoresis.

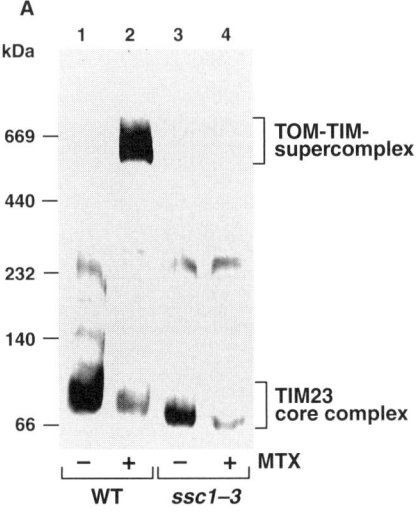
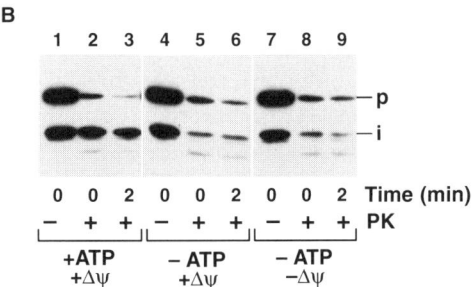

Fig. 4. Generation and analysis of import intermediates. (**A**) Dissection of the TOM–TIM supercomplex on Blue Native gel electrophoresis (BN-PAGE). Wild-type (WT; lanes 1 and 2) and mtHsp70 mutant (*ssc1-3*; lanes 3 and 4) mitochondria were incubated with pre-$b_2(167)_\Delta$-DHFR in the presence (lanes 2 and 4) or absence (lanes 1 and 3) of 5 µM methotrexate (MTX). Addition of MTX resulted in the accumulation of the preprotein in a location spanning both mitochondrial membranes. Mitochondria were extracted with the mild detergent digitonin and protein complexes resolved by BN-PAGE under native conditions. Analysis by Western blot using anti-Tim23 antibodies indicate the formation of the TOM–TIM23 supercomplex at a size of approx 600 kDa only in the presence of MTX (compare lanes 1 and 2). The temperature-sensitive mtHsp70-mutant *ssc1-3* is not able to support the completion of the translocation at nonpermissive conditions. Hence, the supercomplex is not observed in *ssc1-3* mutant mitochondria (see lanes 3 and 4) (**B**) Assessment of the inward-directed import force exerted by mtHsp70. Mitochondria were incubated in the presence of pre-$b_2(167)_\Delta$-DHFR and 5 µM MTX for 10 min. A sample was taken directly after the incubation (lanes 1, 4, and 7). Another sample was taken directly after proteinase K (PK) treatment (lanes 2, 5, and 8). A third sample was taken after a 2-min incubation following PK

1. Dilute MTX stock 200-fold (f.c.: 50 mM) in import buffer w/o BSA.
2. Incubate the appropriate amount of preprotein with 5 μM MTX for 10 min at 25°C.
3. Set up an import reaction (*see* **Subheading 3.4.1.**) for the accumulation; add creatine kinase, and creatine phosphate to a concentration of 200 μM, and 20 mM, respectively. Add 5 μM MTX.
4. Initiate accumulation by addition of a 10-fold excess of MTX-coupled precursor (240 ng, corresponding to 6.25 pmol in the case of $b_2(167)_\Delta$-DHFR), and incubate for 15 min at 25°C.
5. Stop reaction by addition of 1x inhibitor mix, and move samples on ice.
6. Pellet mitochondria by centrifugation at 12,000g for 10 min.
7. Wash mitochondria by addition of 100 μL SEM buffer and centrifugation at 12,000g for 8 min.
8. Solubilize mitochondrial pellet in 55 μL ice-cold digitonine buffer by pipeting up and down carefully with a 20 to 200-μL-pipet tip for 15 to 20 times. Try to avoid digitonine foam. Keep on ice for 15 min.
9. Spin down unsolubilized material by centrifugation at 20,000g for 15 min.
10. Prepare fresh tubes with 5 μL 10x loading dye, to which 45 μL of solubilized material (supernatants) is added after centrifugation. Take 5 μL of remaining supernatants each, and analyse on SDS-PAGE in parallel.
11. Load samples on a 6 to 16.5% BN gel (*see* **Table 1** and **Note 11**). Cover samples carefully with 1x cathode buffer containing 0.2% Coomassie blue G250. Gel run is performed in a cooled chamber (4°C). Keep the voltage at 100 to 150 V for passage through the stack gel, afterward it can be risen up to 600 V. The current should be limited to 15 mA during the entire gel run.
12. After the proteins have migrated into the seperation gel, the Coomassie-containing cathode buffer is replaced by the one without dye (excess Coomassie might interfere with subsequent Western transfer of the proteins).
13. Before Western transfer, the gel is incubated in 1x SDS-PAGE running buffer for 5 min (facilitation of protein transfer).
14. Decorate BN-membrane for Tim23 using TBS(T), supplemented with 5% milk powder (*see* **Note 12**). The SDS Western membrane is decorated against anti-mouse DHFR.

Fig. 4. (Continued) treatment (lanes 3, 6, and 9). In the presence of a membrane potential (+Δψ) and high ATP levels in the matrix (+ATP), an active import motor rendered 20 to 30% of the accumulated intermediates resistant to protease treatment (lanes 2 and 3). Depletion of the membrane potential (-Δψ) or removal of ATP (-ATP) inactivated the translocation machinery resulting in a rapid and complete loss of protease rcsistance.

Table 1
Pipeting Scheme for Blue Native Gel Preparation

% Acrylamide	6	8	10	13	16.5	Stack Gel
3x gel buffer (mL)	3	3	3	3	3	2.5
37% Acrylamide (mL)	1.07	1.46	1.82	2.35	3.05	0.6
Glycerol (mL)	–	–	1.8	1.8	1.8	
Water (mL)	4.888	4.507	2.347	1.817	1.117	4.367
10% APS (µL)	38	38	38	38	38	30
TEMED (µL)	3.8	3.8	3.8	3.8	3.8	3

APS, ammonium persulfate solution; TEMED, tetramethyl-ethylenediamin

3.7. Assessment of the Inward-Directed Translocation Force

MTX-stabilized cytochrome b_2-DHFR fusion protein is accumulated in mitochondria *(9,11)*.

The interaction of the precursor protein inserted into the import channels with the functional import motor is required to generate the matrix-directed import force driving polypeptide movement and unfolding *(16)*. In case of MTX-stabilized $b_2(\triangle)$-DHFR translocation intermediates, the protease-resistant DHFR domain is pulled tightly against the outer mitochondrial membrane by the activity of the import motor. The close apposition of stable DHFR and the outer membrane prevents the access of external proteases to the preprotein translocation intermediates. However, resistance against proteases is achieved only in the case of an active ongoing translocation reaction. Therefore, the quantification of the protease resistance of translocation intermediates is an indirect measure for the import driving activity of the translocation machinery. In combination with the protease treatment, the membrane potential and the ATP conditions can be manipulated to assess the role of the mitochondrial energy sources and the activity of the mtHsp70 import motor in the import process (*see* **Fig. 4B**).

1. Thaw 55 µL (550 µg) mitochondria and 21 µL ATP on ice. Prepare 45 µL 0.2 M NADH and keep on ice.
2. Prepare a standard import master mix with a total volume of 2,100 µL on ice: 21 µL ATP, 42 µL NADH, 21 µL 1 M potassium buffer, pH 7.2, and add up to 2,100 µL import buffer minus the volume of mitochondrial suspension and MTX added. Add 52.5 µL mitochondria, and 5 µM MTX (21 µL from 100x stock solution in import buffer). Mix carefully by inverting the tube and spinning at low speed for 2 to 3 s.
3. Divide the master mix into two portions of 1,000 µL each.

4. Dilute the precursor protein in import buffer w/o BSA to a concentration of 0.7 pmol/μL in a total volume of 230 μL. Add 5 μM MTX, and incubate for 5 min on ice.
5. Start accumulation of translocation intermediates: add 110 μL of precursor to 1,000 μL import master mix, and incubate for 10 min at 25°C. Place on ice afterward.
6. Depletion of ATP: add 11 μL apyrase (f.c.: 0,01 U/μL) and 11 μL oligomycin (f.c.: 20 μM) to one of the 1,000 μL import reactions. The other is treated as mock with 11 μL ethanol. Incubate for 15 min on ice.
7. Re-isolate mitochondria by centrifugation at 12,000g for 10 min.
8. Resuspend mitochondrial pellet carefully in 1100 μL buffer R1 (w/o ATP) or 1,100 μL buffer R2 (w/ ATP), respectively.
9. Divide each of the samples in 2× 500 μL to achieve the following incubation conditions:

    A: +△ψ / -ATP
    B: -△ψ / -ATP
    C: +△ψ / +ATP
    D: -△ψ / +ATP

10. Start incubation of samples A through D at 25°C, take samples 1 and 2 (100 μL each of samples A–D, corresponding to 25 μg mitochondria) immediately, and put on ice.
11. Dissipate the membrane potential in mitochondria of samples B and D by addition of 5 μL inhibitor mix, and mix carefully by flicking the tubes.
12. Add 5 μl ethanol to samples A and C (*mock*) and likewise mix.
13. Sample 3 is taken after 2 min, and sample 4 after 15 min (100 μL each). Keep all samples on ice.
14. Subsequent PK-treatment is performed by addition of 2 μL PK (f.c. 50 μg/mL) to samples 2, 3, and 4 and incubate for 15 min on ice.
15. Stop PK treatment by addition of 1 mM PMSF (*see* **Note 2**); for equal treatment of all samples add PMSF to non-PK-treated samples as well. Keep samples on ice for further 10 min, pellet mitochondria by centrifugation at 12,000g for 10 min.
16. Remove supernatants and wash mitochondrial pellet by addition of 150 μL SEM buffer, containing 1 mM PMSF, and centrifugation at 12,000g for 8 min.
17. Remove supernatants and denature mitochondria in 1x Laemmli buffer.
18. Run samples on a polyacrylamide gel with a resolving gel concentration of 10 to 12% polyacrylamide.
19. Transfer your samples to a PVDF Western blot membrane, and immuno-decorate with antibodies directed against mouse-DHFR.

4. Notes

1. Always prepare a fresh 0.5 M stock solution, and add DTT just prior to use.
2. PMSF addition should always be followed by immediate mixing, otherwise it precipitates. In aqueous solution, PMSF has a half-life of about 30 min. Thus, always

add it just prior to use. Alternatively, isopropanol stocks, which are stable for up to 6 mo, can be used.
3. The SNAP protein requires Zn for folding.
4. Depending on the preprotein, and the mitochondria used, rising the K^+ concentration up to 200 mM might help to increase import fidelity
5. It is more convenient to use already recrystallized digitonine that can be directly dissolved in water (e.g., Calbiochem). Warming up the solution significantly increases the solubility of digitonine. Pellet digitonine precipitates that occur during storage and use the supernatant only in the experiment.
6. Alternatively, 2 mM NADH plus 2 mM ATP might be added. Wild-type mitochondria usually contain sufficient internal substrates (i.e. the addition of substrates will not further increase the fluorescence quenching).
7. Fluorescence intensity will not reach starting (maximal) value again, as you will always have unspecific fluorescence quenching to a certain extent.
8. For preparation of a master mix, it is useful to always use 5 to 10% in excess of the volumes needed.
9. To maintain the full import-competence, do not vortex mitochondria at all before performing the import reaction as this might result in damages of the membranes.
10. Other detergents such as dodecylmaltosite or Triton X-100 may also be suitable.
11. Prepare the BN gel always in advance and keep it at 4°C until it is loaded.
12. Addition of sodium azide to a final concentration of 0.02% (from a 100x stock) and storage at –20°C is useful to preserve the antibody solution for subsequent experiments. Handle sodium azide with care as it is highly toxic.

References

1. Wiedemann, N., Frazier, A. E., and Pfanner, N. (2004) The protein import machinery of mitochondria. *J. Biol. Chem.* **279**, 14473–14476.
2. Strub, A., Lim, J. H., Pfanner, N., and Voos, W. (2000) The mitochondrial protein import motor. *Biol. Chem.* **381**, 943–949.
3. Voos, W., and Röttgers, K. (2002) Molecular chaperones as essential mediators of mitochondrial biogenesis. *Biochim. Biophys. Acta.* **1592**, 51–62.
4. Ryan, M. T., Voos, W., and Pfanner, N. (2001) Assaying protein import into mitochondria. *Methods Cell Biol.* **65**, 189–215.
5. Pena, A., Uribe, S., Pardo, J. P., and Borbolla, M. (1984) The use of a cyanine dye in measuring membrane potential in yeast. *Arch. Biochem. Biophys.* **231**, 217–225.
6. Gärtner, F., Voos, W., Querol, A., et al. (1995) Mitochondrial import of subunit Va of cytochrome-c oxidase characterized with yeast mutants. *J. Biol. Chem.* **270**, 3788–3795.
7. Lim, J. H., Martin, F., Guiard, B., Pfanner, N., and Voos, W. (2001) The mitochondrial Hsp70-dependent import system actively unfolds preproteins and shortens the lag phase of translocation. *EMBO J.* **20**, 941–950.

8. Koll, H., Guiard, B., Rassow, J., et al. (1992) Antifolding activity of hsp60 couples protein import into the mitochondrial matrix with export to the intermembrane space. *Cell.* **68**, 1163–1175.
9. Dekker, P. J., Martin, F., Maarse, A. C., et al. (1997) The Tim core complex defines the number of mitochondrial translocation contact sites and can hold arrested preproteins in the absence of matrix Hsp70-Tim44. *EMBO J.* **16**, 5408–5419.
10. Krayl, M., Guiard, B., Paal, K., and Voos, W. (2006) Fluorescence-mediated analysis of mitochondrial preprotein import in vitro. *Anal. Biochem.* **355**, 81–89.
11. Chacinska, A., Rehling, P., Guiard, B., et al. (2003) Mitochondrial translocation contact sites: separation of dynamic and stabilizing elements in formation of a TOM–TIM-preprotein supercomplex. *EMBO J.* **22**, 5370–5381.
12. Krayl, M., Lim, J. H., Martin, F., Guiard, B., and Voos, W. (2007) A cooperative action of the ATP-dependent import motor complex and the inner membrane potential drives mitochondrial preprotein import. *Mol. Cell. Biol.* **27**, 411–425.
13. Schägger, H. (2001) Blue-native gels to isolate protein complexes from mitochondria. *Methods Cell Biol.* **65**, 231–244.
14. Eilers, M., and Schatz, G. (1986) Binding of a specific ligand inhibits import of a purified precursor protein into mitochondria. *Nature* **322**, 228–232.
15. Rassow, J., Guiard, B., Wienhues, U., Herzog, V., Hartl, F.-U., and Neupert, W. (1989) Translocation arrest by reversible folding of a precursor protein imported into mitochondria. A means to quantitate translocation contact sites. *J. Cell Biol.* **109**, 1421–1428.
16. Voisine, C., Craig, E. A., Zufall, N., von Ahsen, O., Pfanner, N., and Voos, W. (1999) The protein import motor of mitochondria: unfolding and trapping of preproteins are distinct and separable functions of matrix Hsp70. *Cell* **97**, 565–574.

6

In Vitro Import of Proteins Into Isolated Mitochondria

Karl Bihlmaier, Melanie Bien, and Johannes M. Herrmann

Summary

Import of proteins is of vital importance for the biogenesis of mitochondria. The vast majority of mitochondrial proteins is encoded within the nuclear genome and translocated into various mitochondrial compartments after translation in the cytosol as preproteins. Even in rather primitive eukaryotes like yeasts, these are 700 to 1,000 different proteins, whereas only a handful of proteins is encoded in the mitochondrial DNA. *In vitro* import studies are important tools to understand import mechanisms and pathways. Using isolated mitochondria and radioactively labeled precursor proteins, it was possible to identify several import machineries and pathways consisting of a large number of components during the last decades.

Key words: Cell-free protein synthesis; *in vitro*; import; mitochondria; protein translocation; *Saccharomyces cerevisiae*.

1. Introduction

Mitochondria are complex double membrane-bounded organelles that exhibit a huge variety of functions. In addition to their role in cellular respiration, mitochondria participate in biochemical key processes like synthesis of metabolites, lipid metabolism, free radical production, apoptosis, and metal homeostasis. Depending on the organism, mitochondria contain about 500 to 2,000 different proteins. In the yeast *Saccharomyces cerevisiae*, only 8 of these proteins are encoded by mitochondrial DNA. All remaining proteins are encoded in the nucleus and synthesized on cytosolic ribosomes. Following the synthesis as precursor proteins (preproteins) they have to be imported into the organelle (for review *see* **refs.** *1–5*). Mitochondria can be divided into four functionally specialized subcompartments: the outer membrane, the intermembrane space (IMS), the inner membrane, and the matrix. Precursor proteins access one of these different compartments as their final location by the use of specific targeting signals. Targeting signals are recognized by receptors on the surface of the organelle. These receptors are part of the translocase of the outer membrane (TOM) complex. This complex appears to interact with all mitochondrial preproteins and either facilitates their insertion into the outer membrane or their translocation through the outer membrane into the IMS.

Most of the preproteins destined to the inner membrane or the matrix contain cleavable N-terminal targeting sequences, so-called presequences that are hydrophobic on one face and positively charged on the other. These sequences can be predicted on the basis of the primary sequence of proteins by the use of specific algorithms like MITOPRED *(6)*, MitoProt II *(7)* or Predotar *(8)*. Presequences are recognized at the level of the inner membrane by the translocase of the inner membrane (TIM)23 complex. This translocase mediates the translocation across and the insertion into the inner membrane using adenosine triphosphate (ATP) and the membrane potential as an energy source. In the matrix, the presequences are proteolytically removed by the mitochondrial processing peptidase (MPP) and the proteins fold into their native structures by assistance of mitochondrial chaperones.

Some preproteins that are destined for the inner mitochondrial membrane or the IMS contain bipartite targeting sequences. These sequences consist of an N-terminal matrix-targeting signal followed by a hydrophobic stretch which directs the proteins into the inner membrane. Once inserted, the precursors may remain anchored in the inner membrane or are released into the IMS after a further processing event.

Another group of inner membrane proteins is sorted by a second TIM machinery, the TIM22 complex. Substrates of the TIM22 pathway lack presequences and contain targeting information that is scattered throughout the sequence of the mature polypeptide. These proteins, such as members of the metabolite carrier family, cross the TOM channel, are ushered through the IMS by specific chaperones and inserted in the inner membrane by the TIM22 complex in a membrane potential-dependent manner.

Some soluble IMS proteins are imported independently of ATP hydrolysis or the membrane potential. The import of these proteins seems to be driven by the use of cysteine oxidation, which is facilitated by two specialized IMS components, Mia40 and Erv1 *(9)*.

To investigate the different protein import machineries and procedures, *in vitro* studies are an adequate tool. To perform such experiments, preproteins labeled *in vitro* with [^{35}S] can be imported into isolated mitochondria. These assays reveal characteristics and requirements of protein import and where highly effective in the past to characterize the import process of mitochondrial preproteins. In this chapter, a standard protocol for an *in vitro* protein import experiment into isolated mitochondria is provided.

2. Materials

2.1. Synthesis of Preproteins

2.1.1. In Vitro Transcription

1. DNA template: take purified plasmid (1 μg/μL) containing the cloned gene for the mitochondrial precursor protein downstream of the SP6 promotor (*see* **Note 1**).
2. 10x salts buffer: 400 mM HEPES/KOH, pH 7.4, 60 mM Mg acetate, 20 mM spermidine (*see* **Note 2**). Sterilize by filtration and store at −20°C.
3. Premix buffer: 400 μL 10x salts buffer, 8 μL 50 mg/mL essentially fatty acid-free bovine serum albumin (BSA), 40 μL 1 M dithiothreitol (DTT) dissolved in H_2O, 20 μL 0.1 M ATP, 20 μL 0.1 M cytidine triphosphate (CTP), 20 μL 0.1 M uridine triphosphtae (UTP), 4 μL 0.1 M guanosine triphosphate (GTP), 2,690 μL H_2O. Sterilize by filtration and store aliquots at −20°C.
4. RNasin ribonuclease inhibitor (40 U/μL). Store aliquots at −20°C.
5. Methylguanine cap ($m^7G[5']p]pp[5']G$), 2.5 mM final concentration: add 483 μL sterile water to 25 A_{250} units of $m^7G(5')ppp(5')G$ (*see* **Note 3**). Store aliquots at −20°C.
6. SP6 RNA polymerase (25 U/μL). Store at −20°C.
7. LiCl: prepare a 10 M solution in H_2O and store aliquots at −20°C.
8. 70 % (v/v) ethanol, store at −20°C.

2.1.2. In Vitro Translation

1. Take commercially available rabbit reticulocyte lysate. Store at −80°C (*see* **Note 4**).
2. Amino acid mixture, minus methionine (*see* **Note 5**), 1 mM. Store at −20°C.
3. RNasin ribonuclease inhibitor.
4. Mg acetate: prepare a 15 mM solution in water and store at −20°C.
5. Take RNA that was produced in the *in vitro* transcription reaction.
6. [^{35}S]-labeled methionine, specific activity 1,175 Ci/mmol, concentration 10 mCi/mL. Store in aliquots at −80°C (*see* **Note 6**).
7. Cold methionine for the chase: 58 mM in water. Store at −20°C.
8. Sucrose: 1.5 M solution in water. Store at −20°C.

2.2. In Vitro Import Into Isolated Mitochondria

1. 2x Import buffer: 100 mM HEPES/KOH, pH 7.2, 6% fatty acid-free BSA, 1 M sorbitol, 160 mM KCl, 20 mM Mg acetate, 4 mM KH_2PO_4, 2 mM $MnCl_2$ (see Note 7). Store at –20°C.
2. ATP: prepare a 0.2 M solution in water and adjust the pH with KOH to 7.0. This is an 100x stock. Store aliquots at –20°C.
3. Nicotinamide adenine dinucleotide (NADH): 0.2 M solution in water. This is a 100x stock. Keep this in single-use aliquots and store them at –20°C.
4. Valinomycin: make a 0.1 mM solution in ethanol. This is a 100x stock. Store at –20°C (see Note 8).
5. Oligomycin: make a 2 mM solution dissolved in ethanol. This is a 100x stock. Store at –20°C (see Note 9).
6. Proteinase K (PK): make a 10 mg/mL solution of PK dissolved in water. Store in small aliquots at –20°C (see Note 10).
7. Phenylmethylsulfonyl fluoride (PMSF): always prepare a fresh 0.2 M solution of PMSF dissolved in ethanol.
8. SH buffer: 0.6 M sorbitol, 20 mM Hepes/KOH, pH 7.2. Store in 50 mL aliquots at –20°C.
9. SH/KCl buffer: 0.6 M sorbitol, 80 mM KCl, 20 mM Hepes/KOH, pH 7.2.
10. [^{35}S]-labeled mitochondrial precursor protein prepared in the *in vitro* translation reaction. Keep in single-use aliquots of 12 μL at –80°C (see Note 11).
11. Isolated yeast mitochondria at 10 mg/mL kept in single-use aliquots at –80°C (see Note 12).

3. Methods

Radiolabeled preproteins can be easily synthesized in cell-free transcription systems *(10)*. It should be noted that the protein concentrations obtained by this method are very low and in most cases are below the nanomolar range *(11)*. However, because the proteins are typically associated by chaperones they are often highly import competent. This high import competence, and the simple production of these radiolabeled proteins made this method the standard procedure for the generation of preproteins. A protocol for the *in vitro* synthesis of preproteins and a standard import experiment are presented in this chapter. Alternatively, larger concentrations of preproteins (so-called "chemical amounts") can be heterologously expressed in bacteria, purified, and used for the *in vitro* import studies (see Note 13). In order to allow their import, the preproteins need to be unfolded; for example by denaturing them with urea or guanidinium hydrochloride *(12)*. The imported preproteins can be detected by Western blotting which, however, is not very sensitive and difficult to quantify with satisfying reliability. In any case, larger epitope tags on the proteins should be avoided because they may interfere with protein import. Recombinant preproteins can be purified from metabolically radiolabeled bacteria that allows their sensitive and reliable quantification by autoradiography *(13)*.

In general, preproteins are incubated with purified mitochondria. A protocol for the purification of mitochondria from yeast cells is given in Chapter 7 of this book. Following TOM, the preproteins become inaccessible to externally added protease. Moreover, the maturation of preproteins by mitochondrial processing peptidases can be easily detected by sodium dodecyl sulfate-polyacrylamide gel electrophoresis (SDS-PAGE) so that the import reaction can be followed experimentally.

3.1. Synthesis of Preproteins

3.1.1. In Vitro Transcription

1. To make 50 μL of the transcription reaction, mix 30 μL of the premix solution, 2.2 μL RNasin, 2.5 μL m^7G-cap, 0.8 μL SP6 RNA polymerase, and 15 μL of the DNA template (*see* **Note 14**).
2. Incubate 1 h at 37°C.
3. Add 5 μL LiCl and 150 μL ethanol (–20°C) in order to pellet the RNA.
4. Incubate 15 min at –20°C.
5. Spin down the RNA at 37,000g and 4°C for 30 min.
6. Wash the RNA pellet with 70% ethanol (–20°C).
7. Spin down the RNA at 37,000g and 4°C for 5 min.
8. Remove ethanol completely at room temperature. Resuspend the RNA pellet in 30 μL H_2O and add 0.8 μl RNasin.
9. Make single-use aliquots and store at –80°C (*see* **Note 15**).

3.1.2. In Vitro Translation

1. To make an *in vitro* translation reaction, mix 140 μL rabbit reticulocyte lysate, 4 μL of the amino acid mixture minus methionine, 4 μL RNasin, 16 μL [^{35}S]-Met and 30 μL RNA from the transcription reaction.
2. Incubate 1 h at 30°C.
3. Chase the reaction by adding 8 μL of cold methionine.
4. Incubate 5 min at 30°C.
5. Spin down the ribosomes and aggregated proteins by centrifugation at 100,000g and 2°C for 30 min.
6. Make single-use aliquots (12 μL) of the lysate. Freeze aliquots in liquid nitrogen and store them at –80°C.
7. To assess the translation product, analyze 1 μL of the lysate by SDS-PAGE and autoradiography (*see* **Note 16**).

3.2. In Vitro *Import Into Isolated Mitochondria*

The standard import protocol monitors the translocation of preproteins across the outer mitochondrial membrane by assessing their resistance against externally added proteases. In addition, the dependency of the import reaction on ATP or the membrane potential can be tested (*see* **Notes 17–19**). To localize the imported proteins after the import reaction, the mitochondria can be converted to mitoplasts (i.e., mitochondria from which the outer membrane were ruptured by hypotonic swelling). Upon addition of protease, proteins that expose domains into the IMS can be degraded by proteases, whereas proteins of the matrix cannot.

1. Prepare six tubes with 30 μL 2x import buffer.
2. Add in tubes 1 to 3: 19.8 μL H_2O, 0.6 μL ATP, 0.6 μL NADH and 6 μL of mitochondria (*see* **Note 20**).
3. Add in tubes 4 to 6: 20.5 μL H_2O, 0.5 μL valinomycin and 6 μL of mitochondria.
4. Preincubate all tubes for 2 min at 25°C (*see* **Note 21**).
5. Add 3 μL of the radioactive protein lysate to all six tubes (*see* **Note 22**).
6. Incubate 20 min at 25°C.
7. Add 540 μL SH buffer to tubes 1, 2, 4, and 5. Add 540 μL Hepes/KOH pH 7.4 to tubes 3 and 6. Add 5 μL of PK to tubes 2, 3, 5, and 6 (*see* **Note 23**).
8. Incubate 30 min on ice.
9. Add 6 μL of PMSF to all samples in order to stop the digestion.
10. Spin down the mitochondria by centrifugation at 18,000g for 10 min at 4°C.

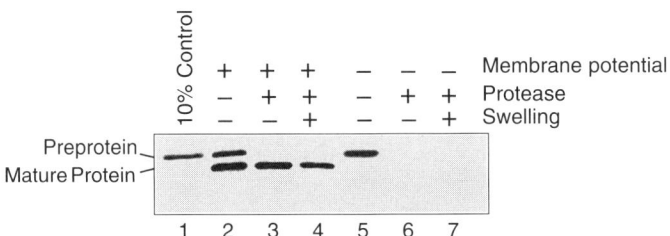

Fig. 1. Import of preSu9-DHFR into yeast mitochondria. A precursor protein consisting of the residues 1–69 of subunit 9 of the ATPase of *Neurospora crassa* fused to mouse dihydrofolate reductase was synthesized in reticulocyte lysate in the presence of [^{35}S]methionine. The radiolabeled preprotein was incubated with isolated yeast mitochondria in the presence of ATP and NADH (lanes 2–4) or valinomycine (lancs 5–7) for 20 min at 25°C. The mitochondria were post-incubated on ice in the presence of proteinase K under non-swelling or swelling conditions as indicated. Protease treatment was stopped by the addition of PMSF. The mitochondria were reisolated and washed. Mitochondrial proteins were separated by SDS-PAGE and detected by autoradiography. Lane 1 shows an equivalent of 10% of the radiolabeled preprotein used per import reaction.

11. Discard the supernatant and wash the mitochondrial pellets with 300 µL SH/KCl solution and 3 µL PMSF.
12. Again, spin down the mitochondria by centrifugation, 18,000g, 10 min, and 4°C.
13. Resuspend the mitochondria in 30 µL reducing sample buffer. Boil the samples for 5 min at 95°C (*see* **Note 24**).
14. The samples are analyzed by SDS-PAGE and fixed on nitrocellulose membranes by blotting. Imported proteins are detected by autoradiography (*see* **Note 25**). An example of such an import experiment is provided in **Fig. 1**.

4. Notes

1. We usually clone open reading frames into pGEM vectors (Promega) downstream the SP6 RNA polymerase transcription initiation site. This generally produces transcripts more efficiently than T3 or T7 RNA polymerases. However, the latter enzymes were successfully used as well. It is recommended to inspect the sequence of the mature protein for the presence of methionine residues. To improve the labeling efficiency, additional ATG codons might have to be inserted into the DNA sequence or at the 3′-end of the gene during the cloning procedure.
2. Spermidine is made from a 200 mM stock solution. Spermidine may deaminate. Thus, make a small stock solution, keep it at –20°C, and prepare new solutions frequently.
3. m7G(5′)ppp(5′)G is an mRNA cap analog, which is used during the *in vitro* transcription reactions in order to yield RNA capped at the 5′-end. This cap is required for satisfactory *in vitro* translations. m7G(5′)ppp(5′)G is shipped in packages with 25 A_{250} units.
4. Avoid multiple freezing and thawing of the lysate. This strongly reduces the translation efficiency. We usually do not freeze the same aliquot more than twice. Lysate is thawed very rapidly by hand warming and subsequently placed on ice.
5. The use of radiolabeled cysteine is possible as well. If cysteine is used, the chase is performed with cold cysteine instead of methionine. We also use radiolabeled methionine–cysteine mixes. This can enhance the labeling efficiency.
6. Refer to the general safety precautions when dealing with radioactivity.
7. We prepare the 2x import buffer with all the ingredients except BSA, dissolve them in water and then adjust the pH with KOH. BSA is added after adjustment of the pH. The purpose of BSA in the buffer is stabilizing some precursor proteins. Although other protocols suggest the usage of 3% (w/v) BSA, this is generally not necessary. In fact, the majority of precursors we tested do not need the addition of BSA to be imported, or are even imported more efficiently without BSA.
8. Valinomycin forms ionophores that dissipate the inner membrane potential.
9. Oligomycin is a potent inhibitor of the ATP synthase.
10. Keep PK in single-use aliquots, and avoid multiple thawing/freezing procedures as this abolishes the enzymatic activity.

11. When we perform a series of import experiments, we keep the remnants of the preprotein aliquots at –20°C and use them mixed with a new aliquot the next day.
12. We keep mitochondria frozen at –80°C in single-use aliquots of 30 or 50 μL. It is recommended to thaw aliquots that are needed quickly by handwarming and place them immediately on ice. Use them as soon as possible after thawing because the quality of mitochondria strongly reduces if not kept at –80°C. The preparation of mitochondria is described in Chapter 7 of this volume.
13. The low amounts of *in vitro* synthesized proteins are generally too small to saturate mitochondrial import sites. When a saturation of mitochondrial import sites is necessary in the experiment, the precursor protein has to be present in excess compared to the available translocases. For this purpose, mitochondrial precursors are heterologously expressed in bacteria and purified as recombinant proteins. It should be noted, however, that this was successfully done for only a few precursors *(12)*.
14. The water used for RNA synthesis should be sterile and of purest quality. Additionally, the yield of RNA can be increased by linearization of the DNA template prior to transcription. This is particularly advisable when T3 or T7 polymerases are used instead of the SP6 polymerase.
15. RNA produced in this way can be stable for more than 1 yr.
16. Lower molecular-weight translation products are often observed in addition to the desired full-length product. This is the result of the initiation of translation at internal start codons that can be suppressed by increasing the concentration of Mg^{2+} and K^+ ions during the translation reaction. However, in most cases an internal initiation is irrelevant since N-terminally truncated proteins lack their mitochondrial targeting signal and will not even bind to receptors of the outer mitochondrial membrane. Sometimes it is difficult to produce full-length proteins. This is frequently observed when the translation products are either very large or hydrophobic. If so, RNA and salt concentrations during the translation reaction have to be optimized. In addition, the plasmid preps should be controlled. Of course, combined transcription–translation reaction systems can be tested, since their yield is usually higher.
17. The import kinetics of a particular preprotein can be assessed using both wild-type mitochondria and mitochondria isolated from mutated yeast strains. This can be used to dissect the import pathway of a preprotein. With this method, one can as well compare established import routes with effects of potential new import components.
18. Described here is a basic protocol for protein import into isolated mitochondria, which can be modified in many ways. For this, we refer to the methods described in the original literature. For example, the effect of reducing agents like DTT on the import reaction can be tested which strongly interferes with the Erv1-mediated translocation of proteins into the IMS.
19. If an import reaction is strongly ATP-dependend, it may be necessary to add an ATP-regenerating system. Therefore, prepare creatine phosphate (CP; 1 M solution in water, which is a 100x stock and stored in aliquots at –20°C) and creatine kinase (CK: 10 mg/mL solution in water, which is a 100x stock and

stored in single use aliquots at −20°C). Add to the import reaction 7.5 μL ATP as well as 15 μL NADH and 6 μL of CP and CK, respectively. This regenerates enough ATP within the mitochondria to keep ATP-driven import running.
20. Mix the components well but avoid vortexing since this breaks the mitochondrial membranes. Be careful when pipeting mitochondria and be as gentle as possible.
21. Import reactions are usually performed at 25°C. Because some precursors are, at this temperature, imported very rapidly, the temperature can be decreased to 12°C in order to observe the import in a linear time range.
22. During imports, we use the term *lysate* for the preprotein within the translation buffer containing the reticulocyte lysate. We normally use 2–5 % (v/v) of lysate in the import reaction, but this is eventually modified since the quality of the lysate is always different. The amount of lysate depends on its quality.
23. Here, the successful translocation across the outer membrane is assessed. Preproteins that are attached to the surface of the mitochondria are still accessible to the added PK and degraded. Preproteins that successfully entered the mitochondria are protease-protected. Trypsin may be used instead of PK, but because of weaker specificity, PK is the preferred protease.
24. Boil all samples immediately at 95°C. Proteases are never stopped completely by the addition of PMSF. Remaining active PK will degrade all proteins very rapidly upon opening of the membranes in the SDS sample buffer.
25. We always load a 10% sample of the amount of lysate used in the import experiment in the first lane during SDS-PAGE analysis. This controls the status of the lysate in each reaction. In addition, the efficiency of the import reaction can be estimated. In the end, we blot the gels to nitrocellulose membranes after SDS-PAGE. One can fix the proteins within the gel and dry the gels, followed by autoradiography. Because blotting allows subsequent immunodecoration, additional control experiments on the quality of the mitochondria and the amount of different import proteins may be assessed.

Acknowledgments

This work was supported by grants of the Fonds der Chemischen Industrie to KB and of the Deutsche Forschungsgemeinschaft (He2803/2-3 and SFB 594 TP B05) and the Stiftung Rheinland-Pfalz für Innovation to JMH.

References

1. Herrmann, J. M., and Hell, K. (2005) Chopped, trapped or tacked - protein translocation into the IMS of mitochondria. *Trends Biochem. Sci.* **30**, 205–212.
2. Koehler, C. M. (2004) New developments in mitochondrial assembly. *Annu. Rev. Cell Dev. Biol.* **20**, 309–335.

3. Pfanner, N., and Geissler, A. (2001) Versatility of the mitochondrial protein import machinery. *Nat. Rev. Mol. Cell Biol.* **2**, 339–349.
4. Neupert, W., and Herrmann, J. M. (2007) Translocation of proteins into mitochondria *Annu. Rev. Biochem.* In press.
5. Endo, T., and Kohda, D. (2002) Functions of outer membrane receptors in mitochondrial protein import. *Biochim. Biophys. Acta.* **1592**, 3–14.
6. Guda, C., Fahy, E., and Subramaniam, S. (2004) MITOPRED: a genome-scale method for prediction of nucleus-encoded mitochondrial proteins. *Bioinformatics.* **20**, 1785–1794.
7. Claros, M. G. (1995) MitoProt, a Macintosh application for studying mitochondrial proteins. *Comput. Appl. Biosci.* **11**, 441–447.
8. Small, I., Peeters, N., Legeai, F., and Lurin, C. (2004) Predotar: a tool for rapidly screening proteomes for N-terminal targeting sequences. *Proteomics.* **4**, 1581–1590.
9. Mesecke, N., Terziyska, N., Kozany, C., et al. (2005) A disulfide relay system in the intermembrane space of mitochondria that mediates protein import. *Cell.* **121**, 1059–1069.
10. Pelham, H. R. B., and Jackson, R. J. (1976) An efficient mRNA-dependent translation system from reticulocyte lysates. *Eur. J. Biochem.* **67**, 247–256.
11. Rassow, J., Guiard, B., Wienhues, U., Herzog, V., Hartl, F.-U., and Neupert, W. (1989) Translocation arrest by reversible folding of a precursor protein imported into mitochondria. A means to quantitate translocation contact sites. *J. Cell. Biol.* **109**, 1421–1428.
12. Becker, K., Guiard, B., Rassow, J., Söllner, T., and Pfanner, N. (1992) Targeting of a chemically pure preprotein to mitochondria does not require the addition of a cytosolic signal recognition factor. *J. Biol. Chem.* **267**, 5637–5643.
13. Lutz, T., Neupert, W., and Herrmann, J. M. (2003) Import of small Tim proteins into the mitochondrial intermembrane space. *EMBO J.* **22**, 4400–4408.

7

Synthesis and Sorting of Mitochondrial Translation Products

Heike Bauerschmitt, Soledad Funes, and Johannes M. Herrmann

Summary

Mitochondria are essential organelles of eukaryotic cells. The biogenesis of mitochondria depends on the coordinated function of two separate genetic systems: one in the nucleus and one in the organelle. The study of mitochondria requires the analysis of both genetic systems and their protein products. In this chapter, we focus on the translation and sorting of mitochondrially encoded proteins into the mitochondrial inner membrane in the baker's yeast *Saccharomyces cerevisiae*. The starting point is the labeling of these proteins, followed by some of the methods developed to investigate their topology and membrane incorporation. The methods described here can be applied also to the study of other aspects of organelle biogenesis such as folding, assembly, and degradation of proteins.

Key words: Mitochondria; OXPHOS; ribosomes; SDS-PAGE; translation; [^{35}S]methionine.

1. Introduction

Mitochondria evolved from bacteria that established an intracellular symbiosis with ancestral cells. Over evolutionary time, most of the genes contributed by the endosymbiont were lost, presumably in order to eliminate redundant processes. Most of the residual genes were transferred from the organellar genome to the nucleus. As a consequence, present-day mitochondria are composed essentially from components that are encoded in the nucleus. However, mitochondria still retain small genomes encoding mainly or exclusively membrane components of the respiratory chain as well as the ribosomal and transfer RNAs required for their translation *(1,2)*. The biogenesis of mitochondria requires the coordinated action of both nuclear and mitochondrial genomes. The mitochondrial genome of *Saccharomyces cerevisiae* is a linear molecule of 85,779 bp (*3*). It contains the genes that encode seven core subunits of the oxidative phosphorylation system (OXPHOS) complexes (cytochrome *b* of the bc_1 complex; the subunits Cox1, Cox2, and Cox3 of the cytochrome *c* oxidase; and the subunits Atp6, Atp8 and Atp9 of the F_o sector of the adenosine triphosphate [ATP] synthase), 1 ribosomal subunit (Var1), 2 rRNAs, and 24 tRNAs. The cotranslational membrane insertion of the seven OXPHOS subunits into the mitochondrial inner membrane is one of the fundamental steps during organellar biogenesis: These seven proteins constitute the membrane cores of the enzymes of the respiratory chain and act as anchor points for the further assembly of the OXPHOS chain.

The radioactive labeling of the mitochondrial translation products is a powerful tool to analyze mitochondrial protein synthesis *per se* or to study the membrane insertion, folding, assembly, or degradation of mitochondrially encoded proteins. Depending on the question to be addressed, translation reactions in complete cells (*in vivo*) or in isolated mitochondria (*in organello*) can be used *(4–7)*. In general, *in vivo* reactions are used when the presence of an active cellular cytosol and nuclear encoded proteins is needed. Moreover, *in vivo* labeling can be easily performed with a large set of strains as it does not require the time-consuming isolation of mitochondria. On the other hand, the *in organello* labeling allows the subfractionation of mitochondria after the labeling in order to monitor the intra-mitochondrial location of the synthesized translation products.

In this chapter, we describe the basic protocols for both techniques, as well as methods to elucidate the protein topology and the membrane association of the studied proteins. For all the experiments *in organello* described within the following pages, an obligate prerequisite is the isolation of intact mitochondria for which a standard protocol is also provided. The isolated mitochondria can be stored and remain functional over several months. Following disruption of the outer membrane and protease treatment, the topology of newly synthesized proteins can be easily verified. Moreover, using fractionation steps like the alkaline extraction of proteins with carbonate the integration of translation products into the inner membrane can be directly assessed. The synthesized proteins are typically visualized by autoradiography after SDS-PAGE. Immunoblotting is mainly employed to control subfractionation and protease protection experiments.

2. Materials
2.1. Growth of Yeast

Amounts are given for 1 L of medium. Sterilize each of the media listed here by autoclaving 20 min at 121°C (*see* **Note 1**).

1. Yeast peptone (YP) medium: 10 g bacto-yeast extract, 20 g bacto-peptone, distilled water ad 1,000 mL. Adjust the pH to 5.5.
2. Lactate medium: 3 g bacto-yeast extract, 1 g KH_2PO_4, 1 g NH_4Cl, 0.5 g $CaCl_2 \cdot 2H_2O$, 0.5 g NaCl, 0.6 g $MgSO_4 \cdot H_2O$, 3 mg $FeCl_3$ (or 0.3 mL from a 1% w/v stock solution), 22 mL 90 % lactic acid (2 % v/v final concentration), distilled water ad 1,000 mL. Adjust the pH to 5.5 with KOH.
3. 5X synthetic minimal medium stock (5X S): 8.5 g yeast nitrogen base (without ammonium sulfate, without amino acids), 25 g ammonium sulfate, distilled water ad 1,000 mL.
4. 5 X synthetic lactate medium stock (5 X SLac): 8.5 g yeast nitrogen base (without ammonium sulfate, without amino acids), 25 g ammonium sulfate, 110 mL 90% lactic acid, distilled water ad 1,000 mL. Adjust the pH to 5.5 with KOH.
5. Glucose stock solution: 400 g glucose, distilled water ad 1,000 mL.
6. Galactose stock solution: 300 g galactose, distilled water ad 1,000 mL.
7. Raffinose stock solution: 20 g raffinose, distilled water ad 100 mL.
8. Auxotrophic markers stock solutions: Adenine sulfate 200 mg in 100 mL of water; uracil 200 mg in 100 mL of water; L-tryptophan 1 g in 100 mL of water, L-histidine-HCl 1 g in 100 mL of water; L-leucine 1 g in 100 mL of water; L-lysine-HCl 1 g in 100 mL of water (*see* **Note 2**).
9. Sterile water.

2.2. Isolation of Mitochondria by Differential Centrifugation

1. MP1 buffer: 100 mM Tris-base (do not adjust the pH), 10 mM dithiothreitol (DTT; add just before use from a 1 M stock solution).
2. Sorbitol buffer: 1.2 M sorbitol.
3. MP2 buffer: 1.2 M sorbitol, 20 mM phosphate buffer, pH 7.4.
4. Zymolyase 20T.
5. MP3 buffer: 0.6 M sorbitol, 10 mM Tris-HCl, pH 7.4, 1 mM EDTA, 0.2 % (w/v) fatty acid-free bovine serum albumin (BSA; essentially fatty acid free), 1 mM phenylmethylsulphonyl fluoride (PMSF; *see* **Note 3**). Chill buffer on ice before use.
6. SH buffer: 0.6 M sorbitol, 20 mM Hepes, pH 7.4.
7. SEH buffer: 0.6 M sorbitol, 1 mM EDTA, 20 mM Hepes, pH 7.4.
8. Glass dounce homogenator (tight-fitting glass pistil).
9. Liquid nitrogen.

2.3. In Organello *Labeling of Mitochondrial-Encoded Proteins*

1. 1.5X *in organello* translation buffer: 375 μL of 2.4 M sorbitol, 225 μL of 1 M KCl, 22.5 μL of 1 M potassium phosphate buffer, pH 7.4, 19 μL of 1 M magnesium sulfate, 45 μL of 100 mg/mL BSA (essentially fatty acid free; *see* **Note 4**),

30 μL of 200 mM ATP, pH 7.0, 15 μL of 50 mM guanosine triphosphate (GTP), 9.1 μL of amino acid stock solution (2 mg/mL of all proteinogenic amino acids except tyrosine, cysteine, and methionine), 10 μL of 10 mM cysteine, 18.2 μL of 1 mg/mL tyrosine, 1.7 mg α-ketoglutarate, and 3.5 mg phosphoenolpyruvate. Add distilled water to a final volume of 1 mL.
3. 10 mg/mL Pyruvate kinase solution. This solution is added to the 1.5 X *in organello* translation buffer at a final concentration of 0.1 % (v/v) immediately before use. The stock solution is stored at 4°C.
4. [^{35}S] methionine, 10 mCi/mL, approx 1,000 Ci/μmole.
5. Cold methionine stock solution: Dissolve 0.298 g of methionine in 10 mL of water for a 200 mM solution. Divide the solution into 0.5 mL aliquots and store them at –20°C.
6. Puromycin stock solution: 1 mg of puromycin is dissolved in 1 mL of water and stored at –20°C.
7. Sample buffer: 2% (w/v) sodium dodecyl sulfate (SDS), 5% (v/v) β-mercaptoethanol, 10% (v/v) glycerol, 0.02% (w/v) bromophenol blue, 60 mM Tris-HCl, pH 6.8. Store the stock solution in aliquots at –20°C.

2.4. Disruption of the Mitochondrial Outer Membrane by Osmotic Swelling and Proteinase K Digestion

1. Swelling buffer: 20 mM Hepes, pH 7.4
2. 10 mg/mL proteinase K (PK).

2.5. Carbonate Extraction of Peripherally Associated Membrane Proteins

1. Sodium carbonate solution: Na_2CO_3 is dissolved at 100 mM in water and prechilled in ice. Prepare this buffer freshly and immediately before use.
2. Trichloroacetic acid (TCA) stock solution: Dissolve 72 g of TCA in 100 mL of water. Protect from light by storing the solution in an amber bottle.
3. Cold acetone: Prechill acetone at –20°C.

2.6. Membrane Flotation Analysis

1. Extraction buffer: 100 mM Na_2CO_3, 4,5 M urea, 1 mM β-mercaptoethanol, 1 mM PMSF.
2. 2 M Sucrose buffer: 2 M sucrose, 100 mM Na_2CO_3, 4,5 M urea.
3. 1.4 M Sucrose buffer: 1.4 M sucrose, 100 mM Na_2CO_3, 4,5 M urea.
4. 0.25 M Sucrose buffer: 0.25 M sucrose, 100 mM Na_2CO_3, 4,5 M urea.
5. 11 x 60 mM Ultra-Clear Centrifuge Tubes (Beckman Instruments Inc.).

2.7. In Vivo Labeling of Mitochondrial-Encoded Proteins

1. Synthetic medium: Prepare 1X synthetic medium from the stock solutions described in **Subheading 2.1**. Choose the appropriate carbon source and the auxotrophic markers in order to prepare the right medium for the yeast strain in use.
2. Cycloheximide stock solution: Cycloheximide is dissolved at 7.5 mg/mL in water. This stock is prepared right before use.
3. Lysis buffer: 1.8 M NaOH, 1.0 M β-mercaptoethanol, 1 mM PMSF.

2.8. Sodium Dodecyl Sulfate-Polyacrylamide Gel Electrophoresis (SDS-PAGE)

2.8.1. Gel Casting

1. Acrylamide stock solution: 30% (w/v) acrylamide, 0.2% (w/v) bisacrylamide (*see* **Note 5**). Store at 4°C.
2. 1.875 M Tris-HCl, pH 8.8.
3. 0.6 M Tris-HCl, pH 6.8.
4. 10% (w/v) SDS stock solution.
5. 10% (w/v) Ammonium persulfate (APS) stock solution. Prepare fresh before use.
6. N,N,N',N'-Tetramethylethylenediamine (TEMED).
7. Isopropanol (technical grade).

2.8.2. Gel Running

1. Running buffer: 1 g/L SDS, 14.4 g/L glycine, 3 g/L Tris-base. Do not adjust pH.

2.9. Coomassie Blue Staining of SDS Gels

1. Coomassie staining solution: 0.3% (w/v) Coomassie brilliant blue R, 50% (v/v) methanol, 10% (v/v) acetic acid.
2. Destaining solution: 30% (v/v) methanol, 7% (v/v) acetic acid.

2.10. Protein Transfer and Immunodecoration

2.10.1. Semi-Dry Protein Transfer

1. Transfer buffer: 20 mM Tris-base, 150 mM glycine, 0.08% SDS, 20% (v/v) methanol. Do not adjust pH.
2. Ponceau S red staining solution: 0.2% (w/v) Ponceau S, 3% (w/v) TCA.

2.10.2. Immunodecoration

1. Tris-buffered saline (TBS) buffer: 150 mM NaCl, 10 mM Tris-HCl, pH 7.5.
2. Blocking solution: 5% (w/v) milk powder in TBS buffer.

3. Primary antibody dilution: Dilute the purified antibody or serum against the protein of interest in a range of 1:250 to 1:5,000 (according to the quality of the antibody) in blocking solution. Store this primary antibody dilution either at −20°C or at 4°C (depending on the stability of the antibody).
4. Secondary antibody: Anti-rabbit, anti-mouse or anti-rat immunoglobulin (Ig)G (depending on the first antibody used) conjugated to horseradish peroxidase (HRP).
5. Secondary antibody dilution: Dilute the antibody right before use following the recommendations of the supplier. Usually dilutions of 1:3,000 to 1:10,000 are recommended.
6. Enhanced chemiluminescent (ECL) reagent—solution I: 220 μL p-cumaric acid (from a 15 mg/mL stock in dimethyl sulfoxide [DMSO]; stored in aliquots at −20°C), 0.5 mL Luminol (from a 44 mg/mL stock in DMSO, stored at −20°C) in a total volume of 50 mL 100 mM Tris-HCl, pH 8.5. Store this solution isolated from light at 4°C.
7. ECL reagent—solution II: 100 mM Tris-HCl pH 8.5, 0.035% (v/v) H_2O_2. Store this solution isolated from light at 4°C. This solution is stable up to 1 wk.

3. Methods
3.1. Growth of Yeast

Yeast cells are grown in Erlenmeyer flasks in liquid cultures under constant agitation (∼ 140 rpm). The optimal growth temperature for *S. cerevisiae* is 30°C, however, temperature-sensitive mutants should be grown at 25°C or 15°C. The growth rate depends notably on the yeast strain. Therefore, it is recommended to measure the generation time for each strain before inoculating the big culture that will be used for mitochondria isolation (*see* **Note 1**).

1. Inoculate 50 mL of culture with the strain of interest.
2. Follow the growth of the strain over several generations in the same conditions. Take care that they should always be kept in logarithmic growth phase (OD_{600} should never exceed 2.0).
3. Measure regularly the OD_{600} of the culture, and calculate the generation time of the strain by using the following formula:

$$\text{generation time} = (t^* \log 2) / \log(OD_{600}\text{final} / OD_{600}\text{initial})$$

where *t* is the time of growth in hours.
4. Inoculate the final culture for the isolation of mitochondria considering the generation time and expecting an OD_{600} of 1.5 at the time of harvesting. The cultures must be in the logarithmic growth phase at the moment of harvesting. The amount of culture inoculated depends exclusively on the experimental needs, usually from 10 L of a culture with an OD_{600} of 1.5, approx 20 g of cells (wet weight) are obtained. The mitochondrial yield from this amount of cells is approx 20 mg.

3.2. Isolation of Mitochondria by Differential Centrifugation

1. Grow the desired amount of yeasts to an OD_{600} of 0.8 to 2.0.
2. Collect the cells by centrifugation at 3,000g for 5 min and determine the wet weight.
3. Resuspend the pellet in 100 mL dH_2O and centrifuge at 3,000 g for 5 min.
4. Resuspend the cells in MP1 buffer (2 mL/g of wet weight) and incubate the suspension for 10 min at 30°C under agitation (~140 rpm).
5. Collect the cells by centrifugation at 2,000g for 5 min. Discard the supernatant and wash the cell pellet with sorbitol buffer (5 mL/g of wet weight).
6. Collect the cells again by centrifugation at 2,000g for 5 min. Discard the supernatant.
7. Resuspend the cells in MP2 buffer (6.7 mL/g of wet weight).
8. Add the Zymolyase 20T (2 mg/g of wet weight) to the suspension and incubate under agitation (~140 rpm) for 20 to 40 min at 30°C until spheroplasts have formed (*see* **Note 6**).
9. All following steps will be performed on ice using ice-cold buffers, keeping the samples always on ice and centrifugation steps performed at 4°C.
10. Harvest the sphaeroplasts by centrifugation at 2,000g for 5 min.
11. Resuspend the pellet in buffer MP3 (6.7 mL/g of wet weight) by gently shaking or stirring with a pipet.
12. Transfer the suspension to a 50 mL dounce homogenizer (tight-fitting glass pistil) and homogenize with 15 strokes.
13. Dilute the resulting homogenate with MP3 buffer (again 6.7 mL/g of wet weight).
14. Centrifuge the homogenate at 2,000g for 5 min to spin down cell debris. Keep the supernatant and transfer it to a fresh centrifugation bucket.
15. Centrifuge 12 min at 17,000g for 12 min. Discard the supernatant.
16. Resuspend the pellet carefully in 10 mL of SH or SEH buffer.
17. Centrifuge 12 min at 17,000g for 12 min. Discard the supernatant.
18. Resuspend the pellet in 0.5 mL of SH or SEH buffer.
19. Take an aliquot to determine protein concentration.
20. Dilute the mitochondrial suspension to a final concentration of 10 mg/mL in SH buffer.
21. Make aliquots of 30 to 50 µL, immediately snap-freeze in liquid nitrogen. Store at –80°C.

3.3. In Organello Labeling of Mitochondrial-Encoded Proteins

1. Before starting the labeling, finish the freshly prepared 1.5X *in organello* translation buffer by the addition of 0.1% (v/v) pyruvate kinase.
2. Thaw 200 µg of isolated mitochondria of the required strains (*see* **Subheading 3.2**. They should be in a volume of 20 µL, *see* **Note 7**).
3. Mix each of the samples with 40 µL of the 1.5X *in organello* translation buffer.
4. Incubate the reaction for 5 min at 30°C.

5. Add 0.5 μL of [^{35}S]methionine to start the labeling of newly synthesized proteins.
6. Incubate the samples for 15 min at 30°C (*see* **Note 8**).
7. Finish the labeling by addition of 10 μL cold methionine and/or 30 μL of the puromycin stock solution (*see* **Note 9**).
8. Incubate the reaction for 5 min at 30°C.
9. Stop the translation reaction by addition of 1 mL cold SH buffer.

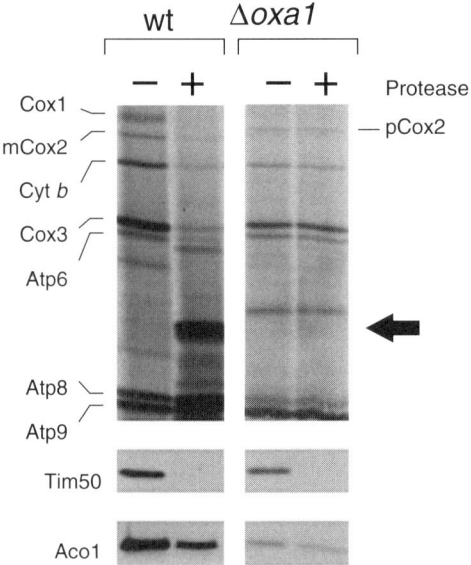

Fig. 1. Labeling of mitochondrial-encoded proteins and analysis of the insertion into the inner membrane. Mitochondria were isolated form wild-type (wt) or insertion-deficient *oxa1* mutant cells. The outer membrane was ruptured by hypotonic swelling as described in **Subheading 3.4**. Translation products were radiolabeled for 15 min at 30°C. After termination of the labeling reaction the mitochondria were incubated without or with proteinase K on ice, reisolated, washed and dissolved in sample buffer. Proteins were resolved by SDS-PAGE and visualized by autoradiography (upper panel) or immunoblotting. The blots were immunodecorated with antibodies against Tim50 and aconitase (Aco1) as markers for the intermembrane space and the matrix, respectively (*see* **Subheading 3.10.**). In wild-type mitochondria the vast majority of the newly synthesized proteins are degraded by the protease giving rise to a characteristic degradation fragment (indicated by an arrow). In contrast, in *oxa1* mitochondria *(8)* the translation products accumulate in the matrix and are not accessible to protease treatment. Thus, all translation products remain stable in the matrix during the procedure. In addition, the protein Cox2 accumulates in its precursor form (pCox2) because the N-terminus of the protein did not reach its processing peptidase Imp1 in the intermembrane space.

10. Collect the mitochondria by centrifugation for 10 min at 20,000g at 4°C. Discard the supernatant.
11. Resuspend the pellet in 50 μL of sample buffer and subject half of all samples to SDS-PAGE (*see* **Subheading 3.8.**) and analyze the samples by autoradiography. **Figure 1** shows an example of how the seven translation products that are components of the OXPHOS appear after a labeling reaction.

3.4. Disruption of the Mitochondrial Outer Membrane by Osmotic Swelling and Proteinase K Digestion (see Note 10)

1. Take 20 μL of mitochondria (200 μg) and dilute them with 180 μL of swelling buffer.
2. Incubate the samples 30 min on ice.
3. Collect the samples by centrifugation for 10 min at 20,000g at 4°C.
4. Resuspend the samples in 20 μL of SH buffer.
5. Treat the samples as if they were mitochondria and follow **steps 1** through **6** of **Subheading 3.3.**
6. Dilute the samples in 1 mL of SH buffer.
7. Split the samples in two (500 μL each). Treat one half from each sample with 5 μL of PK and mock treat the other half.
8. Incubate the samples 30 min on ice.
9. Add 5 μL of PMSF and mix carefully.
10. Centrifuge the samples 10 min at 20,000g at 4°C. Discard the supernatant carefully.
11. Resuspend the pellet in 500 μL of SH buffer.
12. Centrifuge the samples 10 min at 20,000g at 4°C. Discard the supernatant carefully.
13. Resuspend the pellet in 25 μL of sample buffer and heat the samples for 30 to 60 s at 96°C (*see* **Note 11**). Subject the samples to SDS-PAGE (*see* **Subheading 3.8.**).

Examples for the use of this method are shown in **Figs. 1** and **2**.

3.5. Carbonate Extraction of Peripherally Associated Membrane Proteins

1. Label 500 μg of mitochondria following the protocol described in **Subheading 3.3.** up to **step 6**.
2. Take out the equivalent volume to 10%. Collect mitochondria by centrifugation for 10 min at 20,000g and resuspend them in 20 μL of sample buffer. Store at −20°C up to **step 16**. This sample is used as a reference for the protein content of the sample before the treatment.
3. Add 1 mL of cold SH buffer to the residual translation reaction and centrifuge 10 min at 20,000g at 4°C. Discard the supernatant.

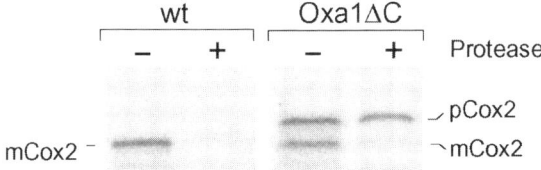

Fig. 2. Analysis of the insertion and topology of Cox2 in yeast mitochondria. Translation products were radiolabeled in wild-type (wt) and Oxa1△C mutant **(9)** mitochondria as described in **Fig. 1**. The samples were lysed in 25 µL of 1% (w/v) SDS, 50 mM Tris-HCl, pH 7.5, cleared by centrifugation for 10 min at 16,000g, and diluted in 1 mL 0.1% Triton X-100, 300 mM KCl, 5 mM EDTA, 10 mM Tris-HCl, pH 7.5. The radiolabeled subunit 2 of cytochrome oxidase was isolated by immunoprecipitation with Cox2-specific antibodies, resolved by SDS-PAGE and visualized by autoradiography. In wild-type mitochondria, Cox2 accumulates in its mature form (mCox2) indicating that its N-terminus was translocated across the inner membrane and processed by the intermembrane space protease Imp1. After complete insertion into the membrane, Cox2 can be completely degraded by protease from the intermembrane space side. In mitochondria lacking the C-terminal domain of Oxa1 (Oxa1△C), a fraction of Cox2 accumulates in the matrix in its precursor form (pCox2) which is not accessible to protease.

4. Resuspend the mitochondrial pellet in 0.5 mL of freshly prepared cold sodium carbonate solution.
5. Incubate for 30 min at 4°C under constant vigorous shaking.
6. Separate the disrupted membranes from the soluble fraction by centrifugation for 20 min at 90,000g and 4°C.
7. Collect the supernatant and transfer it to a new tube and keep it on ice up to **step 9**.
8. Resuspend the pellets in 25 µL of sample buffer. Do not boil the samples. Store them at –20°C until **step 16**.
9. Take the samples from **step 7** and fill up to a final volume of 1 mL with water.
10. Add 300 µL of the TCA stock solution and mix. Incubate the supernatant samples for 30 min at –20°C to precipitate proteins.
11. Centrifuge the samples for 30 min at 35,000g at 4°C. Discard the supernatant.
12. Wash the pellets with 500 µL of cold acetone.
13. Centrifuge the samples for 30 min at 20,000g at 4°C. Discard the supernatant.
14. Let the pellets dry for 5 to 10 min at room temperature.
15. Resuspend the pellets by vigorous shaking in 25 µL of sample buffer. Do not boil the samples (*see* **Note 11**).
16. Analyze all samples by SDS-PAGE (*see* **Subheading 3.8.**) and Western blotting (*see* **Subheading 3.10.**).

3.6. Membrane Flotation Analysis

1. Label 500 μg of mitochondria following the protocol described in **Subheading 3.3.** up to **step 7**.
2. Take out the equivalent volume to 10%. Collect mitochondria by centrifugation for 10 min at 20,000g and resuspend them in 20 μL of sample buffer. Store at –20°C until **step 15**. This sample is used as a reference for the protein content of the sample before the treatment.
3. Add 1 mL of cold SH buffer to the residual labeling reaction.
4. Centrifuge 10 min at 20,000g at 4°C. Discard the supernatant.
5. Resuspend the mitochondrial pellet carefully in 200 μL of extraction buffer.
6. Incubate for 30 min at 4°C under constant vigorous shaking.
7. Mix the sample with 800 μL of 2 M sucrose buffer (*see* **Note 12**) to get a final concentration of sucrose of 1.6 M.
8. Make a step gradient in SW60 Ultra-Clear Centrifuge Tubes (11 × 60 mM; total volume approx 4 mL): First, carefully pour the sample into the tube. Then, slowly and carefully overlay with 2 mL of 1.4 M sucrose buffer. Always stay at the tube wall and close to the meniscus of the solutions while pouring, to avoid disturbing the steps of the gradient! Finally, fill the tube to the very top with 0.25 M sucrose buffer (approx 1 mL volume).
9. Mark the position of each of the two-phase interfaces on the tube.
10. Place the tubes carefully inside of the buckets of the SW60 rotor.
11. Balance all samples to ±0.010 g.
12. Centrifuge for 15 h at 165,000g at 4°C.
13. Collect the upper half of the gradient as the "flotated membranes fraction." Make sure to collect all membranes in this sample. Never go deep into the gradient with the pipet tip, stay at the meniscus!
14. Collect the lower part of the gradient as "soluble protein fraction."
15. Fill both sets of samples to 1 mL with water.
16. Follow **steps 10** to **15** of **Subheading 3.5**.
17. Analyze all samples by SDS-PAGE (*see* **Subheading 3.8.**) and Western blotting (*see* **Subheading 3.10.**).

See **Fig. 3** for an example in which this method was used.

3.7. In Vivo *Labeling of Mitochondrial-Encoded Proteins*

1. Prepare 25 mL overnight cultures of the desired strains in the appropriate synthetic medium. Include the appropriate auxotrophic markers and carbon source as indicated above (*see* **Subheading 3.1.**).
2. Take cells equivalent to 1 mL of an OD_{600} of 1.0.
3. Collect the cells by centrifugation for 5 min at 20,000g.
4. Resuspend the cell pellets in 1 mL of the same synthetic media (to get a cell suspension of 1 OD/mL).
5. Incubate the cell solution for 10 min at 30°C.

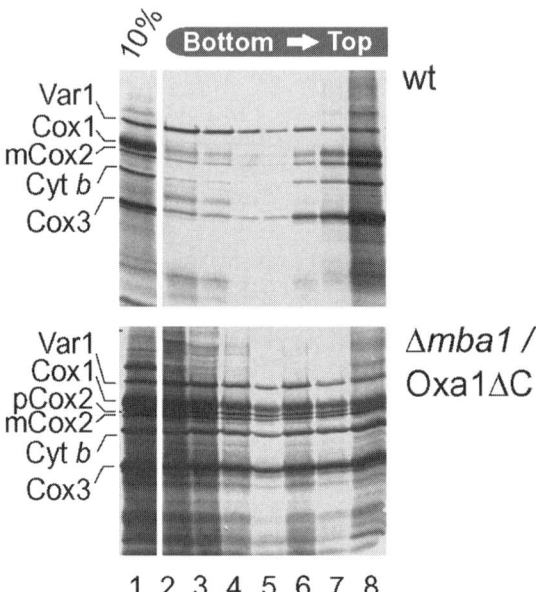

Fig. 3. Extraction of membrane-associated translation products. Translation products were radiolabeled in isolated mitochondria of wild type (wt) and ∆*mba1*/Oxa1∆C *(10)* mutants. The sample was treated with carbonate/urea buffer as described in **Subheading 3.6.** and membrane-embedded proteins were floated by centrifugation in sucrose gradients. The gradients were fractionated. The containing proteins were isolated by precipitation with trichloroacetic acid, separated by SDS-PAGE and visualized by autoradiography. In wild type mitochondria, most translation products migrate to the top of the gradient. Only one protein stays at the bottom which is the only non-membrane proteinVar1. In ∆*mba1*/Oxa1∆C mitochondria, the insertion of protein is impaired and even many of the hydrophobic translation products stay at the bottom of the gradient.

6. Add 50 µL of freshly prepared cycloheximide solution to block cytosolic protein translation.
7. Mix gently and incubate the samples for 5 min at 30°C.
8. Start the labeling of newly synthesized mitochondrial proteins by the addition of 5 µL of [^{35}S]methionine to each sample.
9. Incubate the samples under constant agitation at 30°C.
10. Stop the reaction after 30 min by the addition of 10 µL of cold methionine stock solution and 30 µL of puromycin stock solution.
11. Mix gently and incubate the samples another 2 min at 30°C.
12. Centrifuge the samples 10 min at 20,000*g* and discard the supernatant.
13. Resuspend the cell pellets in 500 µL of water.
14. Spin the cells down again 10 min at 20,000*g*. Discard the supernatant.
15. Repeat this washing step once and resuspend the pellets in 500 µL of water.

16. Add 75 μL of lysis buffer to each sample, mix well and incubate 10 min on ice.
17. Add 100 μL of the TCA stock solution and incubate the samples for 30 min at −20°C to precipitate all proteins.
18. Follow **steps 12** to **16** of **Subheading 3.5**.

3.8. SDS-PAGE

3.8.1. Gel Casting

Gels are cast between a pair of glass plates separated by two plastic spacers. We normally use gels consisting of a running gel of 90 × 150 × 1 mm and a stacking gel of 10 × 150 × 1 mm. Because of their very hydrophobic nature, many mitochondrially encoded proteins do not migrate at their expected size. Thus, the running behavior of these proteins can vary considerably depending on the specific gel system and acrylamide composition. In principle, different sizes and gel systems can be used, but in our hands the clearest separation of the mitochondrial encoded proteins is obtained with 16% SDS-PAGE gels.

1. Wash the two glass plates, the spacers, and the comb carefully with water and ethanol.
2. Place the spacers between both glass plates. Fix them tightly with metal or plastic clamps and place them into a gel pouring tray.
3. Prepare the base gel (20%) by mixing 6.7 mL of the 30% acrylamide stock solution, 2.0 mL of 1.875 M Tris-HCl, pH 8.8, 100 μL of the SDS stock solution, 1.1 mL of water, 50 μL of APS stock solution, and 25 μL of TEMED. Pour the base gel about 1.5 to 2 cm high into the gel tray so that it completely covers the bottom of the glass plates.
4. After the base gel has polymerized, prepare the running gel by mixing 9.0 mL of the 30% acrylamide stock solution, 3.5 mL of 1.875 M Tris-HCl, pH 8.8, 167 μL of the SDS stock solution, 4.2 mL of water, 100 μL of APS stock solution, and 10 μL of TEMED. Pour the running gel mixture between the two glass plates, until approx 0.5 cm below the lower end of the comb. Overlay the polymerizing running gel mixture with 1 mL of isopropanol.
5. Prepare the stacking gel by mixing 830 μL of the 30% acrylamide stock solution, 500 μL of 0.6 M Tris-HCl, pH 6.8, 50 μL of the SDS stock solution, 3.6 mL of water, 25 μL of APS stock solution, and 5 μL of TEMED.
6. When the running gel has polymerized, wash off the isopropanol with water and dry the gel carefully. Pour the stacking gel and insert the comb carefully. Wait until the polymerization is completed. Gels can be used immediately or stored for several days at 4°C inside of a plastic bag covered with paper towels soaked in water.

3.8.2. Gel Running

1. Remove the comb from the gel and put it into a vertical electrophoresis chamber.
2. Fix the gel with at least two metal or plastic clamps.
3. Fill the cathode and the anode buffer reservoir with running buffer. Make sure that the pockets of the gel are all filled and covered with buffer and that the chamber is not leaking!
4. Load the samples and a protein standard on the gel.
5. Electrophoresis is performed at constant 25 mA until the bromophenol blue front has reached the bottom gel. After electrophoresis, the gel can be either stained (*see* **Subheading 3.9.**) or the proteins can be transferred to a nitrocellulose membrane (*see* **Subheading 3.10.**).

3.9. Coomassie Blue Staining of SDS Gels

1. Stop the electrophoresis and discard the running buffer.
2. Disassemble the glass plates carefully and remove the stacking gel.
3. Transfer the gel to a tray filled with Coomassie staining solution. Incubate it for 30 to 60 min at room temperature under gentle agitation.
4. Pour off the Coomassie staining solution, rinse the gel with water and incubate it in destaining solution under gentle agitation (*see* **Note 13**).
5. Remove the used destaining solution as soon as it has turned deep blue (*see* **Note 14**).
6. Fill the tray with fresh destaining solution and incubate again as described above.
7. Repeat **steps 5** and **6** until the background is completely destained.
8. For gel drying, rinse two cellophane papers with water. Place the gel on one cellophane sheet and sandwich it with the second piece. Remove all air bubbles between the cellophane sheets and seal the edges of the sheets with plastic frames or clamps. Air dry the gel over night or use a gel drying unit.

3.10. Protein Transfer and Immunodecoration

3.10.1. Semi-Dry Protein Transfer

1. Prepare a semi-dry blot chamber unit. Wet the lower plate with transfer buffer.
2. Prepare four Whatmann papers and one nitrocellulose membrane of sizes similar to that of the gel and soak them with transfer buffer. Put two Whatmann papers onto the lower plate of the blot chamber. Remove carefully any air bubbles between the plate and the Whatmann papers.
3. Place the nitrocellulose membrane onto the Whatmann papers and remove air bubbles.
4. Disassemble the glass plates from the gel and transfer the gel to a tray filled with blotting buffer for at least 1 min. Then, place the gel on the nitrocellulose

membrane. Cover the gel with two additional pieces of Whatmann paper. Remove air bubbles.
5. Wet the upper plate of the blotting chamber. Close the blotting chamber. Place a weight of about 500 g on top of the blotting chamber and apply a constant current of approx 200 mA for 90 min.
6. Disassemble the blotting chamber. Remove the nitrocellulose membrane, rinse it with water and stain it with Ponceau S red staining solution for 10 min under constant agitation. Wash the membrane with water to remove background staining. Dry the membrane completely. Drying can be accelerated by putting the blot under a red light bulb or by using a hair drier.

3.10.2. Immunodecoration

1. Mark all the proteins of the protein standard on the membrane with a pen. There is no need to do this when a prestained protein standard is used.
2. When the decoration of several proteins within one membrane is needed and the precise running behavior of each of them is known, it is possible to cut the membrane in stripes to decorate each of them independently with the right antibody.
3. Place the membrane(s) in a plastic box and cover them with blocking solution.
4. Incubate at room temperature for at least 1 h under constant gentle agitation.
5. Remove the blocking solution and overlay with the primary antibody dilution. Incubate at room temperature for at least 2 h under gentle agitation (*see* **Note 15**).
6. Remove the primary antibody dilution and wash the membrane(s) three times for 10 min each in TBS solution. Incubate as described above. Usually the primary antibody dilutions can be used more than one time, be careful not to discard this solution after use!
7. Discard the TBS solution and overlay with the secondary antibody dilution. Incubate at room temperature for 1 h under gentle agitation.
8. Wash the membrane(s) as described in **step 6**.
9. For the detection of the signals, place the stripes inside of a transparent plastic bag. Cover them with equal amounts of ECL solution I and ECL solution II (use approx 2 mL each per 10 x 15 cm of membrane). Make sure that the solutions are well mixed, and press out completely the remaining liquid. Expose films for different times according to the intensity of the signal they produce.

3.11. Analysis of Mitochondrial-Encoded Protein Insertion

By exposure of films or phosphoimager screens onto the dried gels or the nitrocellulose membranes, the radiolabeled mitochondrial translation products can be detected. The eight translation products of yeast mitochondria can typically be identified by their specific running behavior and appearance on gels:

Var1 is one of the subunits of the mitochondrial ribosome and the only hydrophilic protein encoded by the mitochondrial genome of yeast. It runs as a sharp band of about 45 kDa. In contrast to most of the other translation products, the migration of Var1 is not significantly influenced by the gel composition. However, the size of the Var1 protein varies significantly between different yeast strains. It therefore received its name (Protein of **Var**iable size **1**).

Cox1 is the largest subunit of the cytochrome oxidase. It can be easily identified because it runs as a fuzzy, broad band. The synthesis of this protein depends considerably on the strain, and many petite mutants fail to produce this protein.

Cox2 is produced as a precursor protein with a small amino-terminal extension that is cleaved after the insertion of the protein into the inner membrane. The precursor and the mature forms of the protein run as sharp bands below the Cox1 signal. The maturation normally occurs cotranslationally and therefore, in mitochondria from wild-type cells, only the mature form of Cox2 can be seen.

Cytochrome *b* is the only subunit of the bc_1 complex that is mitochondrially encoded. It is one of the strongest signals observed after the labeling reaction.

Cox3 and **Atp6** both migrate usually very close to each other and, on many gel systems, both proteins cannot be distinguished.

Atp8 and **Atp9**. These two very small proteins migrate very fast. On our gel system, they run very close to the front of the gel.

During the analysis of mitochondrial translation, it is possible to see that behind the signal of the fully synthesized proteins exists a smeary background that covers the entire line of an SDS-PAGE gel. This "smear" corresponds to the nascent polypeptides that are being synthesized by mitochondrial ribosomes.

4. Notes

1. The appropriate carbon source depends on the strains and its ability to grow on nonfermentable conditions. When working with strains that are unable to respire, the use of galactose is preferred in comparison to glucose, as glucose-repression of many mitochondrial genes leads to poorly developed mitochondria and consequently a rather poor labeling. When working with strains that are able to respire lactic acid is preferred. Glucose, galactose, and lactic acid are added to a final concentration of 2% (w/v). Raffinose can be used at a final concentration of 1% (w/v).
2. Auxotrophic markers (amino acids or nucleotides) might be omitted to select for plasmids. L-Tryptophan turns into a brown solution during the autoclaving process; alternatively it can be sterilized by filtration through a 0.22 μm filter. L-Tryptophan and L-histidine HCl are photosensible, therefore the stock solutions must be kept protected from light.
3. PMSF is dissolved at 200 mM in ethanol. Prepare freshly before use. PMSF inhibits all serine proteases irreversibly and is therefore highly toxic. Appropriate safety measurements must be taken by handling it.

4. BSA normally helps to stabilize the mitochondria during the labeling. Its presence is, however, not absolutely required. The conditions for the *in organello* labeling procedure were initially established by Poyton and co-workers *(7)*. In the original publication, the influence of the buffer composition on the labeling reaction is well documented.
5. Please note that acrylamide is neurotoxic when unpolymerized. Care should be taken to avoid any skin contact.
6. To test for sphaeroplast formation, add 50 µL cells to 2 mL H_2O or sorbitol buffer. The suspension in H_2O should clear up as an indication that the cells are being lysed in this hypotonic solution. This effect can be also detected by measuring the OD_{600} and comparing the optical densities from both suspensions.
7. The concentration of the mitochondria is not very critical, and may be adjusted to the specific conditions of the experiment.
8. Under these conditions, mitochondria translate proteins efficiently for about 30 to 60 min. At longer incubation times, the translation levels off. Alternatively, the labeling reaction can be performed at different temperatures from about 10°C to 40°C with an optimum at 25°C to 35°C.
9. As soon as an excess of cold methionine is added, the translation reaction proceeds without labeling of the products. This reduces the amount of radiolabeled incompleted nascent chains and clears the background between the bands of the full-length translation products.
10. Hypo-osmotic treatment of mitochondria often leads to a decrease in the labeling efficiency. Nevertheless, the disruption of the outer membrane should be done before the labeling reaction, since mitochondria are almost completely resistant to swelling after incubation at high ATP conditions like that used for the labeling reaction.
11. Boiling of the samples must be avoided. Incubation of highly hydrophobic proteins at high temperatures leads to the formation of protein aggregates which do not enter the SDS gels. After boiling for several minutes, especially the signals of Cox1, Cox3, and cytochrome *b* disappear. However, very short boiling steps (between 30 and 45 s) can be used to completely inactivate proteases (e.g., when mitochondria were proteolytically treated after the translation reaction).
12. 2.4 M sucrose is a very dense solution and tends to settle. Therefore, all sucrose containing buffers should always be remixed immediately before use.
13. Make sure that your SDS gel is really swimming in staining or destaining solution. Attachment to the bottom leads to uneven staining and destaining, respectively.
14. Used destaining solution can be recovered by filtering through an active carbon filter. The recovered solution is stored at room temperature and can be used up to several times for destaining SDS gels.
15. Alternatively, this incubation can also be done as an overnight step, especially, if the signal of the antibody is weak. In this case, the blots should be transferred to 4°C.

Acknowledgments

This work was supported by grants of the Deutsche Forschungsgemeinschaft (He2803/2-4 and SFB594 TP B05) and the Stiftung Rheinland Pfalz für Innovation to JMH.

References

1. Burger, G., Gray, M. W., and Lang, B. F. (2003) Mitochondrial genomes: anything goes. *Trends Genet.* **19**, 709–716.
2. Kurland, C. G., and Andersson, S. G. E. (2000) Origin and evolution of the mitochondrial proteome. Microbiol. *Mol. Biol. Rev.* **64**, 786–820.
3. Borst, P., and Grivell, L. A. (1978) The mitochondrial genome of yeast. *Cell.* **15**, 705–723.
4. Tzagoloff, A., Akai, A., and Needleman, R. B. (1975) Assembly of the mitochondrial membrane system. Characterization of nuclear mutants of Saccharomyces cerevisiae with defects in mitochondrial ATPase and respiratory enzymes *J. Biol. Chem.* **250**, 8228–8235.
5. Groot, G. S., Rouslin, W., and Schatz, G. (1972) Promitochondria of anaerobically grown yeast. VI. Effect of oxygen on promitochondrial protein synthesis *J. Biol. Chem.* **247**, 1735–1742.
6. Westermann, B., Herrmann, J. M., and Neupert, W. (2001) Analysis of mitochondrial translation products in vivo and in organello in yeast. *Methods Cell Biol.* **65**, 429–438.
7. McKee, E. E., and Poyton, R. O. (1984) Mitochondrial gene expression in Saccharomyces cerevisiae. I. Optimal conditions for protein synthesis in isolated mitochondria. *J. Biol. Chem.* **259**, 9320–9331.
8. Hell, K., Herrmann, J., Pratje, E., Neupert, W., and Stuart, R. A. (1997) Oxa1p mediates the export of the N- and C-termini of pCoxII from the mitochondrial matrix to the intermembrane space. *FEBS Lett.* **418**, 367–370.
9. Szyrach, G., Ott, M., Bonnefoy, N., Neupert, W., and Herrmann, J. M. (2003) Ribosome binding to the Oxa1 complex facilitates cotranslational protein insertion in mitochondria. *EMBO J.* **22**, 6448–6457.
10. Ott, M., Prestele, M., Bauerschmitt, H., Funes, S., Bonnefoy, N., and Herrmann, J. M. (2006) Mba1, a membrane-associated ribosome receptor in mitochondria. *EMBO J.* **25**, 1603–1610.

8

In Vivo Labeling and Analysis of Mitochondrial Translation Products in Budding and in Fission Yeasts

Karine Gouget, Fulvia Verde, and Antoni Barrientos

Summary

Mitochondrial biogenesis requires the contribution of two genomes and of two compartmentalized protein synthesis systems (nuclear and mitochondrial). Mitochondrial protein synthesis is unique on many respects, including the use of a genetic code with deviations from the universal code, the use of a restricted number of transfer RNAs, and because of the large number of nuclear encoded factors involved in assembly of the mitochondrial biosynthetic apparatus. The mitochondrial biosynthetic apparatus is involved in the actual synthesis of a handful of proteins encoded in the mitochondrial DNA. The budding yeast *Saccharomyces cerevisiae* and the fission yeast *Schizosaccharomyces pombe* are excellent models to identify and study factors required for mitochondrial translation. For that purpose, *in vivo* mitochondrial protein synthesis, following the incorporation of a radiolabeled precursor into the newly synthesized mitochondrial encoded products, is a relatively simple technique that has been extensively used. Although variations of this technique are well established for studies in *S. cerevisiae*, they have not been optimized yet for studies in *S. pombe*. In this chapter, we present an easy, fast and reliable method to *in vivo* radiolabel mitochondrial translation products from this fission yeast.

Key words: Budding yeast; fission yeast; *in vivo* protein synthesis; mitochondria; mitochondrial protein labeling; mitochondrial translation; *Schizosaccharomyces pombe*; *Saccharomyces cerevisiae*.

1. Introduction

Mitochondria are cellular organelles that host many intermediary metabolism reactions as well as the electron transport chain and oxidative phosphorylation system (OXPHOS) pathways required for the aerobic synthesis of adenosine triphosphate (ATP). This ancient organelle, believed to be a descendant of an earlier aerobic prokaryote *(1)*, has retained only a vestige of its original genetic information, most of which has been transferred to the nucleus of the host cell. The maintenance and propagation of functional mitochondria in yeast and higher eukaryotic cells is governed by 30 to 40 genes resident in the mitochondrial DNA (mtDNA) and approx 500 to 600 genes located in the chromosomal DNA.

Mitochondria are semi-autonomous organelles containing their own independent translational machinery. More than 100 nuclear genes code for components of the mitochondrial protein synthetic system, required to synthesize a handful of proteins. Most nuclear genes dedicated to maintaining mitochondrial translation encode ribosomal proteins, aminoacyl-tRNA synthetases, and initiation, elongation, and termination factors *(2,3)*. Additional gene products identified through mutant screens in yeast code for proteins that function in biogenesis of the translational machinery rather than in translation itself *(4–6)*.

Expression of the OXPHOS components is under the control of the two separate genomes. The OXPHOS system consists of five multimeric complexes (I–V), coenzyme Q and cytochrome *c*, which in concert act to conserve the energy of respiratory substrates in the form of a proton-motive gradient that is further converted into chemical energy during ATP synthesis. Only a small portion of the structural subunits forming the respiratory complexes are encoded on the mtDNA. For example, in humans, only 13 out of more than 90 subunits are encoded by the mitochondrial genome. This number is reduced to 7 in the yeasts *Saccharomyces cerevisiae* and *Schizosaccharomyces pombe* (**Fig. 1**) whose mitochondrial respiratory chain do not contain a multimeric complex I. These seven proteins include three cytochrome *c* oxidase subunits, one subunit of the bc_1 complex and three subunits of the F_0F_1 mitochondrial ATPase. The mtDNA also universally encode two ribosomal RNAs and a set of tRNAs. Some tRNAs can be imported from the cytoplasm in some plants and fungi *(7)* but in most cases they are all encoded in the mtDNA, in a smaller number than cytosolic tRNAs (i.e., 22 mammalian mt-tRNAs, 25 mt-tRNAs in both *S. cerevisiae* and *S. pombe*), since a single tRNA is able to read several codons *(8)*. Additionally, the mtDNA of yeasts and some plants also encodes for some

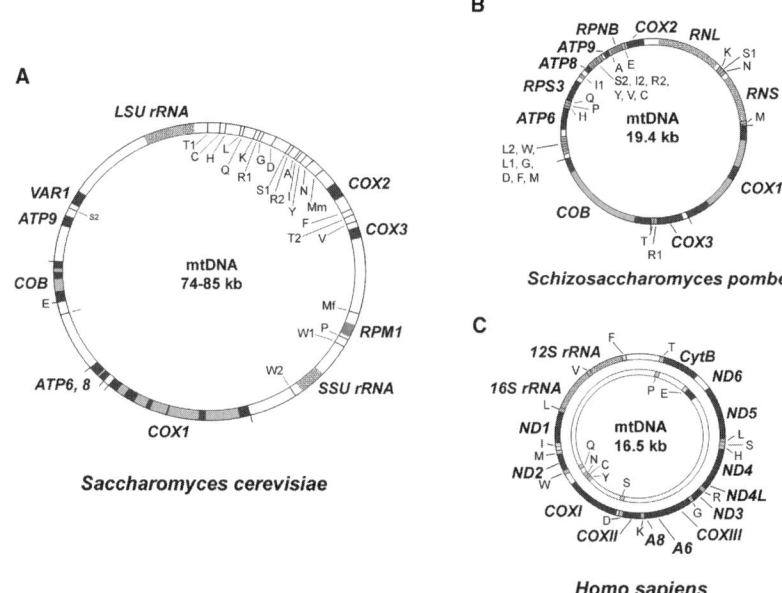

Fig. 1. Gene maps of the mtDNA genomes of *Saccharomyces cerevisiae*, *Schizosaccharomyces pombe*, and *Homo sapiens*. (**A**) *S. cerevisiae* mtDNA molecule (modified from **ref. 20**). Although the major form of mtDNA in *S. cerevisiae* is linear *(21)*, it has been represented in its circular form to facilitate the comparison with the other genomes presented in this figure. The genes *COX1*, *COX2*, and *COX3* encode subunits 1, 2, and 3 of cytochrome *c* oxidase; *ATP6*, *ATP8*, and *ATP9* encode subunits 6, 8, and 9 of ATPase; *COB* encode cytochrome *b*; *SSU rRNA* encodes the small ribosomal RNA; *LSU rRNA* encodes the large ribosomal RNA; and *VAR1* encodes a mito-ribosomal protein of the small subunit. (**B**) *S. pombe* mtDNA molecule (modified from **ref. 10**). *COX1*, *COX2*, and *COX3* encode subunits 1, 2, and 3 of cytochrome *c* oxidase; *ATP6*, *ATP8*, and *ATP9* encode subunits 6, 8, and 9 of ATPase; *COB* encode cytochrome *b*; *RNS* encodes the small ribosomal RNA; *RNL* encodes the large ribosomal RNA; *RPS3* encodes a mito-ribosomal protein of the small subunit, and *RNPB* encodes the RNA component of mt RNase P. (**C**) Human mtDNA molecule (modified from **ref. 22**). The genes *ND1*, *ND2*, *ND3*, *ND4*, *ND4L*, *ND5*, and *ND6* encode subunits 1, 2, 3, 4, 4L, 5, and 6 of NADH dehydrogenase or complex I; *COX1*, *COX2*, and *COX3* encode subunits 1, 2, and 3 of cytochrome *c* oxidase; *A6* and *A8* encode subunits 6 and 8 of ATPase; *CYTB* encode cytochrome *b*; *12S rRNA* encodes the small ribosomal RNA; and *16S rRNA* encodes the large ribosomal RNA. In the three schemes, the genes encoding for tRNAs are represented by the one-letter designation of their cognate amino acid or for this letter followed by a number when there is more than one tRNA recognizing the same amino acid.

mitochondrial ribosome proteins, including Var1 in the case of *S. cerevisiae* *(9)* and Rps3 in the case of *S. pombe* *(10)*.

The budding yeast *S. cerevisiae* has been classically used as a model organism to study mitochondrial biogenesis. Its metabolic properties as a facultative anaerobe allows for an easy selection of respiratory deficient mutants, for example with lesions in elements of the mitochondrial protein synthesis apparatus. However, care has to be taken with the manipulation of translational mutants in *S. cerevisiae* because they have a tendency to lose their mtDNA *(11)*. Even if this organism is a very useful tool to understanding many aspects of mitochondrial protein synthesis of higher eukaryotes including human, their mtDNA organization and gene expression present some differences with respect to more complex systems *(12)*. As an alternative, the fission yeast *S. pombe* mimics human cells more closely than *S. cerevisiae* in various aspects of mitochondrial physiology. For example, *S. pombe* is an obligatory aerobe and cannot survive the loss of mtDNA *(13)*. Its mtDNA is compact and organized in a way similar to human mtDNA *(14)*, and it does not contain messenger-specific translation activators as in *S. cerevisiae* *(15)* but it contains general translational factors with human counterparts *(12)*. The combination of studies in both model systems is definitely important to identify all the players involved in mitochondrial translation and their specific roles (**Fig. 2**).

In the absence of a true *in vitro* mitochondrial translation system, researchers have approached the study of mitochondrial translation by following the synthesis of mtDNA-encoded products either *in organello*, in isolated mitochondria, or *in vivo* in whole cells, the latter being the focus of this chapter. *In vivo* labeling of mitochondrial translation products in *S. cerevisiae* is a very well established technique in our laboratory and others. It consists of following the incorporation of a radiolabeled precursor, usually [^{35}S]-methionine into the newly synthesized mitochondrial proteins in whole cells in the presence of an inhibitor of cytoplasmic protein synthesis, usually cycloheximide or emetine. However, little is known concerning protein synthesis rates in *S. pombe* and the appropriate techniques to explore it *in vivo* have not been fully developed.

In this chapter we explain the similarities and differences between the classical protocol used routinely in *S. cerevisiae* *(16)* for *in vivo* mitochondrial protein synthesis, and the protocol proposed and tested for *S. pombe*. The main points of this technique are (a) the culture conditions that allow a good expression of mitochondrial genes, (b) the silencing of the cytoplasmic translational apparatus followed by the radiolabeling of mtDNA-encoded proteins with [^{35}S]-methionine, and (c) the separation of these proteins by sodium dodecyl sulfate-polyacrylamide gel electrophoresis (SDS-PAGE) followed by blotting to a nitrocellulose membrane and signal detection.

2. Materials
2.1. Strains and Culture Media
2.1.1. Strains

1. S. cerevisiae wild-type strain: W303 (MAT-A; ade2-1; his3-1,15; leu2-3,112; trp1-1; ura3-1).
2. S. pombe wild-type strain 972 h.

2.1.2. Media for Growth of S. cerevisiae

The compositions of the growth media for S. cerevisiae have been described elsewhere (11) and are reported here:

1. Complete galactose containing solid media (YPGal): yeast extract 10 g/L, bactopeptone 20 g/L, galactose 20 g/L, agar 20 g/L. All culture media is prepared in distilled water. Media is autoclaved during 20 min at 20 lb per square inch prior use. The media must be taken out from the autoclave promptly to avoid excessive burning of the sugars. Store the plates at 4°C.

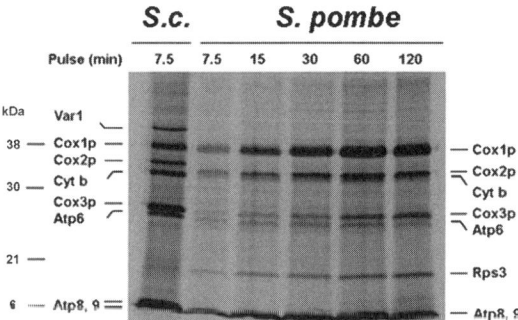

Fig. 2. Kinetics of in vivo [^{35}S]-methionine incorporation into newly synthesized mtDNA-encoded products in wild-type Schizosaccharomyces pombe cells in the presence of cycloheximide. S. cerevisiae and S. pombe wild-type cells were grown and labeled for the indicated amount of time as described in the present article. Saturation of [^{35}S]-methionine incorporation in S. pombe mitochondrial proteins is reached after pulses of 30 min. S. cerevisiae mitochondrial products were labeled and run as a control of the protein separation pattern. Notice that the ribosomal proteins Var1p (in S. cerevisiae) and Rps3 (in S. pombe) have different electrophoretic mobility due to their significantly different molecular weight (47 vs 27 kDa).

2. Minimum galactose-containing liquid media (WO-Gal): 6.7 g/L yeast nitrogen base-containing ammonium sulphate, 20 g/L galactose, prototrophic requirements (e.g., specific amino acids and nucleobases) supplemented as required for each strain (*see* **Note 1**). The pH should be equal or above 6.0. Autoclave as described above and store at room temperature.

2.1.3. Media for Growth of S. pombe

Media compositions for growth of *S. pombe* has been described elsewhere (*see* **ref. 17** and http://www-rcf.usc.edu/~forsburg/media.html) and are reported here:

1. Yeast extract with supplements (YES): 5 g/L yeast extract, 30 g/L glucose, supplements: 225 mg/L adenine, histidine, leucine, uracil, and lysine hydrochloride. Solid media is made by adding 2% Difco Bacto Agar.
2. Edinburgh Minimal medium (EMM): 3 g/L potassium hydrogen phthalate, 2.2 g/L Na_2HPO_4, 5 g/l NH_4Cl, 20 g/L glucose, 20 mL/L salts (52.5 g/L $MgCl_2.6H_2O$, 0.735 g/L $CaCl_2.2H_2O$, 50 g/L KCl, 2 g/L Na_2SO_4), 1 mL/L vitamins (1 g/L pantothenic acid, 10 g/L nicotinic acid, 10 g/L inositol, 10 mg/L biotin), 0.1 mL/L minerals (5 g/L boric acid, 4 g/L $MnSO_4$, 4 g/L $ZnSO_4.7H_2O$, 2 g/L $FeCl_2.6H_2O$, 0.4 g/L molybdic acid, 1 g/L KI, 0.4 g/L $CuSO_4.5H_2O$, 10 g/L citric acid). Solid media is made by adding 2% Difco Bacto Agar. Autoclave as described above. Medium can be stored at room temperature.
3. Minimum liquid medium (EMM liquid) with galactose: EMM medium (without glucose) is supplemented with 20 g/L galactose, 1 g/L glucose (*see* **Note 2**). Autoclave as described above. Medium can be stored at room temperature.

2.2. In Vivo *Radiolabeling of mtDNA-Encoded Products*

1. Reaction buffer for *S. cerevisiae*: 40 mM potassium phosphate buffer, pH 6.0, 20 g/L galactose. Store at room temperature and manipulate in sterile conditions.
2. Reaction buffer for *S. pombe*: 40 mM potassium phosphate buffer, pH 6.0, 20 g/L galactose, 1 g/L glucose (*see* **Note 3**). Store at room temperature and manipulate in sterile conditions.
3. Cycloheximide solution for *S. cerevisiae* : 10 mg/mL cycloheximide in water. It has to be freshly prepared and the solubilization of the antibiotic must be complete (*see* **Note 4**).
4. Cycloheximide solution for *S. pombe* : 10 mg/mL cycloheximide in reaction buffer. It also has to be prepared freshly and without any crystals.
5. Radiolabeled [^{35}S]-methionine at 15 mCi/mL. Store at −80°C in a restricted area (*see* **Note 5**). Always manipulate radioactive material in accordance to your institution's guidelines.
6. Cell solubilization buffer: 1.8 M NaOH, 1 M β-mercaptoethanol, 0.01 M phenylmethylsulfonylfluoride (PMSF). This solution must be prepared freshly.

7. Trichloracetic acid solution (TCA): 50% (v/v) TCA in water. Store at room temperature.
8. Tris-base solution: 0.5 M Tris-base (do not adjust the pH; it should be of approx 11). Store at room temperature.
9. Gel sample buffer: 2% (w/v) SDS, 10% (v/v) glycerol, 60 mM Tris-HCl, pH 6.8, 2.5% (v/v) β-mercaptoethanol, 0.02% (w/v) bromophenol blue. Store at room temperature.

2.3. Separation of the Labeled Newly Synthesized mtDNA-Encoded Products by Sodium Dodecyl Sulfate-Polyacrylamide Gel Electrophoresis (SDS-PAGE)

1. Acrylamide solution: acrylamide 30%, *bis*-acrylamide 0.2%. Before polymerization, acrylamide is highly neurotoxic. Handle with care and store in a dark bottle at 4°C.
2. Ammonium persulfate solution (APS): prepare 10% solution in water and store at 4°C.
3. Stacking gel: 5% (v/v) acrylamide solution, 60 mM Tris-HCl, pH 6.8, 0.1% (w/v) SDS, 0.05% (v/v) APS, 0.1% (v/v) *N,N,N,N'*-Tetramethyl-ethylenediamine (TEMED).
4. Separating gel: 17.5% acrylamide (v/v, as above), 375 mM Tris-HCl, pH 8.8, 0.1% (w/v) SDS, 0.05% (v/v) APS, 0.035% (v/v) TEMED.
5. Running buffer (5X): 30 g/L Tris-HCl, 144 g/L glycine, 5 g/L SDS. Store at room temperature. From this stock, prepare a 1X buffer by diluting the 5X stock in distilled water and adjusting the pH to exactly 8.3 with HCl (*see* **Note 6**).
6. Transfer buffer (1X): 200 mM glycine, 25 mM Tris-HCl, 20% (v/v) methanol. Store at room temperature.

3. Methods

The method proposed below for the analysis of *in vivo* synthesized mtDNA products in *S. pombe* is an adaptation of a standard procedure commonly used for *S. cerevisiae* (*16*) in our laboratory and others (*12,18,19*). The kinetics of the labeling is slower in the wild-type *S. pombe* strain used (maximum labeling after pulses of 30 min; **Fig. 2**) than in a wild-type *S. cerevisiae* strain used in the assay (maximum labeling after 15 min; *see* **refs.** *18,19*), although some inter-assay variability was observed. This simple technique has been very useful in the identification and study of the role of specific proteins in mitochondrial translation in both yeast species (*5,12,19*).

3.1. Culture Conditions for the Growth of Yeast Strains

Two days before the date of the protein synthesis experiment perform the following steps:

1. Prepare a fresh culture of yeast *S. cerevisiae* cells in complete media containing galactose, a carbon source that do not induce repression of the expression of mitochondrial proteins in this yeast. Incubate the culture O/N at 30°C.
2. Start a preculture of *S. pombe* cells, from a fresh plate, in EMM medium. Incubate the culture O/N at 32°C.

One day before the experiment:

1. For *S. cerevisiae*, inoculate enough cells from the fresh solid culture into 10 mL of minimum liquid medium containing galactose and supplemented with the prototrophic requirements to obtain a confluence of OD^{600} between 1 and 3 after overnight incubation (16–18 h) at 30°C under constant shaking.
2. For *S. pombe*, inoculate enough cells from the preculture into 50 mL of minimum liquid medium containing 2% galactose and 0.1 % of glucose. Grow cells exponentially (not exceeding densities of 1×10^7 cells/mL) for at least eight generations at 32°C under constant shaking.

On the day of the experiment:

Estimate the confluence of the cultures by measuring spectrophotometrically their OD at 600 nm as stated previously. Do not allow the cell culture to reach high densities to avoid loss of mitochondrial protein synthesis efficiency in overgrown cultures.

3.2. In Vivo *Radiolabeling of mtDNA-Encoded Products*

1. Transfer the equivalent of 0.6 OD of budding yeasts and 1.5 OD of fission yeasts (approx 10^7 and 2.5×10^7 cells, respectively) into microtubes with safe lock, due to further manipulation with radioactive isotope.
2. Pellet cells by centrifugation at 12,000g for 30 s and wash them in 500 µL of their specific reaction buffer.
3. Pellet all cells by centrifugation as above.
4. Resuspend the cells in 500 µL of reaction buffer supplemented with cycloheximide to inhibit cytoplasmic protein synthesis. Efficient inhibition of cytoplasmic ribosomes is readily accomplished by incubating *S. cerevisiae* cells in a buffer containing 0.2 mg/mL cycloheximide during 2.5 min prior to adding the labeled precursor. *S. pombe* cells require concentrations of 10 mg/mL and preincubation of 15 min (*see* **Note 7**).
5. Following preincubation with cycloheximide, directly add the radiolabeled precursor to the cell suspension maintaining the same buffer. Add 3 µL (*S. cerevisiae*) and 5 µL (*S. pombe*) of [^{35}S]-methionine with care and appropriate protections toward radiation hazard (*see* **Note 8**). Mix by vortexing.

6. Allow incorporation of [^{35}S]-methionine into the newly synthesized mitochondrial products by the convenient amounts of time depending on the objective of the experiment (*see* **Note 9**).
7. Pellet the cells by centrifugation at 10,000*g* for 1 min and resuspend them in 75 μL of cell solubilization buffer to stop the labeling reaction. Homogenize by vortexing and immediately add 500 μL of H$_2$O to dilute the cell solubilization buffer and 575 μL of 50% TCA to precipitate all the proteins after 5 min incubation on ice.
8. Pellet the TCA precipitated proteins by centrifugation at 10,000*g* for 5 min.
9. Wash the pellet once with 1.5 mL of Tris-base to neutralize and eliminate traces of TCA and once with distilled water to wash out traces of Tris-base.
10. Resuspend the final washed pellet in 25 μL of gel sample buffer. The solution must remain blue (*see* **Note 10**).
11. The samples can be used immediately for separation in a PAGE electrophoresis system or kept in a restricted area at –80°C until use.

3.3. Separation of the Labeled Newly Synthesized mtDNA-Encoded Products by SDS-PAGE

1. Prepare a 17.5% SDS-PAGE separating gel, overlaying it with isopropanol to obtain a flat gel surface after polymerization, which takes approx 30 to 45 min. After polymerization of the running gel, discard the isopropanol, wash out its traces with abundant distilled water, carefully dry the gel surface with a piece of filter paper, and overlay a 5% SDS-PAGE stacking gel. Insert the appropriate comb and wait approx 30 min for the stacking gel to polymerize (*see* **Note 11**).
2. Mount the gel on the electrophoresis apparatus and add the 1X running buffer to both anode and cathode chambers. Check that the upper chamber is not leaking, and for the absence of air bubbles between the gel and the buffer in the lower chamber. Remove the comb of the gel and rinse the wells twice with running buffer. Load the samples. If you have empty wells, fill them with 25 μL of 1X sample buffer.
3. Connect the chamber to a power supply and run the gel at 40 mA for approx 3 h or until the loading dye is just about running off the bottom (*see* **Note 12**).
4. Disassemble the electrophoretic glass plate sandwich, remove the stacking gel and discard it in the appropriate container for disposal of radioactive materials.
5. Blot the proteins onto a nitrocellulose membrane using a standard Western blotting technique using either a wet or a semi-dry blotting apparatus.
6. After finishing the transfer, carefully eliminate any traces of PAGE from the nitrocellulose membrane.
7. Allow the membrane to air-dry completely and expose it to x-ray film overnight prior to development (final results shown in **Fig. 2**; *see* **Note 13**).

4. Notes

1. Frequently the so-called "wild-type" *S. cerevisiae* strains carry mutations in several genes coding for factors involved in the metabolism of specific amino acids (frequently affecting the synthesis of a combination of the following: leucine, tryptophan, histidine, and methionine) or nucleobases (commonly uracil and adenine). These amino acids and nucleobases must be exogenously supplemented in the culture media. The wild-type strains of *S. pombe* do not contain usually this kind of mutations and do not require specific supplements.
2. *S. pombe* cells are not able to metabolize galactose efficiently and do grow at a poor rate in media containing exclusively this carbon source. However *S. pombe* cells can grow at higher rates if the galactose containing media is supplemented with a small amount (0.1%) of glucose *(12)*.
3. The reaction buffers are supplemented with the same carbon sources used for the growth of the different yeasts.
4. Cycloheximide can be difficult to dissolve. To obtain a complete solubilization warm the solution at 60°C for a few minutes and then to cool it down on ice.
5. Some providers of [^{35}S]-methionine add a colored inactive reagent in the solution. We recommend it for safety reasons because any leak or solution drop can be easily localized.
6. A pH of 8.3 is crucial for a correct separation of translation products, both with *S. cerevisiae* and *S. pombe* because the mitochondrial-encoded proteins are highly hydrophobic and do not separate in SDS-PAGE gels exclusively by their size. Such a pH 8.3 is necessary to allow for a good separation of specifically Var1p from Cox1p and Cox2p. With the system described here it is possible to obtain a good separation of the precursor and mature forms of Cox2p, which differ in 15 amino acids.
7. The concentration of cycloheximide in the reaction buffer is crucial to obtain an efficient inhibition of cytoplasmic protein synthesis. *S. pombe* requires cycloheximide concentrations 50 times higher than do *S. cerevisiae*. Although the reasons for that are not clear at the moment, it could be possible that the drug is less permeable to *S. pombe* cells or that this fission yeast has more efficient multidrug-resistant systems than *S. cerevisiae*.
8. The amount of [^{35}S]-methionine used in the *S. pombe* protocol is higher than the one used for labeling in *S. cerevisiae* because the number of cells used in the experiment is three times higher. Please note that you are now dealing with radioactive material and that you need to wear appropriate protections.
9. The time of the pulses must be adjusted depending on the objective of the experiment. Usually in the experimental conditions explained in this chapter the incorporation of [^{35}S]-methionine into the newly synthesized product is saturated after 15 min in *S. cerevisiae* and after 30 min (**Fig. 2**) in *S. pombe*.
10. If the sample turns yellow after addition of the sample buffer, it indicates that traces of acid are still present in the sample. In that event, neutralize the sample by adding a couple of microliters of diluted Tris-base.

11. We use a 14.5 × 17-cm glass plates with 1.5 mm spacers. A 4-cm stacking gel with 2.5-cm deep comb allows us to easily load up to 40 to 60 μL of sample if needed, although we usually resuspend our samples in a volume of 30 μL.
12. The gel of the sizes proposed in **Note 11** usually takes 3 h to run at constant amperage of 0.04. Do not use higher amperages to avoid overheating the gel and altering the migration pattern.
13. Depending on the labeling efficiency it can be necessary to expose a film from several hours (4–5 h can be enough) to a couple of days until the desired intensity of the signal is obtained.

Acknowledgments

This research was supported by National Institutes of Health Research Grant GM071775A (to A.B.), a Research Grant from the Muscular Dystrophy Association (to A.B.), and the National Science Foundation Grant MCB-0344798 (to F.V.)

References

1. Margulis, L. (1975) Symbiotic theory of the origin of eukaryotic organelles; criteria for proof *Symp. Soc. Exp. Biol.*, 21–38.
2. Pel, H. J., and Grivell, L. A. (1994) Protein synthesis in mitochondria *Mol. Biol. Rep.* **19**, 183–94.
3. Towpik, J. (2005) Regulation of mitochondrial translation in yeast. *Cell Mol. Biol. Lett.* **10**, 571–594.
4. Sirum-Connolly, K., and Mason, T. L. (1995) The role of nucleotide modifications in the yeast mitochondrial ribosome *Nucleic Acids Symp. Ser.*, 73–75.
5. Barrientos, A., Korr, D., Barwell, K. J., Sjulsen, C., Gajewski, C. D., Manfredi, G., Ackerman, S., and Tzagoloff, A. (2003) *MTG1* codes for a conserved protein required for mitochondrial translation *Mol. Biol. Cell* **14**, 2292–2302.
6. Datta, K., Fuentes, J. L., and Maddock, J. R. (2005) The yeast GTPase Mtg2p is required for mitochondrial translation and partially suppresses an rRNA methyltransferase mutant, mrm2 *Mol. Biol. Cell* **16**, 954–963.
7. Bhattacharyya, S. N., and Adhya, S. (2004) The complexity of mitochondrial tRNA import *RNA Biol.* **1**, 84–88.
8. Fukuhara, H., and Bolotin-Fukuhara, M. (1976) Deletion mapping of mitochondrial transfer RNA genes in *Saccharomyces cerevisiae* by means of cytoplasmic petite mutants *Mol. Gen. Genet.* **145**, 7–17.
9. Terpstra, P., Zanders, E., and Butow, R. A. (1979) The association of var1 with the 38 S mitochondrial ribosomal subunit in yeast *J. Biol. Chem.* **254**, 12653–12661.
10. Bullerwell, C. E., Leigh, J., Forget, L., and Lang, B. F. (2003) A comparison of three fission yeast mitochondrial genomes *Nucleic Acids Res.* **31**, 759–768.

11. Myers, A. M., Pape, L. K., and Tzagoloff, A. (1985) Mitochondrial protein synthesis is required for maintenance of intact mitochondrial genomes in *Saccharomyces cerevisiae EMBO J.* **4**, 2087–2092.
12. Chiron, S., Suleau, A., and Bonnefoy, N. (2005) Mitochondrial translation: elongation factor tu is essential in fission yeast and depends on an exchange factor conserved in humans but not in budding yeast *Genetics* **169**, 1891–1901.
13. Haffter, P., and Fox, T. D. (1992) Nuclear mutations in the petite-negative yeast *Schizosaccharomyces pombe* allow growth of cells lacking mitochondrial DNA *Genetics* **131**, 255–260.
14. Schafer, B. (2003) Genetic conservation versus variability in mitochondria: the architecture of the mitochondrial genome in the petite-negative yeast *Schizosaccharomyces pombe Curr. Genet.* **43**, 311–326.
15. Costanzo, M. C., Bonnefoy, N., Williams, E. H., Clark-Walker, G. D., and Fox, T. D. (2000) Highly diverged homologs of *Saccharomyces cerevisiae* mitochondrial mRNA-specific translational activators have orthologous functions in other budding yeasts *Genetics* **154**, 999–1012.
16. Zambrano, A., Fontanesi, F., Solans, A., et al. (2007) Aberrant Translation of Cytochrome *c* Oxidase Subunit 1 mRNA Species in the Absence of Mss51p in the Yeast *Saccharomyces cerevisiae Mol. Biol. Cell* **18**, 523–535.
17. Moreno, S., Klar, A., and Nurse, P. (1991) Molecular genetic analysis of the fission yeast *Schizosaccharomyces pombe Methods Enzymol.* **194**, 795–823.
18. Herrmann, J. M., Stuart, R. A., Craig, E. A., and Neupert, W. (1994) Mitochondrial heat shock protein 70, a molecular chaperone for proteins encoded by mitochondrial DNA *J. Cell Biol.* **127**, 893–902.
19. Barrientos, A., Korr, D., and Tzagoloff, A. (2002) Shy1p is necessary for full expression of mitochondrial *COX1* in the yeast model of Leigh's syndrome *EMBO J.* **21**, 43–52.
20. Grivell, L. A. (1995) Nucleo-mitochondrial interactions in mitochondrial gene expression *Crit. Rev. Biochem. Mol. Biol.* **30**, 121–164.
21. Nosek, J., Tomaska, L., Fukuhara, H., Suyama, Y., and Kovac, L. (1998) Linear mitochondrial genomes: 30 years down the line *Trends Genet.* **14**, 184–188.
22. Rotig, A., and Munnich, A. (2003) Genetic features of mitochondrial respiratory chain disorders *J. Am. Soc. Nephrol.* **14**, 2995–3007.

9

Exploring Protein–Protein Interactions Involving Newly Synthesized Mitochondrial DNA-Encoded Proteins

Darryl Horn, Flavia Fontanesi, and Antoni Barrientos

Summary

Biogenesis of the mitochondrial respiratory chain enzymes involves the coordinated action of the mitochondrial and nuclear genomes. As a matter of fact, the structural subunits forming these multimeric enzymes are encoded in both genomes. In addition, the assistance of nuclear encoded factors, termed *assembly factors*, is necessary to allow for the expression of the mitochondrial DNA-encoded subunits and to facilitate their maturation, membrane insertion, and further assembly into the corresponding enzymatic complex. These processes involve transient interactions among the newly synthesized mitochondrial products and specific assembly factors. The identification and characterization of these interactions can be achieved by the method described here, consisting of pulling down tagged versions of the assembly factors immediately after radiolabeling the mitochondrial translation products in isolated mitochondria, and analyzing the radiolabeled pulled-down material.

Key words: GST pull-down; *in organello* protein synthesis; mitochondria; mitochondrial biogenesis; mitochondrial translation; *Saccharomyces cerevisiae*.

1. Introduction

The mitochondrial respiratory chain and oxidative phosphorylation system (OXPHOS) consist of five multimeric enzymes embedded into the inner mitochondrial membrane. Biogenesis of these enzymes involves the coordinated action of the nuclear and mitochondrial genomes. In all eukaryotes, only a handful of proteins are encoded in the mitochondrial DNA (mtDNA), most of them subunits of the different OXPHOS enzyme complexes with the exception of mito-ribosomal proteins encoded in the yeast and plant mtDNAs *(1)*. In the yeast *Saccharomyces cerevisiae*, eight proteins are encoded in the mitochondrial genome, the mito-ribosomal protein Var1, three cytochrome *c* oxidase subunits, one subunit of the bc_1 complex and three subunits of the F_0F_1 mitochondrial ATPase. Biogenesis of the mtDNA-encoded proteins and their assembly in their correspondent multimeric complexes require the availability of their nuclear-encoded structural partners and the assistance of a large number of additional nuclear-encoded products *(2)*. These proteins act at all levels of the biogenesis and assembly process of the mitochondrial-encoded subunits, including their synthesis in the mitochondrial matrix *(3,4)*, insertion in the inner mitochondrial membrane *(5,6)*, addition of prosthetic groups if required and assembly into the multimeric enzyme to which they belong *(7–9)*. Through their biogenetic pathways, the newly synthesized mtDNA-encoded proteins transiently interact with specific biogenetic factors (**Fig. 1A**). The identification of these interacting factors and the characterization of these interactions are both important to understand the process of mitochondrial biogenesis and its regulation.

Because the lack of a true *in vitro* mitochondrial protein synthesis system, this process and the fate of mtDNA-encoded proteins during and after their synthesis have been traditionally approached either *in vivo* or *in organello* *(10)*. In vivo, whole cells are poisoned with cytoplasmic protein synthesis inhibitors prior to the addition of a labeled precursor, usually [^{35}S]-methionine. In this system, the study of protein–protein interactions require further purification of a mitochondrial-enriched fraction once the synthesis has been completed. A more convenient alternative consists of labeling the mtDNA-encoded products *in organello* in previously purified, intact mitochondria and readily proceed to extract the newly synthesized products in the presence of salt and detergents under native conditions in which the interactions of the newly synthesized proteins with their interacting biogenetic factors are preserved. In this system, subsequent pull-down experiments allow for the identification of the interacting partners.

In recent years, a number of techniques have been developed to detect protein–protein interactions. The use of tagged proteins, which can be easily purified, is the basis of these techniques. Here we describe a method to examine the interactions between newly synthesized mitochondrial proteins and their biogenetic factors in the ycast *S. cerevisiae* (**Fig. 1B**). Briefly, the overall method consists of (a) constructing yeast strains expressing tagged versions of nuclear-encoded protein candidates to interact with newly synthesized mtDNA-encoded product/s, (b) purifying mitochondria competent for protein synthesis, (c) *in organello* radiolabeling of mitochondrial protein synthesis products, and (d) protein pull-down assays.

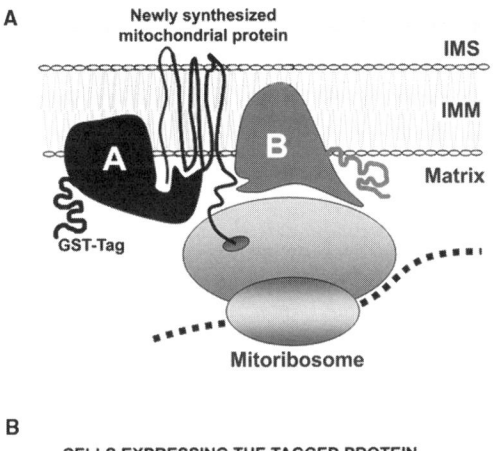

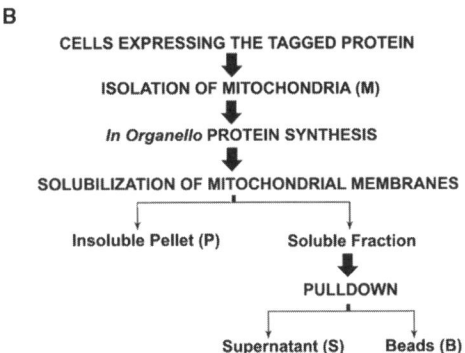

Fig. 1. Exploring protein–protein interactions involving newly synthesized mitochondrial DNA (mtDNA)-encoded proteins. (**A**) Nuclear-encoded mitochondrial proteins (represented by A and B) interact with newly synthesized mtDNA-encoded proteins to facilitate their synthesis, maturation, insertion into the inner mitochondrial membrane and further assembly into multimeric respiratory complexes. (**B**) Diagram depicting overall scheme of the protocol followed to exploring protein–protein interactions involving newly synthesized mtDNA-encoded proteins including constructing tagged protein expressing strain, isolation of mitochondria, *in organello* protein synthesis, and pull-down assay. Letters in bold and in parentheses indicate samples taken for analysis on SDS-PAGE.

The success of the methodology is exemplified with the recent observation by our group that newly synthesized Cox1p interacts with Mss51p, a cytochrome *c* oxidase (COX) subunit 1 (*COX1*) mRNA specific translational activator (**Fig. 2A,B**) that also plays an important role in regulation of COX assembly in the yeast *S. cerevisiae* (*11*).

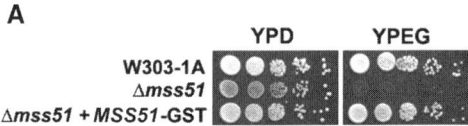
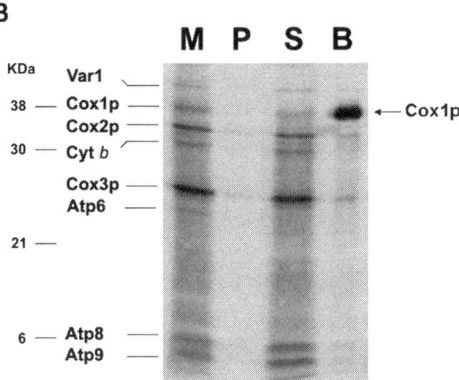

Fig. 2. Mss51p interacts with newly synthesized Cox1p. (**A**) Serial dilutions of the haploid wild-type strain W303, a null mutant of *mss51* and the same null mutant with a chromosomally integrated plasmid expressing Mss51p-GST were spotted on YPD and YPEG plates and incubated at 30°C for 2.5 d. (**B**) Mitochondria were prepared from an *mss51* null mutant with a chromosomally integrated plasmid expressing Mss51p-GST. Mitochondria were labeled with [^{35}S]-methionine for 30 min and extracted with 1% lauryl maltoside, 1 M KCl, and 1 mM PMSF. The extract was clarified by centrifugation at 50,000g for 30 min and incubated with glutathione–sepharose beads for 4 h at 4°C. After centrifugation at 500g for 5 min, the supernatant was collected and the beads were washed three times with PBS. Mitochondria (M) corresponding to 2 mg protein, equivalent volumes of the membrane pellet (P) after lauryl maltoside extraction and of the supernatant from the glutathione–sepharose beads (S) were separated on a 17.5% polyacrylamide gel by SDS-PAGE as described in the text. The amount of washed beads (B), however, corresponded to approx 50 mg of the starting mitochondria. The mitochondrial translation products and the size of molecular-weight markers are identified in the margin.

2. Materials

2.1. Generating Strain Expressing Tagged Versions of Protein Candidates to Interact With Newly Synthesized mtDNA-Encoded Product/s

2.1.1. Cloning

1. Primers for polymerase chain reaction (PCR) amplification.
2. High-fidelity polymerase and buffers.

3. DNA gel extraction kit.
4. Required DNA restriction enzymes and the corresponding digestion buffers.
5. T4 DNA ligase and buffer.

All enzymes must be kept at −20°C.

2.1.2. Yeast Strains and Yeast Growth Media

1. Yeast strain W303-1A (MAT-A *ade2-1 his3-1,15 leu2-3,112 trp1-1 ura3-1*).
2. Minimum medium *(12)*: 0.67% (w/v) yeast nitrogen base without amino acids, 2% glucose.
3. Complete media *(12)* containing glucose (Yeast extract/peptone/dextrose [YPD]: 2% glucose, 1% yeast extract, 2% peptone), or galactose (YPGal: 2% galactose, 1% yeast extract, 2% peptone) or ethanol plus glycerol (YPEG: 2% ethanol, 3% glycerol, 1% yeast extract, 2% peptone) as the carbon sources.

   Agar (20 g/L) is added when required to obtain solid media. All culture media is prepared in distilled water. Media is autoclaved during 20 min at 20 lb per square inch prior use. The media must be taken out from the autoclave promptly to avoid excessive burning of the sugars.

2.1.3. Yeast Transformation Using Lithium Acetate

1. Lithium acetate solution (TEL): 10 mM Tris-HCl, pH 7.5, 1 mM ethylenediamine tetraacetic acid (EDTA), 100 mM lithium acetate. Sterilize and store at room temperature.
2. Polyethylene glycol solution (PEG): 40% polyethylene glycol 4,000 (w/v) in TEL. Sterilize by filtering (do not autoclave). Because PEG easily degenerates, it is advised to prepare the solution freshly for each experiment or to store it at room temperature for a short period of time.
3. DNA carrier solution: 2 mg/mL DNA sodium salt type III from Salmon testes in sterile TE solution (10 mM Tris-HCl, pH 8.0, 1 mM EDTA). Disperse the DNA into solution by passing it up and down repeatedly in a 10 mL pipet and mix vigorously in a magnetic stirrer until it dissolves completely. Aliquot and store at −20°C. Denature the DNA prior using it by boiling an aliquot at 95°C for 5 min and cool it down in ice.

2.1.4. Whole-Genome DNA Extraction

1. Cell solubilization solution (solution A): 50 mM Tris-HCl, 10 mM EDTA pH 8.0, 0.3% β-mercaptoethanol, and 0.5 mg/mL Zymolyase-20T.

2.2. Isolation of Mitochondria for *in organello* Protein Synthesis

1. Complete media containing galactose as the carbon source: 2% galactose, 1% yeast extract, 2% peptone.

2. Cell wall digestion buffer: 1.2 M sorbitol, 20 mM potassium phosphate buffer, pH 7.4, 0.6 mg/mL Zymolyase-20T (see **Note 1**).
3. Homogenization buffer: 10 mM Tris-HCl, pH 7.5, 1 mM EDTA, 0.2% fatty acid-free bovine serum albumin (BSA), 1 mM phenylmethylsulfonyl fluoride (PMSF), and 0.6 M sorbitol.
4. Isoosmotic mitochondrial buffer (SEH Buffer): 0.6 M sorbitol, 1 mM EDTA, and 20 mM Hepes, pH 7.2.

2.3. In organelle *Protein Synthesis*

1. Amino acid stock solution: solubilize 20 mg each of the amino acids alanine, arginine, aspartic acid, asparagine, glutamic acid, glutamine, glycine, histidine, isoleucine, leucine, lysine, phenylalanine, proline, serine, threonine, tryptophan, and valine in 10 mL distilled water. Aliquot in 100 µL portions and store at –80°.
2. 10 mM cysteine: Solubilize 1.2 mg of cysteine in 1mL of distilled water. Aliquot in 20 µL and store at –80°C.
3. 1 mg/mL tyrosine: solubilize 1 mg of tyrosine in 900 µL of distilled water, adjust to pH 7.0 with KOH and add water to a total volume of 1 mL. Aliquot in 20 µL and store at –80°C (see **Note 2**).
4. 1.5X translation buffer (TB): 0.9 M sorbitol, 225 mM KCl, 22.5 mM potassium phosphate buffer, pH 7.2, 30 mM Tris-HCl, pH 7.2, 19 mM MgSO$_4$, 4.5 mg/mL BSA, 6 mM ATP, 0.75 mM GTP, 1.7 mg α-ketoglutarate, 3.5 mg phosphoenolpyruvate, 9.1 µL/mL amino acid stock solution (2 mg/mL of all amino acids except methionine, cysteine, and tyrosine), 0.1 mM cysteine, 18.2 µg/mL tyrosine (see **Note 3**).
5. Washing buffer: 0.6 M sorbitol, 1 mM EDTA, 5 mM cold methionine.
6. Radiolabeled [^{35}S]-methionine at 15 Ci/mL. Always manipulate radioactive material in accordance to your institution's guidelines.

2.4. Pull-Down Assay

1. Glutathione-*S*-transferase (GST) agarose beads (Amersham Biosciences).
2. Phosphate-buffered saline (PBS): 137 mM NaCl, 2.7 mM KCl, 10 mM phosphate buffer, pH 7.4.

2.5. Sodium Dodecyl Sulfate-Polyacrylamide Gel Electrophoresis (SDS-PAGE) for Separation of the Pulled-Down Material

1. 30.2% polyacrylamide solution: 30% acrylamide, 0.2% *bis*-acrylamide and store it in a dark bottle at 4°C.
2. Ammonium persulfate solution (APS): prepare 10% solution in water and keep it at 4°C.

3. Separating PAGE: 17.5% polyacrylamide, 0.376 M Tris-HCl, pH 8.8, 0.1% (w/v) SDS, 0.05 % APS, 0.035 % (v/v) tetramethyl-ethylenediamin (TEMED.)
4. Stacking PAGE: 5% polyacrylamide, 0.06 M Tris-HCl, pH 6.8, 0.1% SDS, 0.05 % APS, 0.1 % (v/v) TEMED.
5. Running buffer 1X: 50 mM Tris, 384 mM glycine, 0.1% SDS. Adjust the pH precisely to 8.3 (with 1 M HCl), which is crucial for a good separation of the proteins (*see* **Note 4**).
6. Gel-loading buffer: 2% (w/v) SDS, 10% glycerol, 60 mM Tris-HCl, pH 6.8, 2.5% (v/v) β-mercaptoethanol, 0.02% (w/v) bromophenol blue.

3. Methods
3.1. Generating Strains Expressing Tagged Proteins

In recent years, a number of techniques have been developed to detect protein–protein interactions. The use of tagged proteins, which can be easily purified, is the base of these techniques. Several tags can be used, including hemagglutin epitope tag (HA), GST, polyhistidine (His), and *Staphylococcus aureus* Protein A. It is necessary to be aware that the tag could interfere with the function of the protein (*see* **Note 5**). For this reason, it is necessary to test whether the tagged protein retains the wild-type function by complementation analysis.

In general, we use recombinant DNA technology to produce yeast-expression vectors that code for polypeptides containing our protein/s of interest linked to the selected tag. The resulting hybrid genes are integrated into the chromosomal DNA of a yeast strain carrying a null mutation of the gene being integrated. Because we are studying factors involved in the biogenesis of the mitochondrial respiratory chain and OXPHOS system, the knockout strain will be respiratory deficient and therefore unable to grow in nonfermentable carbon sources. The expressed tagged protein, if functional, must restore the ability of the knockout strain to grow in respiratory media at a rate equivalent to the wild-type strain.

3.1.1. PCR Amplification of the Disrupting DNA Fragment

The first task in creating a strain expressing a tagged protein is to remove or knockout the wild-type gene/protein. This can be easily accomplished in the yeast *S. cerevisiae* by taking advantage of its ability to perform homologous recombination with very short region of homology (~40 nucleotides; *see* **Note 6**). The gene to be disrupted will be substituted by a disruptor prototrophic marker gene, commonly coding for a factor necessary for the synthesis of an amino acid (i.e., histidine, leucine, or tryptophan), a nucleobase (i.e., adenine or uracyl), or one that confers resistance to an antibiotic (i.e., kanamycin).

1. Design primers to amplify by PCR the disruptor gene flanked by 40 to 60 nucleotides of DNA sequence homologous to the DNA sequence flanking the gene that is to be knocked out. Use commercially available plasmids containing the disruptor gene or wild-type genomic DNA as the template for the PCR reaction.
2. Purify the PCR fragment either directly from the reaction suspension or after separating the DNA on a 0.8% agarose if required to eliminate any unspecific amplification product, by using a commercially available extraction kit.

3.1.2. Yeast Transformation

Transform the yeast cells with the purified PCR fragment. Several methods can be used to transform yeast cells. Here we describe the transformation by lithium acetate/single-stranded DNA carrier/PEG method because it is relatively quick and simple to perform and as well as one of the more efficient.

1. Grow a 10 mL preculture of wild-type cells overnight in complete YPD media at 30°C with constant shaking.
2. Transfer an aliquot of the preculture to a fresh 10 mL flask of YPD to obtain a final OD^{600} of approx 0.1 to 0.2. Grow the cells until the culture reaches a confluence of OD^{600} of approx 0.4–0.5. This is very important because only cells in the mid-logarithmic phase of growth will become significantly competent for transformation.
3. Transfer 1.5 mL of the culture to a microtube and pellet the cells at $1,500g$ for 5 min.
4. Wash the cells in 1 mL of TEL and resuspend them in 0.1 mL of TEL. To this cell suspension, add 5 μL of salmon sperm carrier DNA premixed with 1 to 4 μg of transforming DNA in a volume of less than 15 μL. Incubate for 30 min at 30°C without shaking.
5. Add 0.7 mL of 40% polyethylene glycol (PEG) prepared in TEL (*see* **Note 7**) and mix it by pipeting. Incubate for 30 to 60 min at room temperature without shaking.
6. Heat-shock the cells for 10 min at 42°C.
7. Pellet the cells by centrifugation at $1,500g$ for 1 min and wash them in 0.2 mL of sterile dH_2O. Resuspend the cells in 0.1 mL of sterile dH_2O and plate them on selective solid media lacking the appropriate amino acid, nucleobase or containing the selective antibiotic.

3.1.3. Confirmation of the Disruption

Confirm the selected clones by testing for integration of the selectable marker in the correct locus and consequent disruption of the desired gene. This can be done by amplifying the locus of interest by PCR using genomic DNA as the template.

1. Yeast-genomic DNA is extracted from 1 mL of a 10 mL overnight culture in YPD media.
2. Pellet the cells and wash them once with 0.5 mL of water. Resuspend the cells in 150 μL of solution A and incubate them for 1 h at 37°C to digest the cell wall.

3. Solubilize the cell membranes by adding 20 μL of 10% SDS and vortex.
4. To facilitate the precipitation of the DNA add 100 μL of 8 M ammonium acetate, vortex and incubate the sample at −20°C for 15 min prior to centrifugation for 10 min at 4°C.
5. Transfer 180 μL of the supernatant fraction to a new tube and precipitate the DNA by adding 120 μL of isopropanol and incubating 5 min at room temperature.
6. Pellet the DNA by centrifugation for 15 min at 4°C and wash it once with 300 μL 80% ethanol. Air-dry the pellet of DNA and resuspend it in 30 μL of sterile water.
7. Analyze the genetic locus of interest by PCR using primers that allow for the amplification of fragments of different size in the mutant and in the wild-type strains.

3.1.4. Creation of a Construct Expressing the Tagged Protein

1. Amplify by PCR the gene of interest, the one that was previously disrupted, from genomic DNA from a wild-type strain with primers that include at least 500 bp upstream of the start codon, thus containing the promoter, and a primer that would allow cloning with a fusion protein tag such as GST (*see* **Note 8**).
2. Digest the PCR product and a suitable yeast-integrative vector with the appropriate restriction enzymes.
3. Purify the digested DNA from a 0.8% agarose gel using a DNA gel extraction kit.
4. Ligate the purified DNA fragments at 16°C overnight with T4 DNA ligase, transform the ligation into *Escherichia coli* and plate the bacteria on selective media.
5. Test the transformant clones for the presence of the correct construct by PCR and/or appropriate restriction digestion of isolated plasmid mini-preps.
6. Grow overnight the positive clones and isolate the constructs by standard plasmid DNA maxi-preps.
7. Linearize the construct by digestion with an enzyme that cuts only once and within the selectable prototrophic marker.
8. Transform the linearized construct into the yeast knockout strain by the lithium acetate method as described in **section 3.1.2**.
9. Test the positive clones for plasmid integration by PCR using as the template genomic DNA extracted as described in **section 3.1.3**.
10. Test the clones for the ability of the transformant gene encoding for tagged protein to complement the deletion mutant phenotype (**Fig. 2A**).

3.2. Isolation of Mitochondria Competent for in organello Protein Synthesis

Among the different methods available for isolating high-quality mitochondrial fractions, we have found the following protocol, adapted from the method described by Herrmann et al. *(13)* to be best suited for preparation of mitochondria competent for *in organello* protein synthesis (*see* **Note 9**).

1. Grow the cell cultures overnight in galactose media at 30°C with constant shaking
2. From this confluent culture, inoculate 5 mL (the volume must be empirically calculated for each strain) into 800 mL of galactose media so that the OD^{600} at the time of harvest (usually after overnight incubation) is between 1.6 and 2 (*see* **Notes 10** and **11**).
3. Harvest the cells by centrifugation at 900*g* for 5 min and then wash them once with 250 mL of distilled deionized water.
4. Resuspend the washed cells in 100 mM Tris-HCl pH 8.8, 10 mM DTT using 1mL per 2 g of cells and incubated at 30°C for 10 min with very gentle shaking.
5. Pellet the cells by centrifugation at 2,200*g* for 10 min and washed them with 1.2 M sorbitol using 150 mL per 10 g of cells.
6. Resuspend the washed cells in cell wall digestion buffer using 1mL for every 0.15 g of cells and incubated for 30 to 60 min at 30°C with gentle shaking.
7. After 30 min the cells should be checked for conversion to spheroplasts. This can be done by adding 50 μL of cells to 2 mL of water and a control sample of 50 μL of cells to 2 mL of 1.2 M sorbitol. The water sample should have an absorbance at 600 nm of 10 to 20% reduction as compared to the control if most of the cells have been converted to spheroplasts (*see* **Note 12**).
8. After approx 80% of the cells have been converted to spheroplasts, they are pelleted by centrifugation at 1,250*g* for 5 min.
9. Resuspend the spheroplasts in homogenization buffer using 1 mL per 0.15 g of cells (*see* **Note 13**).
10. Homogenize the spheroplasts in a loose glass/Teflon dounce using 10 strokes (*see* **Note 14**).
11. Centrifuge the homogenate at 1,700*g* for 10 min. Repeat this step with the supernatant fraction to eliminate residual cellular debris.
12. Centrifuge the supernatant fraction at 14,000*g* for 12 min to pellet mitochondria.
13. The pelleted mitochondria are resuspended in 10mL of SEH buffer and centrifuged at 2,200*g* for 5 min to pellet damaged mitochondria.
14. Centrifuge the supernatant fraction containing mitochondria with intact outer membrane at 14,000*g* for 12 min.
15. Resuspend the mitochondrial pellet in a small amount of SEH (200 μL–500 μL) depending on the pellet size. Measure the protein concentration by standard procedures and adjust the mitochondrial suspension to a final concentration of 10 mg/mL. If the mitochondria are not used immediately for *in organello* protein synthesis, prepare aliquots of 50 μL, freeze them by immersion in liquid nitrogen and stored them at –80°C until further use.

3.3. In Organello *Protein Synthesis*

This technique uses the natural translational machinery of the mitochondria to synthesize proteins encoded in the mitochondrial genome (*see* **Notes 15** and **16**).

1. Resuspend 150 μL of 10 mg/mL isolated mitochondria (1.5 mg protein total), freshly prepared or thawed just prior use, in 330 μL of 1.5X translation buffer and add 10 μL of 10 mg/mL pyruvate kinase to maintain the mitochondria metabolically active. Mix by gentle pipeting and incubate for 5 min at 30°C.
2. Add 10 μL of [^{35}S]-methionine and allow for its incorporation into the newly synthesized mtDNA products with pulses of 30 min at 30°C.
3. Stop the labeling reaction by adding 1 mL of washing buffer containing excess of cold methionine.
4. Re-isolate mitochondria by centrifugation at 18,000g for 5 min at 4°C and wash them once with 0.5 mL of the same buffer.
5. Finally, resuspend the pellet in 125 μL of SEH buffer. Aliquot 5 μL of this suspension into 75 μL of 1X gel-loading buffer (**Fig**. **1B [M]**) to keep a sample of total labeled mitochondria. Use the remaining 120 μL of mitochondrial suspension for further analysis in the pull-down assay.

3.4. Pull-Down Assay

The fusion protein is extracted under native conditions and is subsequently purified on the appropriate type of sepharose resin depending on the tag that has been used. The procedure described here is based on the use of GST-tagged proteins, a strategy that has been extensively used in our laboratory. To test if a newly synthesized mtDNA-encoded protein is pulled down together with the tagged protein, the putative complex will be fractionated by SDS-PAGE and transferred to a membrane that will be exposed to an x-ray film in order to detect the radiolabeled proteins.

3.4.1. Mitochondrial Membranes Solubilization and Protein Extraction Under Native Conditions and GST Pull-Down

1. To the 120 μL of labeled mitochondrial suspension add the following reagents in order: 22 mg of KCl, 121 μL of distilled water, 3 μL of 200 mM PMSF, and 30 μL of 10% lauryl maltoside, to solubilize the mitochondrial membranes and extract the newly synthesized proteins, which are cotranslationally inserted into the inner mitochondrial membrane, complexed with their partners of interest (hopefully our GST-tagged protein).
2. Centrifuge the sample at 60,000g for 10 min at 4°C and keep the pellet for further testing. Perform an additional similar centrifugation with the supernatant fraction to completely eliminate any residual insoluble material and reserve the supernatant, which should contain the tagged protein and all labeled newly synthesized mitochondrial proteins.
3. Wash the pellet with 125 μL of PBS and resuspended it in 125 μL SEH. From the resuspended pellet, place an aliquot of 5 μL in 75 μL of gel-loading buffer.

The pellet sample will be used to test that all labeled proteins were effectively extracted (**Fig. 1B [P]**).

4. To the supernatant fraction add 30 μL of GST beads suspension, prewashed with cold PBS, and incubate in an orbital rotor with gently rotation for 4 h at 4°C.
5. Pellet the beads by centrifugation at 500g for 5 min at 4°C (*see* **Note 17**).
6. Reserve the supernatant fraction and place 5 μL into 75 μL of gel-loading buffer to obtaining a sample of the soluble unbound material (**Fig. 1B [S]**).
7. Wash the beads twice with 1 mL of PBS and resuspend them in 75 μL of gel-loading buffer to obtain a sample of the material bound to the beads (**Fig. 1B [B]**).

3.4.2. SDS-PAGE for Separation of the Pulled-Down Material

The four samples (whole mitochondria [M], pellet [P], supernatant [S], beads [B]) are run on a 17.5% SDS-PAGE gel (*see* **Notes 18 and 19**).

1. Prepare a 17.5% SDS-PAGE separating gel, overlay it with isopropanol to obtain a flat gel surface after polymerization, which takes approx 30 to 45 min. After polymerization of the running gel, discard the isopropanol, wash out its traces with abundant distilled water, carefully dry the gel surface with a piece of filter paper and overlay a 5% SDS-PAGE stacking gel. Insert the appropriate comb and wait approx 30 min for the stacking gel to polymerize (*see* **Note 20**).
2. Run the gel at 40 mA for approx three h or until the loading dye is just about running off the bottom.
3. Blot the proteins onto a nitrocellulose membrane using a standard Western blotting technique using either a wet or a semi-dry blotting apparatus.
4. Allow the membrane to dry completely and expose it to x-ray film overnight prior to development (**Fig. 2B**; *see* **Note 21**).

4. Notes

1. Digestion buffers containing Zymolyase-20T must be prepared immediately before using to maintain the maximum activity of the enzyme.
2. Cysteine and tyrosine are very insoluble and stock solutions of these amino acids (10 mM cysteine and 1 mg/mL tyrosine) are better prepared and stored separately. This translation buffer contains the appropriate amounts of each component, previously established empirically *(14)*, to ensure an optimal synthesis of the mtDNA-encoded products.
3. The 1.5X translation buffer is prepared immediately prior use.
4. The mitochondrial-encoded proteins are highly hydrophobic and do not separate in SDS-PAGE gels exclusive by their size. To work with a running buffer of exactly pH 8.3 is necessary to allow for a good separation of specifically Var1p from Cox1p and Cox2p. With the system described here it is possible to obtain a good separation of the precursor and mature forms of Cox2p.

5. It is often necessary to add a flexible peptide fragment between your protein of interest and the tag. This can be accomplished by adding two to three glycine residues at this junction.
6. Depending on the research project, it can be interesting and/or necessary to create double knockout mutants. The first knockout is for the gene that is going to be subsequently tagged and integrated into the genome, and the second knockout allows for the examination of the interaction or loss of affects when the second gene is missing.
7. Because the PEG solution is very viscous, it is fundamental to mix well the cells after adding the solution.
8. Remember to create a fusion protein, thus include a stop codon after the tag and not after the gene. Also make sure the coding sequence corresponding to the tag is in-frame with the gene.
9. No detergent should be used to wash glassware that is used for mitochondrial preparations. Remaining traces of detergent could disrupt the mitochondrial membranes.
10. The confluence of the cell culture at harvest time, measure as OD at 600 nm, is very important. For a good quality and appropriate yield of mitochondrial membranes we should work with cultures at OD_{600} between 1.6 and 2.
11. The yield of mitochondrial proteins is approx 2 mg per liter of initial culture for wild-type strains, and close to a half of that value for respiratory deficient strains. Taking this into account, inoculate as many 800 mL cultures as necessary in order to recover enough mitochondria to proceed with the labeling of the mitochondrial products, purification of our tagged protein and pull-down experiments.
12. The digestion of the yeast cell wall with Zymolyase is a crucial step. Over digestion with Zymolyase could cause not only digestion of the yeast cell wall but also the mitochondrial membranes themselves. On the other hand, under digestion will decrease the yield of mitochondria isolated.
13. All cells subfractionate must be kept on ice at this point and all buffers chilled.
14. Ten strokes with our dounce is optimum for the balance between a good quality and a good quantity of mitochondria. However, the number of strokes must be empirically adjusted for each type of homogenizer.
15. Good quality mitochondria are an absolute requirement for the protein synthesis reaction to be successful.
16. Cycloheximide could be added to the reaction to inhibit protein synthesis of cytoplasmic ribosomes that are attached to the outer mitochondrial membrane and frequently copurify with mitochondria. This is unnecessary if the mitochondria have been properly washed and the preparation is in general of good quality.
17. It is important that the agarose beads are not centrifuged at more than $500g$. Centrifugation at higher speeds can crush the beads and thus release prematurely any proteins that bound to them.
18. mtDNA–encoded proteins are very hydrophobic and do not separate on PAGEs exclusively depending on their molecular weight. The proportion of acrylamide/*bis*-acrylamide and the concentration of SDS in the gel and running buffer, as well as the pH of the running buffer are all parameters that affect the final

resolution of the different labeled proteins. With the conditions described here, all the proteins clearly separate resulting in the migration pattern observed in **Fig. 2**. From top to the bottom in the gels, the proteins migrate in the following order: Var1p, Cox1p, Cox2p, Cyt *b*, Cox3p, Atp6, Atp8, and Atp9.

19. In some experiments, it might be possible to detect a band running between Cox1p and Cox2p. This band corresponds to the precursor form of Cox2p, which has a slightly slower electrophoretic mobility than the mature protein.
20. We use a 14.5 × 17-cm glass plates with 1.5-mm spacers. A 4-cm stacking gel with 2.5-cm deep comb allows us to easily load up to 40 to 60 µL of sample if needed, although we usually resuspend our samples in a volume of 30 µL. This size gel usually takes 3 h to run at constant amperage of 0.04. Do not use higher amperages to avoid overheating the gel and altering the migration pattern.
21. Depending on the labeling efficiency it can be necessary to expose a film for several days until the desired intensity of the signal is obtained.

Acknowledgments

This research was supported by National Institutes of Health Research Grant GM071775A (to A.B.), and a Research Grant from the Muscular Dystrophy Association (to A.B.), and a Telethon-Italy Fellowship GFP05008 (to F.F.)

References

1. Scheffler, I. E. (1999) Mitochondria. Wiley-Liss Eds, New York.
2. Tzagoloff, A., and Dieckmann, C. L. (1990) PET genes of *Saccharomyces cerevisiae*. *Microbiol. Rev.* **54**, 211–225.
3. Mulero, J. J., and Fox, T. D. (1993) Alteration of the Saccharomyces cerevisiae COX2 mRNA 5'–untranslated leader by mitochondrial gene replacement and functional interaction with the translational activator protein PET111. *Mol. Biol. Cell* **4**, 1327–1335.
4. Costanzo, M. C., and Fox, T. D. (1993) Suppression of a defect in the 5' untranslated leader of mitochondrial COX3 mRNA by a mutation affecting an mRNA-specific translational activator protein. *Mol. Cell Biol.* **13**, 4806–4813.
5. Hell, K., Herrmann, J., Pratje, E., Neupert, W., and Stuart, R. A. (1997) Oxa1p mediates the export of the N- and C-termini of pCoxII from the mitochondrial matrix to the intermembrane space. *FEBS Lett.* **418**, 367–370.
6. Hell, K., Neupert, W., and Stuart, R. A. (2001) Oxa1p acts as a general membrane insertion machinery for proteins encoded by mitochondrial DNA. *Embo J.* **20**, 1281–1288.
7. Barros, M. H., Carlson, C. G., Glerum, D. M., and Tzagoloff, A. (2001) Involvement of mitochondrial ferredoxin and Cox15p in hydroxylation of heme O. *FEBS Lett.* **492**, 133–138.

8. Glerum, D. M., Shtanko, A., and Tzagoloff, A. (1996) SCO1 and SCO2 act as high copy suppressors of a mitochondrial copper recruitment defect in *Saccharomyces cerevisiae. J. Biol. Chem.* **271**, 20531–20535.
9. Glerum, D. M., Shtanko, A., and Tzagoloff, A. (1996) Characterization of COX17, a yeast gene involved in copper metabolism and assembly of cytochrome oxidase. *J. Biol. Chem.* **271**, 14504–14509.
10. Barrientos, A. (2002) In vivo and in organello assessment of OXPHOS activities. *Methods* **26**, 307–316.
11. Barrientos, A., Zambrano, A., and Tzagoloff, A. (2004) Mss51p and Cox14p jointly regulate mitochondrial Cox1p expression in *Saccharomyces cerevisiae. EMBO J.* **23**, 3472–3482.
12. Myers, A. M., Pape, L. K., and Tzagoloff, A. (1985) Mitochondrial protein synthesis is required for maintenance of intact mitochondrial genomes in *Saccharomyces cerevisiae. EMBO J.* **4**, 2087–2992.
13. Herrmann, J. M., Stuart, R. A., Craig, E. A., and Neupert, W. (1994) Mitochondrial heat shock protein 70, a molecular chaperone for proteins encoded by mitochondrial DNA. *J. Cell Biol.* **127**, 893–902.
14. McKee, E. E., and Poyton, R. O. (1984) Mitochondrial gene expression in *Saccharomyces cerevisiae*. I. Optimal conditions for protein synthesis in isolated mitochondria. *J. Biol. Chem.* **259**, 9320–9331.

10

Purification of Yeast Membranes and Organelles by Sucrose Density Gradient Centrifugation

Jennifer Chang, Victoria Ruiz, and Ales Vancura

Summary

Many experiments require isolation and purification of membranes and organelles from a cell-free lysate. A combination of differential and sucrose density gradient centrifugation provides adequate separation of most yeast organelles in a single experiment. Yeast cells are converted to spheroplasts and gently lysed under conditions that preserve the integrity of organelles. The total lysate is subjected to differential centrifugation and the resulting membrane pellets are fractionated on density gradients. The method is based on the fact that different membranes contain different ratios of lipid to protein, and thus exhibit different density, allowing them to migrate through the gradient until they reach isopycnic position. The fractionated gradients are analyzed by Western blotting with antibodies that recognize marker proteins specific for individual organelles.

Key words: Cell fractionation; organelles; *Saccharomyces cerevisiae*; sucrose density gradient; Western blotting.

1. Introduction

Purification of membranes and organelles is an integral part of many experimental strategies. It is used for preparation of organelles and membranes in transport experiments *(1)*, as a first step in protein purification *(2)*, as well as a method of choice for determination of the subcellular distribution of a protein *(3–5)*. Additionally, characterization of the nature of interaction of a particular protein with a membrane requires isolation of the organelle. Yeast *Saccharomyces cerevisiae* is not an ideal model organism for biochemical experiments, however, this drawback is more than made up for by the amenability of yeast to genetic manipulation. Although specialized methods exist for purification of different organelles *(6–11)*, separation of yeast membranes and organelles by sucrose density gradient centrifugation provides adequate fractionation of most organelles in a single experiment *(3,4)*. The method takes advantage of the fact that different membranes contain different ratios of lipid to protein and thus exhibit different density. The lysate is prepared by enzymatic removal of the yeast cell wall and gentle mechanical lysis in a glass-teflon homogenizer. The lysate is fractionated into pellet P1 (10,000g), pellet P2 (170,000g), and supernatant fraction (S). The P1 and P2 pellet fractions are placed at the bottom of ultracentrifuge tubes and overlaid with solutions containing different concentrations of sucrose. During the centrifugation (16 h at 170,000g), the membranes migrate upward through the sucrose gradient until they reach isopycnic position in the gradient. The gradients are fractionated, and abundance of different marker proteins is determined by Western blotting.

2. Materials

2.1. Yeast Strains, Growth Media, and Cell Lysis

1. *S. cerevisiae* wild-type strain W303 (*MAT*a, *ade2-1, his3-1,15, leu2-3,112, trp1-1, ura3-1*) or any other strain derived by genetic manipulation from W303.
2. Yeast extract/peptone/dextrose (YPD) media: 1% bacto-yeast extract, 2% bacto-peptone, 2% dextrose.
3. Spheroplast buffer: 40 mM potassium phosphate, pH 7.5, 1.4 M sorbitol.
4. Yeast lytic enzyme (Sigma no. L2524 ; activity > 2,000 U/mg): 10 mg/mL in 10% sucrose, 10 mM Tris-HCl, pH 7.5.
5. Lysis buffer: 30 mM triethanolamine, 1.2 M sorbitol, 1.3 mM ethylenediamine tetraacetic acid (EDTA); pH is adjusted to 7.2 with acetic acid.
6. Protease inhibitors: 1 mg/mL each of leupeptin, aprotinin, and pepstatin. This is a 200X stock.
7. Phenylmethylsulfonyl fluoride (PMSF): 100 mM in ethanol. This is a 100X stock.

2.2. Gradient Preparation

1. Gradient solution A: 55% sucrose, 20 mM Tris-HCl, pH 7.4, 1 mM EDTA, 1 mM PMSF, 0.1% 2-mercaptoethanol, 5 μg/mL each of leupeptin, aprotinin, and pepstatin.
2. Gradient solution B: 20 mM Tris-HCl pH 7.4, 1 mM EDTA, 1 mM PMSF, 0.1% 2-mercaptoethanol, 5 μg/mL each of leupeptin, aprotinin, and pepstatin.

3. Methods

This method separates yeast organelles and membranes by a combination of differential centrifugation and sucrose density gradient centrifugation. Although differential centrifugation separates membranes and organelles according to their differences in size and density, sucrose density gradient centrifugation separates according to the differences in buoyant density. The growth and fractionation of cells can be accomplished in about 3 or 4 d, the analysis of fractions by immunoblotting can take an additional several days, depending on how many marker proteins are assayed. The whole procedure can be divided in several stages. Briefly, cells are grown, converted to spheroplasts, and lysed. Particulate fraction of the cell is separated by differential centrifugation and sucrose density gradient centrifugation, and analyzed by Western blotting.

3.1. Cell Growth and Lysis

1. Yeast culture is grown in 20 mL of YPD media overnight and then transferred to 100 mL of the same media and grown until the next morning. A_{600} of this culture is determined and a 2-L flask containing 300 mL YPD is inoculated with such a volume of this preculture so that the starting A_{600} equals 0.5. This culture is grown until A_{600} equals 2.0 (approximately two cell divisions; *see* **Note 1**). The culture is transferred to 250-mL centrifugation bottles and the bottles are placed in a prechilled rotor and centrifuged 20 min at 2000g. The supernatant is carefully poured off and the pellet is resuspended in water and transferred to 50-mL plastic conical tube. The volume of the suspension is adjusted to 40 mL.
2. The cells are centrifuged 5 min at 2000g (Beckman Avanti centrifuge equipped with a rotor that accommodates 50 mL plastic conical tubes is preferred) and the cell pellet is resuspended in spheroplast buffer (total volume of the suspension is 40 mL). The suspension is centrifuged again (5 min at 2,000g) and resuspended in 20 mL of the spheroplast buffer supplemented with 200 µl of 2-mercaptoethanol. A_{600} is determined and the formation of spheroplasts is initiated with 2 mL of yeast lytic enzyme (10 mg/mL in 10% sucrose, 10 mM Tris-HCl, pH 7.5) and allowed to proceed at room temperature with occasional gentle mixing. A_{600} is determined every 15 min until it drops at least 75% of the original A_{600} (10 µL aliquot of the cell suspension is placed in 1 mL of H_2O in plastic spectrophotometer cuvette and the reading is taken after the A_{600} stabilizes). It usually takes 30 to 60 min. If the spheroplasting does not proceed well, it is possible to add more enzyme.
3. The spheroplasts are centrifuged 5 min at 500g and washed four times with 40 mL of ice-cold spheroplast buffer (because the spheroplasts are very fragile, the centrifugation is always 5 min at 500g and spheroplasts are handled very gently, resuspensions are done with a pipet, vortexing is avoided; *see* **Note 2**).
4. The spheroplasts are resuspended in 10 mL of the lysis buffer containing 1 mM PMSF and 5 µg/mL of each aprotinin, leupeptin, and pepstatin and homogenized

by 10 strokes in a glass-teflon homogenizer (30 mL size; see **Note 3**). The homogenate is transferred to a 50 mL conical tube and centrifuged 3 min at 450g in Avanti centrifuge. The supernatant is removed, pellet is resuspended again in 10 mL of the lysis buffer (containing 1 mM PMSF and 5 µg/mL of each aprotinin, leupeptin, and pepstatin), and homogenized by 20 strokes in a glass-teflon homogenizer (30 mL size). The supernatants are combined (the volume is determined) and transferred to a 30-mL Corex tube and designated total lysate (TL). An aliquot of 100 µL is removed from TL (see **Note 4**).

3.2. Cell Fractionation

1. The TL is centrifuged 10 min at 10,000g in Avanti centrifuge, the supernatant is removed and the pellet is resuspended in 2.5 mL of ice-cold gradient solution A and designated pellet 1 (P1). The supernatant is centrifuged 1 h at 170,000g in ultracentrifuge (75 Ti Beckman rotor; Nalgene thick wall tubes 16 × 76 mm, polycarbonate; no. 3425-1613). The resulting supernatant (S) is removed and the volume is determined. The pellet is resuspended in 2.5 mL of ice-cold gradient solution A and labeled as pellet 2 (P2).
2. P1 and P2 samples in solution A are transferred to glass-teflon homogenizer (5 mL size), manually homogenized with five strokes of glass-teflon homogenizer, and transferred to the bottom of the tube for sucrose density gradient (Nalgene thin wall polypropylene tube, 14 × 89 mm; no. 3410-1489). The tube is gently mixed by vortexing, and 100 µL samples are removed from P1 and P2. Samples (100 µL) of TL, P1, P2, and S are mixed with 100 µL of 2X sample buffer for SDS/PAGE and boiled immediately for 10 min.
3. Sucrose solutions with different densities are prepared in 15 mL plastic conical tubes by mixing gradient solution A and gradient solution B (pipetting is done carefully with 1 mL or 200 µL micropipetor; solution A contains sucrose, is viscous and attention must be paid to fully release the solution from the pipet tip; see **Note 5**). Samples P1 and P2 in the centrifugation tubes are carefully overlaid with these ice-cold sucrose solutions according to the scheme in **Table 1**, starting with solution 1. All the solutions are pipeted slowly along the wall of the tube on the top of the previous solution in such a way that mixing between individual solutions is minimized. The tubes are kept on ice while the gradient is being assembled. The layers of the individual solutions should be clearly visible. The tubes must be handled very carefully to prevent mixing of the layers.
4. The tubes with the sucrose gradients are weighed, balanced with gradient solution B and carefully placed in buckets of Beckman SW41 rotor (prechilled at 4°C; see **Note 6**). The rotor is carefully placed in the centrifuge and centrifuged for 16 h at 170,000g with slow acceleration and deceleration (both are set at 7).
5. Immediately (not more than 15 min) after the centrifuge stops, the rotor is very carefully removed and transported to the lab bench (it is essential to do this slowly and carefully to prevent mixing of the gradient). The tubes are removed and

Table 1
Composition of the Sucrose Gradient

Solution	Sucrose (%)	Volume for 1 gradient tube (mL)	Solution A (mL) (4 gradients)	Solution B (mL) (4 gradients)
1	50.0	1.0	3.636	0.364
2	47.5	1.0	3.456	0.544
3	45.0	1.5	4.920	1.080
4	42.0	1.5	4.560	1.440
5	40.0	1.5	4.360	1.640
6	37.5	1.0	2.728	1.272
7	35.0	1.0	2.544	1.456
8	30.0	1.0	2.180	1.820

placed in ice (into holes in the ice formed by empty tubes) and individual 600 µL fractions are removed with 1 mL micropipetor (the samples are removed along the side of the tube, the pipet tip touches the tube at the top of the liquid). Nineteen samples are removed and the pellet is resuspended in 600 µL of gradient solution A, so the total number of samples from each gradient is 20. The samples should be aliquoted as needed for sucrose and protein assays, and Western blotting.

3.3. Analysis of the Fractions

1. Each fraction is assayed for sucrose concentration by refractometry. This confirms that the gradient of sucrose concentration was maintained in the centrifuge

Table 2
Differential Centrifugation Analysis

Marker protein	P1	P2	S
PM ATPase (plasma membrane)	31	63	3
Vps10p (Golgi)	29	57	8
V-ATPase (vacuoles)	32	27	35
Porin (mitochondria)	84	16	0
Dpm1p (endoplasmic reticulum)	63	29	5

Note Cells (W303-1a) were lysed and subjected to differential centrifugation. The amounts of the marker proteins in the resultant P1 pellet ($10,000g$), P2 pellet ($170,000g$), and soluble fraction (S) were determined by quantitative Western blot analysis. The results are expressed as the percent of each marker protein located in P1, P2, or S, in relation to the level of each protein in the initial total lysate. Each value represents an average of at least five experiments, which did not differ by more than 10%.

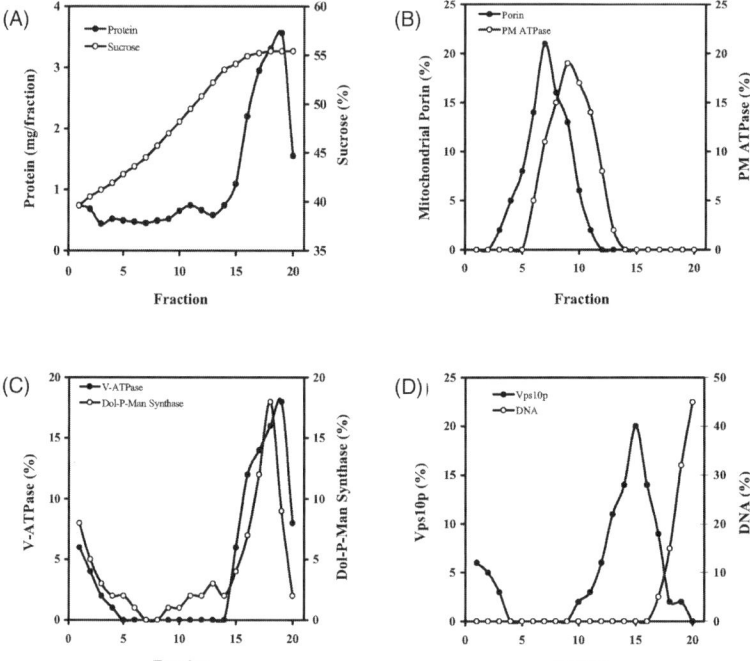

Fig. 1 Sucrose density gradient fractionation of the P2 pellet. The P2 fraction from W303-1a strain was subjected to density gradient centrifugation. The resultant gradient was fractionated from top (fraction 1) to bottom (fraction 20) and assayed for marker proteins by Western immunoblotting. The amounts of marker proteins in each fraction are reported as percentages of the total amounts loaded onto the gradient. (A) protein content and sucrose concentration, (B) porin (mitochondria) and plasma membrane ATPase, (C) V-ATPase (vacuolar membrane) and dolichol-phosphate mannose synthase Dpm1p (endoplasmic reticulum), and (D) Vps10p (Golgi).

tube. We use an inexpensive hand-held Brix refractometer for 35 to 60% sucrose concentration (Fisher Scientific no. 13-946-22).
2. The protein and DNA concentrations in each fraction are also assayed. We use a modified Bradford Coomassie blue binding assay (Pierce no. 23236) and bovine serum albumin (BSA) as a standard for the protein assay. DNA is assayed fluorimetrically with Hoechst dye 33258 *(12)*.
3. The marker proteins in all fractions are assayed by Western blotting. As a marker for vacuoles we use vacuolar H^+-ATPase subunit (assayed with monoclonal antibody [MAb] 10D7, Invitrogen no. A6426), marker for mitochondria is mitochondrial porin (assayed with monoclonal antibody 16G9, Invitrogen no. A6449), marker for endoplasmic reticulum membrane is dolichol-phosphate-mannose synthase Dpm1p (assayed with MAb 5C5, Invitrogen no. A6429), marker for late Golgi is Vps10p (assayed with MAb 18C8, Invitrogen no. A21274), marker for

cytosol is 3-phospho glycerate kinase Pgk1p (assayed with MAb 22C5, Invitrogen no. A6457), marker for plasma membrane is plasma membrane H^+-ATPase (assayed with polyclonal antibody, Santa Cruz Biotechnology no. sc-19389; *see* **Note 7**).
4. Immunoreactivity is measured by enhanced chemiluminiscence using sheep anti-mouse or anti-rabbit immunoglobulin-G conjugated to horseradish peroxidase (GE Biosciences). Images are collected on x-ray film and quantified by densitometry or acquired using Genegnome imaging station (Syngene; *see* **Note 8**). Example of a typical differential and sucrose density gradient centrifugation is provided in **Table 2** and **Fig. 1**.

4. Notes

1. When growing cells transformed with a plasmid, selection medium is used for preculture and subsequently the growth in YPD is limited to only two generations to limit loss of the plasmid.
2. Yeast lytic enzymes very likely contain some proteases that could degrade proteins in the cell lysate. Therefore, it is important to extensively wash the spheroplasts before cell breakage.
3. The lysis buffer contains 1.2 M sorbitol, which prevents osmotic lysis of the spheroplasts and organelles. Cell lysis is achieved by gentle homogenization in a glass-teflon homogenizer. The mixture of aprotinin, leupeptin, and pepstatin can be replaced with a protease inhibitor cocktail optimized for fungal cells (Sigma no. P8215) without any observable decrease in quality of the preparation.
4. Smaller number of preparations can be processed by manual homogenization in the glass-teflon homogenizer. Motor-driven homogenizer is recommended for scaled-up preparations.
5. Dissolution of sucrose in solution A can be facilitated by heating up the solution in a water bath in a microwave oven.
6. The centrifugation tubes have relatively thin walls and must be completely filled to the top with the gradient solutions. Half-filled tubes will collapse during centrifugation.
7. Other useful antibodies for different subcellular fractions are available from Invitrogen. A host of other antibodies against yeast proteins are available from Santa Cruz Biotechnology. Some of them could probably be used for identification of subcellular organelles. Alternatively, the fractions can be also analyzed by measuring activities of enzymes with well-defined subcellular localization: plasma membrane: vanadate-sensitive plasma membrane ATPase *(13)*, endoplasmic reticulum: NADPH-cytochrome *c* reductase *(14)*, mitochondria: cytochrome *c* oxidase *(15)*, Golgi: GDPase *(16)*, vacuole: α-mannosidase *(17)*.
8. Typically, after initial run with a particular antibody, we adjust loading of all samples so that the intensities of the corresponding bands fall within a linear range of response.

References

1. Sanderson, C.M., and Meyer, D.I. (1991) Purification and functional characterization of membranes derived from the rough endoplasmic reticulum of *Saccharomyces cerevisiae*. *J. Biol. Chem.* **266**, 13423–13430.
2. Vancurova, I., Choi, J.H., Lin, H., Kuret, J., and Vancura, A. (1999) Regulation of phosphatidylinositol 4-phosphate 5-kinase from *Schizosaccharomyces pombe* by casein kinase I. *J. Biol. Chem.* **274**, 1147–1155.
3. Goud, B., Salminen, A., Walworth, N.C., and Novick, P.J. (1988). A GTP-binding protein required for secretion rapidly associates with secretory vesicles and the plasma membrane in yeast. *Cell* **53**, 753–768.
4. Vancura, A., Sessler, A., Leichus, B., and Kuret, J. (1994) A prenylation motif is required for plasma membrane localization and biochemical function of casein kinase I in budding yeast. *J. Biol. Chem.* **269**, 19271–19278.
5. Wang, X., Hoekstra, M.F., DeMaggio, A.J., et al. (1996). Prenylated isoforms of yeast casein kinase I, including the novel Yck3p, suppress the gcs1 blockage of cell proliferation from stationary phase. *Mol. Cell. Biol.* **16**, 5375–5385.
6. Whitters, E.A., McGee, T.D., and Bankaitis, V.A. (1994) Purification and characterization of a late Golgi compartment from *Saccharomyces cerevisiae*. *J. Biol. Chem.* **269**, 28106–28117.
7. Bryant, N.J., and Boyd, A. (1995) Localization of a protein A-tagged Kex2 protein to the vacuole of *Saccharomyces cerevisiae* allows rapid purification of vacuolar membranes. *Yeast* **11**, 201–210.
8. Strambio-de-castillia, C., Blobel, G., and Rout, M.P. (1995) Isolation and characterization of nuclear envelopes fronm the yeast *Saccharomyces*. *J. Cell Biol.* **131**, 19–31.
9. Kang, M.S., Young, J.A., and Cabib, E. (1985) Modification of yeast plasma membrane density by concanavalin A attachment. *J. Biol. Chem.* **260**, 12680–12684.
10. Sanderson, C.M., and Meyer, D.I. (1991) Purification and functional characterization of membranes derived from the rough endoplasmic reticulum of *Saccharomyces cerevisiae*. *J. Biol. Chem.* **266**, 13423–13430.
11. Zinser, E., and Daum, G. (1995) Isolation and biochemical characterization of organelles from the yeast, *Saccharomyces cerevisiae*. *Yeast* **11**, 493–536.
12. Cesarone, C.F., Bolognesi, C., and Santi, L. (1979). Improved microfluorimetric DNA determination in biological material using 33258 Hoechst. *Anal. Biochem.* **100**, 188–197.
13. Bowman, B.J., and Slayman, C.W. (1979). The effects of vanadate on the plasma membrane ATPase of Neurospora crassa. *J. Biol. Chem.* **254**, 2928–2934.
14. Kreibich, G., Debey, P., and Sabatini, D.D. (1973). Selective release of content from microsomal vesicles without membrane disassembly. I. Permeability changes induced by low detergent concentrations. *J. Cell Biol.* **58**, 436–462.
15. Mason, T.L., Poyton, R.O., Wharton, D.C., Schatz, G. (1973). Cytochrome c oxidase from bakers' yeast. I. Isolation and properties. *J. Biol. Chem.* **248**, 1346–1354.

16. Abeijon, C., Orlean, P., Robbins, P.W., Hirschberg, C.B. (1989). Topography of glycosylation in yeast: characterization of GDPmannose transport and luminal guanosine diphosphatase activities in Golgi-like vesicles. *Proc. Natl. Acad. Sci. U S A.* **86**, 6935–6939.
17. Walworth, N.C., and Novick, P.J. (1987). Purification and characterization of constitutive secretory vesicles from yeast. *J. Cell Biol.* **105**, 163–174.

11

Microscopic Analysis of Lipid Droplet Metabolism and Dynamics in Yeast

Heimo Wolinski and Sepp D. Kohlwein

Summary

Lipid-associated disorders are a worldwide health concern and major efforts are directed toward understanding—at the molecular levels—mechanisms of lipid storage and degradation, in healthy and diseased states. Yeast is a widely used model organisms to study such processes at the cellular level because of significant functional and structural conservation of the factors involved in lipid metabolism. The focus of this study is on the microscopic investigation of the turnover of lipid droplets that are the intracellular storage compartments for fat, using a combination of green fluorescent protein-tagging and vital dye-labeling techniques. The applications and pitfalls of such techniques in understanding lipid storage and degradation are discussed.

Key words: Fluorescence microscopy; green fluorescent protein; lipolysis; neutral lipid storage; Nile Red; *Saccharomyces cerevisiae*; yeast.

1. Introduction

Lipid droplets (LD) are the predominant fat storage organelle in yeast, plant, and mammalian cells. Although different in size and composition, they serve similar functions in all types of cells, to provide fatty acids, diacylglycerols, sterols and perhaps other lipid precursors or intermediates, and specific proteins important for cellular proliferation and maintenance. Rather than being passive storage depots for fat, LD have only recently been recognized as dynamic and metabolically highly active organelles *(1)*. The discovery of physiologically important and highly regulated enzymatic processes taking place on the surface of these organelles *(2)* have attracted great interest in the basic questions of LD assembly, function and potential malfunction, which may lead to lipid-associated disorders in humans. Surprisingly little is known, however, about the mechanisms that target proteins to the LD and that control LD formation and inheritance *(3)*. Specific and robust targeting mechanisms must exist to ensure specific LD association of proteins in order to spatially separate anabolic and catabolic pathways involved in neutral lipid homeostasis. Furthermore, the interplay between lipid synthesis and further processing that takes place in the endoplasmic reticulum (ER) or in other organelles, for example, peroxisomes *(4)*, and storage in the LD is poorly understood.

Multiple aspects of LD formation and physiology are conserved in yeast, which provides a suitable model system for studying LD biogenesis. Proteomics *(5)* and microscopic screens *(6,7)* have identified numerous LD-associated proteins with functions implicated in lipid metabolism. In particular, microscopic analyses combining green fluorescent protein (GFP)-tagging strategies and vital dye-labeling provide unprecedented insights into the biogenesis and dynamics of this organelle. Such studies have uncovered growth-specific alterations and the impact of inactive fat degradation on neutral lipid pools and LD content *(8)*. The characteristics of the hydrophobic fluorescence probe, Nile Red, that is commonly used to monitor LD, however, requires specific preparation and analysis techniques in order to generate reliable quantitative image data.

Nile Red is widely used for labeling hydrophobic structures, primarily triacylglycerol and steryl ester-rich LD, and phospholipid membranes in a cell *(9)*. The absorption and fluorescence properties of Nile Red are sensitive to the polarity of the environment. In this respect, Nile Red fluorescence is more blue-shifted in hydrophobic lipid environments (neutral lipids), and emits red-shifted fluorescence in the more polar phospholipid membranes. Because of its spectral characteristics and the preferential solubility of the dye in hydrophobic (lipid) solvents, Nile Red is an excellent probe particularly for selective staining of intracellular LD *(10–12)*. The solubility of Nile Red also in phospholipid membranes, with an accompanying red-shift in fluorescence emission, provides an additional advantage for differentiated detection of LD morphology and dynamics, and their interaction with intracellular membranes *(9)*.

2. Materials

2.1. Cell Culture

1. Wild-type strain BY4742 and chromosomally integrated GFP fusion constructs are obtained from Euroscarf (Institute of Microbiology, Johann Wolfgang Goethe-University Frankfurt, Germany) as well as from Invitrogen, Inc. *(6)*.

2. Yeast extract/peptone/dextrose (YPD): Bacto-yeast extract (1% w/v) and Bacto-peptone (2% w/v) are weighed in an appropriate flask and filled with distilled water to 9/10 of the final volume. Autoclave at 121°C for 20 min; add 1/10 volume of 10X glucose solution (2% w/v final concentration) by sterile filtration to the media, mix well and dispense into sterile flasks or glass tubes. For solid media plates, 2% bacto-agar is added to the solution prior to autoclaving.

2.2. Cell Preparation

1. Nile Red (Invitrogen, Inc., Cat. no. N3013): 1 mg/mL in dimethylsulfoxide (DMSO; Merck Inc.). Store 20 µL aliquots of the stock solution in 500 µL reaction tubes at −20°C. Centrifuge Nile Red stock solution briefly in a mini centrifuge prior to cell staining, to remove precipitates that may have formed.
2. Formaldehyde solution min. 37% (Merck Inc.).
3. Redigrad™ (Amersham Biosciences Inc.).
4. 1.5 M NaCl.
5. 50 mM Tris-HCl, pH 7.5.
6. 1 M sorbitol.
7. Agarose electrophoresis grade (Invitrogen Inc.).
8. COREX™ high-speed glass centrifugation tubes (38 mL).

2.3. Microscopy Hardware and Imaging Software

1. Microscope slides 76× 26 mm (Roth Inc.).
2. Cover slips 50× 24 mm no.1 (Menzel Inc.).
3. Leica TCS SP2 confocal microscope with differential interference contrast (DIC; transmission imaging), spectral detection and sequential scanning feature.
4. 100X Oil immersion lens (LEICA HCX PL APO, NA 1.4).
5. LEICA TCS SP2 microscope control and imaging software v2.60.1537.
6. Adobe Photoshop™ CS.
7. Image J, public domain image-processing software (author: Wayne Rasband, National Institute of Mental Health, Bethesda, MD.)

3. Methods

Nile Red is the vital dye of choice for detecting LD in a cell (*9,10*). Upon using Nile Red staining in combination with GFP fluorescence experiments in different stages of cellular growth (*8*), we noted a marked dependence of Nile Red-staining efficiency on the growth phase: Whereas stationary-phase cells (cultivated for 72 h), or numerous defective mutants, readily take up Nile Red and display intense fluorescence of LD, actively growing and dividing cells are typically poorly labeled presumably owing to active pleiotropic drug resistance pumps (PDR; Wolinski and Kohlwein, manuscript in preparation). Furthermore, Nile Red and GFP display substantial spectral overlap that requires optimized image recording, which routinely is not feasible on standard epifluorescence microscopes with GFP, fluorescein isothiocyanate, and rhodamine filter blocks. Third, Nile Red fluorescence is quickly saturated and subject to rapid bleaching, which requires strongly reduced excitation intensities for Nile Red imaging. Additionally, Nile Red fluorescence characteristics change depending on the environment, and display a significant red-shift when interacting with phospholipid membranes (9). Thus, depending on the type of experiment and for quantitative purposes, different protocols are required to accurately determine LD morphology and dynamics in yeast. The labeling efficiency of Nile Red is significantly reduced in actively growing yeast cells and may be very heterogeneous in mutant strains. Thus, exponentially growing cells are preferentially fixed prior to staining with Nile Red. If GFP fusions are to be detected in addition to Nile Red, fixation conditions require optimization to maintain GFP fluorescence in the presence of the fixative. Because detection of both fluorophores requires very different microscope settings in terms of excitation intensity, fixation provides the additional advantage of sequential scanning modes without loss of resolution owing to intracellular movements.

3.1. Cell Cultivation

Growth conditions largely determine Nile Red staining efficiency and GFP expression levels. Thus, great care has to be taken to maintain reproducible growth conditions.

3.1.1. Stationary Phase Cells

1. Cultivate yeast cells in culture flasks containing 5 mL YPD for 12 h at 30°C on a rotary shaker (180 rpm).
2. Inoculate 100 mL fresh YPD in a 500-mL flask with 100 µL of the preculture and cultivate cells for 72 h, to stationary phase.

3.1.2. Logarithmically Growing Cells

1. Cultivate yeast cells in culture flasks containing 5 mL YPD medium for 12 h at 30°C on a rotary shaker (180 rpm).
2. Inoculate 100 mL fresh YPD medium in a 500-mL flask with 100 µL of the preculture and cultivate cells for 72 h, to stationary phase.
3. Inoculate 2 to 10 mL fresh YPD medium with 1/100 volume of stationary-phase culture and cultivate for 4 to 8 h at 30°C on a rotary shaker (180 rpm).

3.2. Separation of Quiescent Yeast Cells Using Density Gradient Centrifugation

This protocol is based on the method described previously *(13)* for purification of quiescent cells from stationary-phase culture. The procedure proves to be extremely valuable for microscopic investigation of cells because of the quantitative elimination of necrotic or apoptotic cells, and stationary-phase cells with large vacuoles. Thus, vital staining and GFP expression patterns turn out very homogeneous in the cell population (**Note 1**).

1. Add 9 mL Redigrad™ and 1 mL sterile 1.5 M NaCl to 38 mL high-speed centrifuge tubes (e.g., COREX™ glass tubes).
2. Establish density gradient by centrifuging tubes at 19.000 g for 15 min at room temperature in a fixed angle rotor.
3. Centrifuge 2 mL of stationary-phase yeast culture (72 h) for 5 min at 1,000 g in a tabletop centrifuge. Aspirate off supernatant.
4. Resuspend cell pellet in 1 mL sterile 50 mM Tris-HCl pH 7.5.
5. Overlay the cell suspension onto the previously formed Redigradd™/NaCl gradient in the high-speed centrifuge tubes.
6. Centrifuge for 1 h at 430 g in a tabletop centrifuge with a swing-out rotor; do not use brakes for slowing down the rotor.
7. Carefully remove the tubes from the centrifuge; avoid shaking.
8. Collect the lower fraction (~1 mL) of quiescent cells with a syringe and transfer into a 15 mL polypropylene tube. Add 500 µL of sterile 50 mM Tris-HCl pH 7.5; vortex briefly. Harvest cells at 1,000 g for 5 min. Wash cells with sterile 50 mM Tris-HCl pH 7.5.

3.3. Nile Red Vital Staining of Unfixed Stationary-Phase Yeast Cells

Nile Red staining efficiency is high in stationary-phase cells, thus fixation with formaldehyde can be omitted under these conditions to obtain reliable LD staining (**Note 2**).

1. Transfer 1 mL of a stationary-phase yeast cell suspension (72 h) to a 1.5-mL reaction tube. Optionally wash the cells twice with 1 mL sterile 50 mM Tris-HCl, pH 7.5.
2. Centrifuge cells at 1,000 g for 1 min in a tabletop centrifuge. Aspirate off supernatant.
3. Resuspend cells in 1 mL sterile 50 mM Tris-HCl, pH 7.5
4. Add 1 µL of the Nile Red stock solution to the cell suspension (final concentration: 1 µg/mL). Vortex briefly.
5. Incubate cells for 20 min at room temperature. Do not close the lid of the reaction tube during staining.
6. Centrifuge cells at 1,000 g for 2 min using a tabletop centrifuge. Aspirate off supernatant and resuspend cells in the remaining liquid.

7. Mount 1 μL of dense cell suspension on a standard microscope slide for short-term analysis or on a microscope slide covered with agarose (*see* **Subheading 3.5.**), for long-term observation (**Note 3**).

3.4. Fixation of GFP or Nile Red-Labeled Yeast Cells

For labeling logarithmically growing or stationary phase cells, and in combination with GFP detection, cells are preferentially fixed with formaldehyde prior to Nile Red staining (**Note 4**, *see also* **Subheading 3.6.**).

1. Transfer 1 mL of yeast culture to a 1.5 mL reaction tube.
2. Centrifuge cells at 1,000 g for 1 min in a tabletop centrifuge. Aspirate off supernatant.
3. Resuspend cells in 500 μL sterile 1 M sorbitol. Incubate for 60 s at room temperature.
4. Add 27 μL of a 37% formaldehyde solution (v/v) to the sample (final concentration 2%, v/v). Incubate the cells for 1 to 5 min at room temperature (**Note 5**). Vortex briefly every 30 s.
5. Centrifuge cells for 2 min at 1,000 g in a tabletop centrifuge.
6. Wash cells three times with 500 μL sterile 50 mM Tris-HCl pH 7.5.
7. Add 0.2 μL of the Nile Red stock solution to the cell suspension (final concentration: 0.4 μg/mL); vortex briefly.
8. Incubate cells for 10 min at room temperature.
9. Centrifuge cells at 1,000 g for 2 min using a tabletop centrifuge. Aspirate off supernatant and resuspend cells in the remaining liquid.
10. Mount 1 μL of dense cell suspension on a standard microscope slide for short-term analysis or on a microscope slide covered with agarose (**Subheading 3.5.**), for long-term observation.

3.5. Immobilization of Labeled Yeast Cells for Long-Term Microscopic Observation

1. Boil 2.5% low-melting temperature agarose in distilled water in a microwave oven (600 W) until the agarose solution is completely clear. Alternatively, use 2% bacto™-agar instead of agarose.
2. Transfer 5 mL of melted agarose into a 15-mL plastic tube and cool to approx 50°C in a water bath.
3. Add 5 μL of the Nile Red stock solution to the agarose solution (final concentration 1 μg/mL) and vortex for 30 s.
4. Place a standard microscope slide (76× 26 mm) on a flat surface and pipet 3 mL of the agarose solution on the microscope slide until it is completely covered.
5. Allow polymerization of the agarose on the slide for 15 min at room temperature.

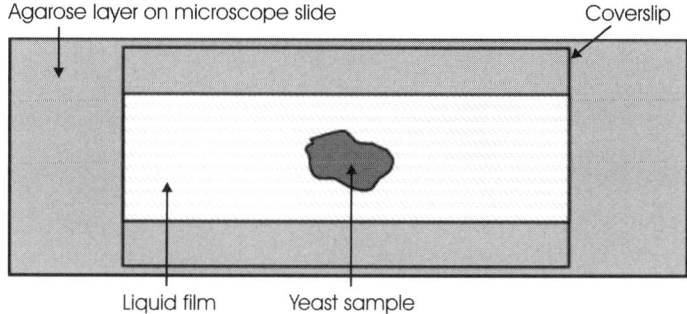

Fig 1. Immobilization of yeast cells using the "agarose sheet" method for long-term microscopic observation. Slides are covered with melted agarose (optionally containing appropriate growth components and dyes). After polymerization, cells are placed in the center of the slide and covered with a cover slip. Cells are spread out in a liquid film as a monolayer right underneath the cover slip, providing least distortion of the laser beam and thus optimum resolution for transmission and fluorescence microscopy. Note that the condensor position of the microscope needs to be adjusted to the appropriate level, to obtain optimal illumination of the specimen. Conditions are appropriate for maintaining cell viability and monitoring cellular growth and division over extended periods of time (>10 h).

6. Add 1 µL of (labeled) yeast cell suspension to the centre of the agarose layer. Mount the cell preparation with a large (50 × 24 mm) cover slip (**Fig. 1**; **Notes 6** and **7**).

3.6. Microscopy

3.6.1. Transmission Microscopy: Differential Interference Contrast

1. Mount cells on standard microscope slides or on agarose-coated slides.
2. Visualize cells using DIC optics. Note that the condensor position needs to be adjusted for cells mounted on agarose-covered slides, owing to the extended distance, for optimized results (**Fig. 2**).

3.6.2. Fluorescence Microscopy of Nile Red-Labeled Cells

Nile Red fluorescence can be excited with both 488 nm (Argon laser) and 543 nm (HeNe laser) light. For specific LD staining, Nile Red is preferentially excited at 488 nm and fluorescence detected at 550 to 575 nm emission. For detection of membranes in addition to LD, Nile Red is preferentially excited with 543 nm and emission detected between 550 and 575 nm for LD, and between 600 and 700 nm for LD plus intracellular membranes. towing to the

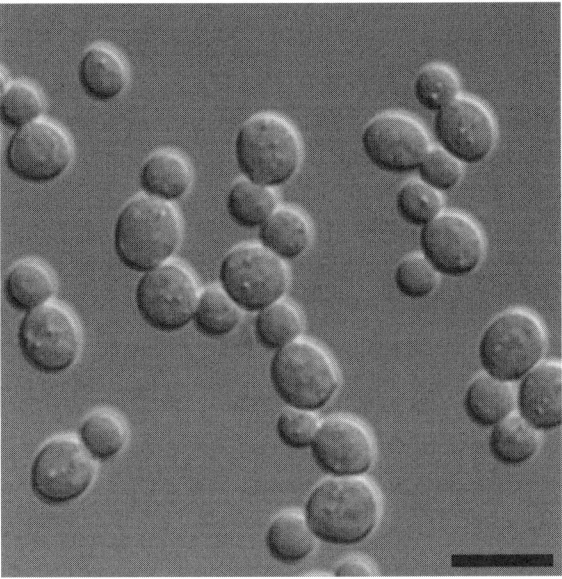

Fig 2. Differential interference contrast image of yeast cells prepared by density gradient centrifugation and subsequent cultivation in complete media for 90 min. Note the lack of any damaged cells and the high level of synchrony of cell division. Bar = 10 μm.

rapid saturation and bleaching characteristics, Nile Red excitation should be performed with reduced laser intensity (1–5% of maximum laser power at 488 nm excitation, and 50% laser power at 543 nm excitation; **Note 8**).

1. Label unfixed (**Subheading 3.3.**) or fixed cells (**Subheading 3.4.**) with Nile Red and mount on agarose-covered slides.
2. Image cells using 543 nm excitation (HeNe laser) at 50% output power, and 550 to 575 nm detection for LD, and 600 to 700 nm detection range for LD plus membrane visualization (**Fig. 3**).

In the absence of Nile Red in the agarose, Nile Red fluorescence is rapidly lost from the cells, unless cells are damaged or dead (**Fig. 4**).

3.6.3. Fluorescence Microscopy of GFP and Nile Red Double-Labeled Cells

GFP and Nile Red fluorescence display significant spectral overlap, which requires careful normalization and standardized settings for double-labeling experiments (**Fig. 5**).

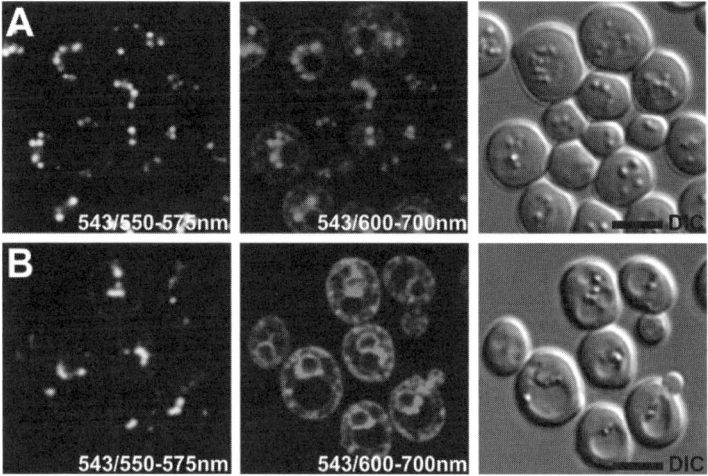

Fig. 3. Nile Red fluorescence of labeled subcellular yeast structures at 543 nm excitation of stationary-phase yeast cells (row A) and late logarithmic phase cells (Row B). Nile Red fluorescence imaged at 550 to 575 nm (left panels) shows strong and specific fluorescence associated with lipid droplets (LD), both in growing and stationary-phase cells. Fluorescence in the 600 to 700 nm emission range (middle panel) displays strong LD labeling and additional, weaker signals in internal and peripheral membranes (presumably endoplasmic reticulum, vacuolar membrane, and plasma membrane). Note that the fluorescence intensity of membranes in late log-phase cells is significantly increased. Right panels: differential interference contrast images. Bar = 5 μm.

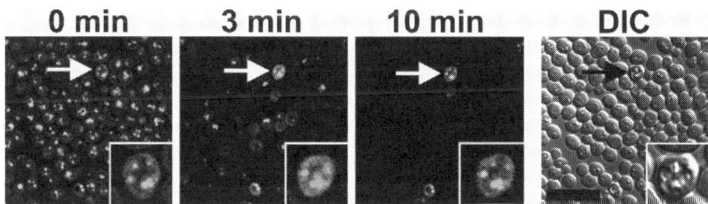

Fig. 4. Time-lapse record showing fast disappearance of Nile Red from stationary-phase yeast cells. Yeast cells were labeled with Nile Red (**Subheading 3.3.**) and mounted on agarose covered slides without Nile Red. Three minutes after mounting, a significant reduction or loss of the Nile Red signal is detectable in most cells. After 10 min, only visibly damaged cells retain fluorescence (white arrow; the corresponding cell is displayed in the lower right corner of each time-lapse image at higher magnification). Nile Red excitation and emission was at 543 nm and 600 to 700 nm, respectively. These data show that Nile Red labeling is reversible, and that Nile Red preferentially stains damaged or dead cells, which limits its use for live cell imaging, in particular of mutant cell cultures. Bar = 20 μm.

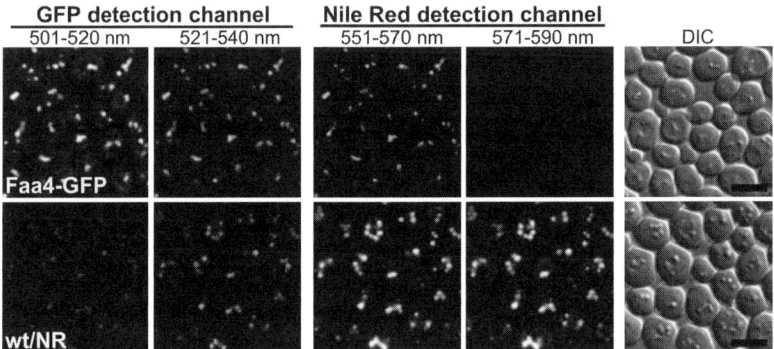

Fig. 5. Spectral overlap of Nile Red and GFP fluorescence emission at 488 nm excitation. Upper row: spectral imaging ("lambda scan") of fluorescence emission of chromosomally integrated Faa4p-GFP. Laser intensity is set to 25% of maximum output power. Maximum GFP fluorescence is observed at 501 to 520 nm, but is also significant in the 521 to 540 nm and 551 to 57 0 nm detection range. Lower row: spectral imaging of Nile Red fluorescence emission. Laser intensity is set to 5% maximum output power. Using 488 nm excitation, strong Nile Red fluorescence is detected at 551 to 570 nm and also in the range of 571 to 590 nm. Significant Nile Red fluorescence appears in the 521 to 540 nm emission range that is typically covered by fluorescein isothiocyanate or GFP filter blocks on standard microscopes. Bar = 5 μm.

1. Mount fixed and Nile Red-labeled cells expressing the GFP construct on standard slides (**Subheading 3.4.**)
2. Detect Nile Red and GFP fluorescence by sequential scanning, using the sequential imaging feature of the LEICA TCS SP2 software
3. Use 488 nm excitation, and 5% laser output power for Nile Red detection, at 571 to 590 nm emission
4. Use 488 nm excitation, and 25% laser output power for GFP detection, at 501 to 520 nm emission.
5. In control experiments, label fixed wild-type cells with Nile Red or fixed GFP-expressing cells, and subject specimens to the same microscopic analyses to carefully monitor spectral crosstalk, in particular Nile Red fluorescence in the GFP detection channel. If the Nile Red concentration was too high, crosstalk of Nile Red fluorescence into the GFP detection channel can be further reduced by three to five scans at full laser power (488 nm excitation), which will efficiently bleach Nile Red fluorescence, prior to monitoring GFP fluorescence (**Fig. 6**).

4. Notes

1. Redigrad™ density gradient centrifugation was previously shown to be largely inert on yeast cell physiology *(13)*. The resulting cell population is very homogeneous in terms of cell size, labeling characteristics, and growth.

Microscopic Analysis of Lipid Droplet Metabolism and Dynamics 161

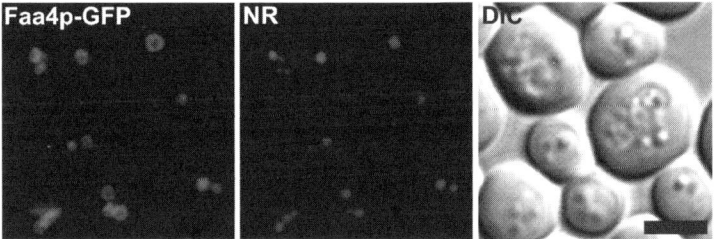

Fig. 6. High-resolution sequential imaging of Faa4p-GFP and Nile Red in fixed, co-labeled yeast cells. The localization of Faa4p-GFP at the periphery of the lipid droplets (LD) as well as the Nile Red-labeled LD "core structures" are clearly resolved. GFP excitation was at 488 nm and emission was detected at 501 to 520 nm (variable emission detection on the LEICA SP2 confocal microscope); Nile Red excitation was at 488 nm and emission was detected at 571 to 590 nm. Scanning was performed sequentially, first detecting Nile Red at 5% laser output power, and subsequent GFP detection at 25% laser power. The Nile Red signal in the 501 to 520 nm emission range can be neglected under these conditions. Bar = 5 μm.

2. Media for yeast cultivation, such as complete media (YPD) containing large quantities of fluorescent metabolites, may cause significant fluorescence signals at the interface between the cover slip and the liquid phase. This phenomenon may interfere with high-resolution imaging of Nile Red-labeled yeast structures. In this case, the cells may have to be washed prior to the staining procedure. However, to avoid significant changes of the cellular physiology and morphology during the staining procedure and washing steps (e.g., the formation of large vacuoles), we prefer to stain the cells directly in the corresponding cultivation media.
3. Nile Red fluorescence disappears from stationary-phase yeast cells, presumably by washout, thus we typically do not wash the cells after staining. Nile Red is practically nonfluorescent in the aqueous environment. Thus, confocal imaging of cells in the presence of excess dye in the media has no significant effect on the background signal.
4. To overcome limitations of the imaging of both GFP and Nile Red-labeled stationary phase yeast cells, that is, photobleaching of Nile Red, spectral overlap of GFP and Nile Red fluorescence emission, or PDR-mediated active export of Nile Red from yeast cells, we apply this short fixation protocol in combination with sequential imaging. In the sequential scan mode, parameters such as laser intensity and emission wavelength are optimized for both GFP and Nile Red. These imaging parameters are initially determined for each fluorophore separately. The experimentally determined parameters are then applied to the imaging of double-labeled fixed yeast samples automatically and in a sequential order.
5. The fixation time depends on the type of GFP fusion and on the growth stage of the cells and has to be determined experimentally. In the authors' laboratory, stationary-phase wild-type cells are fixed for 2 to 5 min and exponentially growing

yeast cells are fixed for a maximum time of 60 s. In general, fixation times should be as short as possible because longer fixation times may influence morphology of LD and GFP fluorescence. Sorbitol treatment efficiently supports GFP integrity and maintenance of its fluorescence, but may interfere with vacuole morphology.
6. After mounting the cover slip the cells are spread out as a monolayer in a liquid film, which should not extend to the horizontal borders of the cover slip, indicative of too much liquid that results in floating of the cells. If a significant reduction of Nile Red fluorescence from yeast cells is observed during microscopic observation the concentration of Nile Red in the agarose solution needs to be increased.
7. To supply the cells with nutrients during long-term microscopic analysis the agarose (agar) can be melted in various yeast growth media. Using additional equipment, such as an objective heater system, this preparation technique also enables cultivation and microscopic analysis of yeast cells directly on the microscope table, over extended periods of time under controlled conditions.
8. We observed rapid photobleaching of Nile Red when applying optimized zoom settings required for high-resolution confocal imaging. However, the rate of Nile Red photobleaching is strongly dependent on the intensity of the illuminating laser source. In this context, setting the laser intensity of the 488 nm argon laser line or the 543 nm HeNe laser line of the confocal microscope system to approx 1 to 5% or 50%, respectively, of the maximum output power significantly reduces Nile Red photobleaching. However, these values may need adjustment as the laser power decreases over time. Owing to the excellent fluorescence properties of Nile Red, high-contrast images with optimum resolution are obtained even at very low excitation intensities that also enable reliable multidimensional imaging of Nile Red-labeled structures.

Acknowledgements

Work in our laboratory is supported by grants from the Austrian Ministry for Education, Science and Culture (GEN-AU program, project GOLD—Genomics of Lipid-associated Disorders), and the Austrian Science Fund, FWF (project SFB Lipotox).

References

1. Beckman, M. (2006) Cell biology. Great balls of fat. *Science* **311**, 1232–1234.
2. Zechner, R., Strauss, J. G., Haemmerle, G., Lass, A., and Zimmermann, R. (2005) Lipolysis: pathway under construction. *Curr. Opin. Lipidol.* **16**, 333–340.
3. Czabany, T., Athenstaedt, K., and Daum. G. (2007) Synthesis, storage and degradation of neutral lipids in yeast. *Biochim. Biophys. Acta* **1771,** 299–309.
4. Binns, D., Januszewski, T., Chen, Y., et al. (2006) An intimate collaboration between peroxisomes and lipid bodies. *J. Cell. Biol.* **173**, 719–731.

5. Athenstaedt, K., Zweytick, D., Jandrositz, A., Kohlwein, S. D., and Daum, G. (1999) Identification and characterization of major lipid particle proteins of the yeast Saccharomyces cerevisiae. *J. Bacteriol.* **181**, 6441–6448.
6. Huh, W. K., Falvo, J. V., Gerke, L. C., et al. (2003) Global analysis of protein localization in budding yeast. *Nature* **425**, 686–691.
7. Natter, K., Leitner, P., Faschinger, A., et al. (2005) The spatial organization of lipid synthesis in the yeast *Saccharomyces cerevisiae* derived from large scale green fluorescent protein tagging and high resolution microscopy. *Mol. Cell. Proteomics.* **4**, 662–672.
8. Kurat, C. F., Natter, K., Petschnigg, J., et al. (2006) Obese yeast: triglyceride lipolysis is functionally conserved from mammals to yeast. *J. Biol. Chem.* **281**, 491–500.
9. Greenspan, P., Mayer, E. P., and Fowler, S. D. (1985) Nile red: a selective fluorescent stain for intracellular lipid droplets. *J. Cell Biol.* **100**, 965–973.
10. Greenspan, P., and Fowler, S. D. (1985) Spectrofluorometric studies of the lipid probe, nile red. *J. Lipid Res.* **26**, 781–789.
11. Fowler, S. D., and Greenspan, P. (1985) Application of Nile red, a fluorescent hydrophobic probe, for the detection of neutral lipid deposits in tissue sections: comparison with oil red O. *J. Histochem. Cytochem.* **33**, 833–836.
12. Sackett, D. L., Wolff, J. (1987) Nile red as a polarity-sensitive fluorescent probe of hydrophobic protein surfaces. *Anal. Biochem.* **167**, 228–234.
13. Allen, C., Buttner, S., Aragon, A. D., et al. (2006) Isolation of quiescent and non-quiescent cells from yeast stationary-phase cultures. *J. Cell Biol.* **174**, 89–100.

12

Use of Bimolecular Fluorescence Complementation in Yeast *Saccharomyces cerevisiae*

Kari-Pekka Skarp, Xueqiang Zhao, Marion Weber, and Jussi Jantti

Summmary

Visualization of protein–protein interactions *in vivo* offers a powerful tool to resolve spatial and temporal aspects of cellular functions. Bimolecular fluorescence complementation (BiFC) makes use of nonfluorescent fragments of green fluorescent protein or its variants that are added as "tags" to target proteins under study. Only upon target protein interaction is a fluorescent protein complex assembled and the site of interaction can be monitored by microscopy. In this chapter, we describe the method and tools for use of BiFC in the yeast *Saccharomyces cerevisiae*.

Key words: Bimolecular fluorescence complementation; EYFP; fluorescence microscopy; GFP; protein interactions; *Saccharomyces cerevisiae*; yeast.

1. Introduction

Proteins carry out a vast number of cellular functions through interactions with one or multiple binding partners. A given protein may display differential interactions depending on its cellular localization. Examples of such differential interactions are proteins that shuttle between nucleus and cytosol having different interaction partners at these two cellular compartments. A central issue in modern cell biology is to reveal such differential interactions and how they are regulated in space and time (e.g., during cell cycle progression or in response to extracellular stimuli).

Identification and analysis of protein interactions by biochemical approaches based on pull-down or immunoprecipitation experiments, or analysis of protein interactions with different variants of the yeast two-hybrid assay, are extremely useful techniques. However, typically none of these techniques necessarily reveal the cellular localization where the protein interaction normally takes place. In the case of immunoprecipitations and pull-down experiments cells need to be broken and the assays are performed under *in vitro* conditions. Such conditions do not favor preservation of weak protein–protein interactions that nevertheless are typical for regulatory molecules. Different two-hybrid interaction assays usually require that the interaction occurs at a predetermined site of the cell such as the nucleus. A complicating feature in these assays is that all cofactors needed for an efficient protein interaction to take place may not be present in this unnatural site of interaction.

Development of green fluorescent protein (GFP) technology has enabled visualization of protein localization and movement dynamics in *in vivo* conditions *(1)*. However, the visualization of the total localization pattern of GFP-tagged target protein does not reveal where within its overall distribution the proteins interact with different binding partners. GFP technology has recently been utilized in two *in vivo* protein interaction methods that can be used to reveal the cellular site of interaction, the fluorescence resonance energy transfer (FRET), and bimolecular fluorescence complementation (BiFC; *2,3*). In FRET, difference in fluorescence signal or its lifetime is detected when the fluorophores are in the same complex *(4)*. In contrast, in BiFC, a fluorescence signal is generated only when nonfluorescent fragments of GFP are brought together by interaction of the target molecules fused to GFP fragments. BiFC is thus potentially very sensitive with low background. The drawback of BiFC is that it does not allow interaction dynamics studies as assembly of the GFP results in significant stabilization of the target protein–GFP complexes *(2,5)*. Recently, BiFC has been shown to function in various cells from bacteria to mammals (for a review *see* **ref. 5**). At the moment, there exists several versions of BiFC that make use of the different spectral properties of variants of GFP. Protein interactions have been visualized with the help of fragments of enhanced yellow fluorescent protein (EYFP) or its variants venus and citrine, cyan fluorescent protein (CYFP) or its variant cerulein and blue fluorescent protein (BFP) and monomeric red fluorescent protein (mRFP) fragments *(6–8)*. Combinations of these variants enable even simultaneous visualization of interaction of several proteins *(9)*.

Yeast *Saccharomyces cerevisiae* offers an attractive model to study spatial and temporal regulation of protein–protein interactions. Molecular biological manipulations are fast and easy and synchronized cell cultures can be obtained. Additionally, for most of the genes, mutations or combination of mutations exist (even for most nonessential genes) that cause a detectable phenotype. Rescue of such phenotypes by the GFP fragment containing target protein enables verification of the *in vivo* functionality of the tagged proteins, an important factor when considering the reliability of the results obtained. In this chapter, we report the method and tools for the use of BiFC system with EYFP fragments in the yeast *S. cerevisiae*.

2. Materials
2.1. Yeast Transformation

1. 10X TE: 100 mM Tris-HCl, 10 mM ethylenediamine tetraacetic acid (EDTA), pH 7.5. Store at room temperature. Working concentration is 1X.
2. 10X LiAc Solution: 1 M LiOAc, pH 7.5 (adjusted with acetic acid). Store at room temperature. Working concentration is 1X.
3. 50% Polyethylene glycol (PEG): 50 g PEG-4000 in 100 mL. Store at room temperature. Working concentration is 40%.
4. Salmon sperm DNA: 10 μg/μL, store in small aliquots at −20°C.
5. Vectors: *See* **Fig. 1**.

2.2. Verification of Expression
2.2.1. Lysate Preparation

1. 2% SDS containing protease inhibitors: add 1 tablet protease inhibitors (Complete, EDTA free [Roche]) to 25 mL of 2% sodium dodecyl sulfate (SDS) in water, store in aliquots at −20°C.
2. Acid-washed glass beads 0.45 mm diameter (Sigma).
3. Laemmli sample buffer: 0.3 M Tris-HCl (pH 6.8), 30% glycerol, 10% SDS, 1.54 M dithiothreitol (DTT), 0.4 mg/mL bromphenol blue. Store in 1 mL aliquots at −20°C.

2.2.2. SDS-PAGE

1. 12% separating gel: 4.65 mL distilled water, 1.25 mL 3 M Tris-HCl (pH 8.8), 0.05 mL 20% SDS, 4 mL acrylamide/*bis*-acrylamide (30%/0.8%), 0.05 mL 10% ammonium persulfate (APS), 0.005 mL tetramethyl-ethylenediamine (TEMED). Mix directly before use. This amount is sufficient for preparing two gels with a spacer thickness of 0.75 mm in Hoefer Multiple Gel Caster (Amersham Bioscience).
2. 4% stacking gel: 4.117 mL distilled water, 0.208 mL 3 M Tris-HCl (pH 6.8), 0.025 mL 20% SDS, 0.67 mL acrylamide/*bis*-acrylamide (30%/0.8%), 0.025 mL 10% APS, 0.005 mL TEMED. Mix directly before use. This amount is sufficient for preparing two gels with a spacer thickness of 0.75 mm in Hoefer Multiple Gel Caster (Amersham Bioscience).
3. 10X running buffer: 30 g Tris-HCl, 144 g glycine, 10 g SDS, add distilled water to 1 L. Store at room temperature. Working concentration is 1X.
4. BCA™ Protein Assay Kit (Pierce).

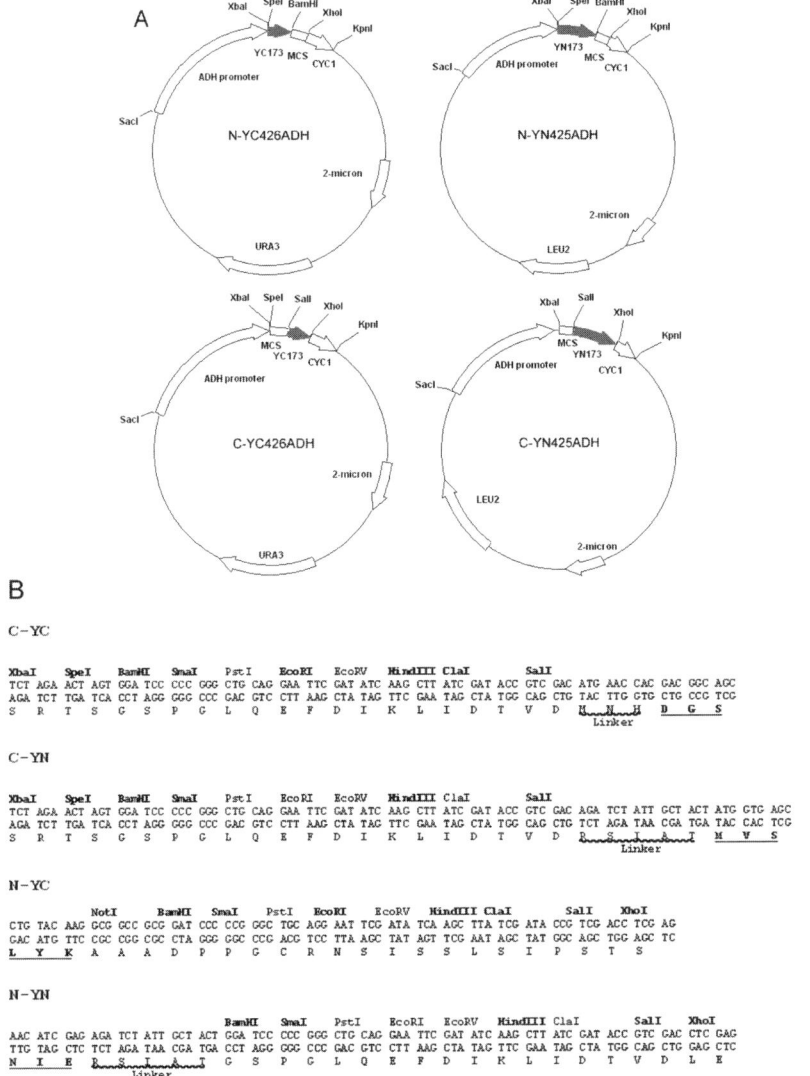

Fig.1. Plasmid maps and sequences of the multiple cloning sites. Plasmids enable tagging of your target protein either at the amino- (N-YC426ADH and N-YN425ADH) or carboxy-(C-YC426ADH and C-YN425ADH) terminus. Underlined bold single-letter amino acid sequence is EYFP sequence. Unique restriction enzyme sites are shown with bold.

2.2.3. Western Blotting

1. Western blotting buffer: 3 g Tris-HCl, 14.4 g glycine, 200 mL methanol, add distilled water to 1 L. Store at +4°C.
2. Ponceau S solution: 100 mg Ponceau, 1 mL acetic acid, add distilled water to 100 mL. Store at room temperature.

3. 10XTBS: 30 g Tris-HCl, 80 g NaCl, 2 g KCl, add distilled water to 1 L. Adjust pH to 7.4 with HCl. Working concentration is 1X and add 500 µL Tween-20 to obtain Tris-buffered saline Tween-20 (TBST). Tween-20 dissolves faster if you add it in to 10X TBST before dilution with water to 1L. Store at room temperature.
4. Enhanced chemiluminescence (ECL) detection reagents: Pierce ECL Western Blotting Substrate kit (or equivalent).
5. Incubation chambers for filters: Plant tissue culture container (PLANTCON®).
6. Secondary antibodies: Horseradish peroxidase conjugated immunoglobulin G specific for your primary antibody.

2.2.4. Microscopy

1. Fixation: 20% paraformaldehyde solution (Electron Microscopy Sciences no. 15713-3).
2. PTFE Printed Slides with six or eight wells (Electron Microscopy Sciences).
3. For preparation of ConA-coated slides: 0.1 µg/mL concanavalin A.
4. Sample mounting for fixed cells: Dissolve 5 g Mowiol in 20 mL Tris-HCl (pH 8.0) overnight. Add 10 mL 100% glycerol and let it mix overnight. Store at $-20°C$.

3. Methods

3.1. Generation of Target Gene–YFP Fusions

The plasmids presented here enable introduction of the target gene in the multiple cloning sites indicated in **Fig. 1**. Plasmids reported here enable high-copy expression of the target protein tagged either at the amino- or carboxy-terminus with EYFP fragments (*see* **Note 1**). The amino terminal (YN) fragment encodes amino acids 1 to 172, whereas the carboxy terminal fragment (YC) encodes amino acids 173 to 238 of EYFP. All plasmids have *ADH1* promoter and *CYC1* terminator and are based on the Mumberg et al. plasmid series *(10)*. Plasmids can be requested from the corresponding author.

3.2. Yeast Transformation

1. Inoculate yeast cells into 5 mL liquid YPD *(11)* medium and grow overnight at 30°C with shaking to approx 2×10^7 cells/mL ($A_{600}=1$).
2. Dilute to OD_{600} 0.1 to 0.2 in 20 mL of fresh, warm YPD and re-grow to 1 to 2×10^7 cells/mL ($OD_{600} <1$). Typically, this takes 4 to 5 h.
3. Harvest the cells by centrifugation at $3,200g$ for 2 min, and wash in 1 volume of sterile water.
4. Pellet the cells by centrifugation at $3\ 200\ g$ for 2 min, pour off the water and resuspend the cells in 1.0 mL 1XTE/LiAc solution made fresh from sterile 10X stocks and transfer the suspension to a 1.5 mL eppendorf tube.

5. Pellet the cells in microfuge at top speed for 5 s and remove the supernatant with a micropipet.
6. Resuspend the cells to 2×10^9 cells/mL with 1XTE/LiAc.
7. Incubate carrier DNA for 10 to 15 minutes at 100°C and place it immediately on ice.
8. Mix 50 μL of yeast cell suspension with 0.5 to 1 μg transforming plasmid DNA and 50 μg single-stranded carrier DNA (salmon sperm DNA) in a 1.5-mL eppendorf tube. The protocol provided here transforms both YN and YC plasmids at the same time. Remember to prepare one negative control that is lacking the plasmid DNA.
9. Add 300 μL of 40% sterile polyethylene glycol (PEG)-4000 in 1XTE/LiAc made fresh from a 50% PEG stock and 10XTE and LiAc stocks. Mix by gentle pipeting until the cell pellet has been completely resuspended.
10. Incubate with agitation for 30 min at 30°C.
11. Add 40 μL tissue-culture grade dimethyl sulfoxide (stored at −70°C in small aliquots), and heat shock at 42°C for 15 min.
12. Spin down in microfuge for 5 s and wash cells gently with 1 mL of sterile water.
13. Spin down again and resuspend the cell pellet in 200 microliter of 1X TE.
14. Plate 50 μL and the rest on SC-leu-ura plates. Grow in an incubator at 30°C. Colonies appear in 2 d. There should be no colonies in the negative control lacking the plasmid DNA.

3.3. Verification of Expression

It is important to verify that your constructs produce a fusion protein of expected size (*see* **Notes 1** and **2**).

3.3.1. Lysate Preparation

1. Inoculate one yeast colony into 20 mL of liquid Sc-ura-leu medium and grow overnight with shaking at 30°C.
2. Dilute cell culture in fresh, warm 20 mL Sc-ura-leu medium to OD_{600} 0.2 and regrow to OD_{600} 0.8–1.0 (*see* **Note 3**).
3. Harvest cells by centrifugation at 3,200*g* for 2 minutes, discard the supernatant.
4. Resuspend the cells in 1 mL of water, transfer into a 1.5 mL microfuge tube and collect by centrifugation.
5. Resuspend the cells in 400 μL 2% SDS containing the protease inhibitors (**Subheading 2.2.1.**), vortex.
6. Add 200 μL of 0.45 mm acid-washed glass beads and heat for 3 min in a 95°C block heater or equivalent.
7. Vortex twice for 90 s with maximum speed.
8. Centrifuge for 5 min full speed in a microfuge and collect the clear lysate in a new tube. Avoid the pellet.

9. Measure the protein concentration by BCA method (BCA™ Protein Assay Kit, Pierce) or equivalent (*see* **Note 4**). Typically a lysate with a protein concentration 8 to 12 µg/µL is obtained.
10. Dilute the lysate to a concentration of 4 µg/µL with 6X Laemmli sample buffer and water to yield 1X sample buffer concentration.

3.3.2. SDS-PAGE

The SDS-polyacrylamide gel electrophoresis (PAGE) protocol has been adjusted to Hoefer SE250 Mighty Small II system (Amersham Biosciences). Apply 20 to 40 µg total protein of every sample to 12% mini SDS-PAGE gel. This percentage of gel gives optimal resolution for 20 to 60 kDa proteins.

1. Heat the samples for 5 min in a 95°C block heater with the lid closers on.
2. Centrifuge the sample in microfuge for 5 min at top speed.
3. Load 5 to 10 µL of sample per well.
4. Run at 10 to 20 mA for each gel until the dye front is at the bottom of the gel.
5. Remove the gel from the running apparatus. Discard the upper gel.

3.3.3. Western Blotting

For electrophoretic transfer of proteins from the gel to nitrocellulose membrane, we use Bio-Rad mini Trans-Blot Cell system.

1. Assemble a sandwich for transfer by prewetting the fiber pads, two Whatmann no. 3 filter papers, and the nitrocellulose membrane (slightly bigger than gel) in Western blotting buffer. The order of constituents in the "sandwich" is as follows:
   a. Pad
   b. Filter papers
   c. Gel
   d. Nitrocellulose membrane
   e. Filter papers
   f. Pad

2. Remove air bubbles by stroking the assembled "sandwich" (e.g., with the flat side of a test tube). Put the "sandwich" into the gel holder cassette and mount the assembled cassette into the electrophoresis blotting module so that the membrane is facing the positive charge.
3. Transfer at 500 mA for 1 h with the cold pack and prechilled buffer or overnight at 50 mM at room temperature toward the plus.
4. When finished, remove the nitrocellulose membrane with forceps (avoid touching the filter without gloves) and put membrane into an incubation chamber (*see* **Note 5**) and cover with Ponceau S solution. Shake for 1 min. The Ponceau S solution can be reused many times.

5. Rinse with distilled water until protein bands appear. If the filter looks fine, continue to the next step.
6. Immerse membrane in 20 mL of blocking buffer (8% nonfat milk powder in TBST) and incubate for 1 h at room temperature with shaking.
7. Pour of the blocking buffer and rinse the membrane once with TBST.
8. Incubate with primary antibody specific for your target protein (see **Note 2**; dilution in TBST) for 60 min at room temperature in 10 mL of the antibody dilution.
9. Remove the primary antibody and wash at least three times for 10 min each with 20 to 30 mL of TBST at room temperature.
10. Incubate with horseradish peroxidase-conjugated secondary antibody (dilution in TBST) for 60 min at room temperature.
11. Remove the secondary antibody and wash at least three times for 10 min each with TBST at room temperature.
12. Detect with Pierce ECL Western Blotting Substrate kit (or equivalent) and autoradiography film as specified by the manufacturer of the kit.

3.4. Microscopy

3.4.1. Cell Growth

1. Inoculate a 10 mL overnight culture of the yeast strain in Sc-ura-leu medium at 30°C.
2. Dilute a 10 mL cultivation to A_{600} 0.2 and grow to A_{600} 0.8 to 1.0. This will take 4 to 5 h.

3.4.2. Fixation

1. Add 20% paraformaldehyde directly in to the cell culture to give 4% final concentration. Paraformaldehyde irritates respiratory tract so it is advisable to do this step in a hood.
2. Fix 10 to 20 min on a shaker at room temperature.
3. Collect the fixed cells in an eppendorf tube by centrifugation at top speed for 5 s.
4. Wash with 1 mL of water for one time and resuspend in 100 µL of water.

3.4.3. Preparation of ConA-Coated Slides

1. Pipet a 50 µL drop of 0.1 µg/mL concanavalin A in the wells of a PTFE-printed microscope slide and let stand for 10 min.
2. Wash a few times with water.

3.4.4. Mounting of Fixed Cells

1. Add 30 µL yeast cells on the slide. To obtain optimal density of cells it is useful to prepare a dilution series of the cells in the ConA-coated wells of a PTFE-printed

microscope slide. This will give you several samples on one slide and possibility to easily obtain pictures from multiple and single cells.
2. Let the cells adhere to the glass for 10 min. Do not let dry.
3. Aspirate away the unbound cells and add 6 μL of the mounting solution on the cells and gradually lay the cover slip down to avoid air bubbles. Lay paper towels over the slide and gently squeeze out the excess of mounting medium.

3.4.5. Mounting of Living Cells

Preferably, use synthetic medium lacking tryptophane (presence of tryptophane can result in background fluorescence).

1. Add 30 μL yeast cells on the slide. To obtain optimal density of cells it is useful to prepare a dilution series of the cells in wells of a PTFE-printed microscope slide.
2. Let the cell adhere to the glass for 10 min and wash excess of cells away. Do not let dry.
3. Aspirate away the unbound cells and add 6 μL of fresh growth medium and lay the cover slip.
4. Seal the edges of the cover slip with nail polish.

3.4.6. Microscopy

The microscope used needs to have a light source and filters optimal for YFP excitation and signal detection. Preferably, 100X oil immersion objectives should be used because of the small size of the cells. However, 63X can also be used. For bright field observation of the cells, the microscope should optimally have DIC (Nomarski) optics. As for any method, appropriate negative controls should be used. When available, one should prepare a negative control of cells expressing one of the binding partners that does not normally interact with the other target protein. In case no such mutant in known, one should use a functionally unrelated protein (of similar size) as the other binding partner. One should also test that no signal is obtained when only the EYFP fragments are expressed in the cells.

4. Notes

1. Care should be taken to maintain the correct reading frame when fusing your target gene with the EYFP fragments. When feasible, it is advisable to test tagging of your target proteins both at the amino- and at the carboxy-terminus. In many cases tags in one combination of orientation give the best signal.
2. In case there are no good antibodies available for your target protein, it is useful to generate a peptide tag (e.g., HA or myc) to your protein by introducing a DNA

sequence for the tag in the oligo nucleotide that are used when cloning your target gene into the plasmids. This will enable verification of the expression level of your target protein and enables also detection of the total distribution of the target proteins (e.g., with indirect immunofluorescence.)
3. Do not grow the culture to stationary phase because the *ADH1* promoter activity decreases gradually when glucose is consumed. We usually collect the cells around A_{600} 1.0.
4. In contrast to many other protein assays, the BCA™ Protein Assay Kit, Pierce (or equivalent) is compatible for SDS concentrations up to 5%.
5. We have found plant cell cultivation chambers (**Subheading 2.2.3**.) handy for membrane staining with Ponceau S and incubation with antibodies. 10 mL of antibody dilution is enough when the lid is on. The antibody dilutions can in most cases be used several times. For storage at +4°C, it is advisable to add (e.g., NaN_3 0.02% w/v). However, do not use NaN_3 for antibodies coupled with horseradish peroxidase as it inhibits peroxidase activity.
6. In order to be confident with the results obtained with BiFC it is advisable to test the *in vivo* functionality of the tagged target genes. For this purpose the ability of the tagged target gene to rescue the phenotype of the conditional or knockout mutants of your gene of interest should be used. Knockout strains can be ordered with very small cost (e.g., from EUROSCARF collection that has deletions for almost all genes in the *S. cerevisiae* genome: http://web.uni-frankfurt.de/fb15/mikro/euroscarf/).

Acknowledgments

Chang-Deng Hu is acknowledged for EYFP fragment-containing plasmids and comments on the manuscript. Marko Kaksonen is thanked for helpful comments on live cell imaging. The work was supported by Academy of Finland grant No. 211171 to J.J.

References

1. Tsien, R. Y. (1998) The green fluorescent protein. *Annu. Rev. Biochem.* **67**, 509–544.
2. Hu, C-D. Chinenov, Y., and Kerppola, T. K. (2002) Visualization of interactions among bZIP and Rel family proteins in living cells using bimolecular fluorescence complementation. *Mol Cell.* **9**, 789–798.
3. Miyawaki A. (2003) Visualization of the spatial and temporal dynamics of intracellular signaling. *Dev Cell.* **4**, 295–305.
4. Tsien, R. Y. Bacskai, B. J., and Adams, S. R. (1993) FRET for studying intracellular signalling. *Trends Cell Biol.* **3**, 242–245.
5. Kerppola, T. K. (2006) Visualization of molecular interactions by fluorescence complementation. *Nat Rev Mol Cell Biol.* **7**, 449–456.

6. Hu, C-D. Grinberg, A., and Kerppola, T. (2005) Visualization of protein interaction in living cells using bimolecular fluorescence complementation (BiFC) analysis. In: *Current Protocols in Cell Biology.* (Bonifacino JS, Dasso M, Harford JB, Lippincott-Schwartz J, Yamada KM., ed). Wiley, Hoboken, NJ: pp. 21.3.1–21.3.21.
7. Shyu, Y.J., Liu, H., Deng, X., and Hu, C-D. (2006) Identification of new fluorescent protein fragments for bimolecular fluorescence complementation analysis under physiological conditions. *Biotechniques.* **40**, 61–66.
8. Jach, G., Pesch, M., Richter, K., Frings, S., and Uhrig, J.F. (2006) An improved mRFP1 adds red to bimolecular fluorescence complementation. *Nat, Methods.* **3**, 597–600.
9. Hu, C-D., and Kerppola, T. K. (2003) Simultaneous visualization of multiple protein interactions in living cells using multicolor fluorescence complementation analysis. *Nat. Biotechnol.* **21**, 539–545.
10. Mumberg, D., Muller, R., and Funk, M. (1995) Yeast vectors for the controlled expression of heterologous proteins in different genetic backgrounds. *Gene.* **156**, 119–122.
11. Sherman, F. (1983) Getting started with yeast. In: *Methods in enzymology* (Fink, G. and Hicks, J. B., eds). Academic Press, New York, NY: pp. 3–20.

II

MAMMALIAN CELLS

13

Using Quantitative Fluorescence Microscopy to Probe Organelle Assembly and Membrane Trafficking

Brian Storrie, Tregei Starr, and Kimberly Forsten-Williams

Summary

With current light microscopy and laboratory-level computational capability, many questions in organelle assembly and membrane trafficking that were once treated in a qualitative manner can now be treated quantitatively. We present here an overview of the principles involved in doing quantitative fluorescence microscopy. We illustrate these with examples drawn from our work with the Golgi apparatus and endosomes in cultured mammalian cells. The principles themselves can be applied to any system.

Key words: Confocal fluorescence microscopy; deconvolution; fluorescence microscopy; Golgi apparatus; quantitation.

1. Introduction

Light microscopy (LM) is the most direct of all cell biology techniques where the adage "seeing is believing" is true but often not the whole story. It is the oldest technique dating from the early 17th century. Indeed, the term *cell* comes from Robert Hooke and his light microscopic observations of thin slices of cork in the mid-17th century. Clearly, the physical properties of light have not changed over the last 400 yr, although our ability to shape optical components and to precision-engineer microscope stands has greatly improved. However, Hooke would readily recognize today's commonly used microscope stand consisting of a single objective, sample stage, light source, and focusing mechanism for what it is. Even the confocal microscope with its pinholes to eliminate out-of-focus light would be readily recognized as a microscope. What would not be recognized or readily comprehended would be the detector and adjacent computer and monitor, critical components of the modern microscopy system. Today, imaging is a digital technique in which the final image itself exists in computer memory. In fact, with the laser scanning confocal microscope (LSCM), there is no image recognizable to the human mind outside of the computer. In our digital world, an image is truly a data array consisting of intensity values with associated spatial coordinates. The image can be processed and corrected for out-of-focus light by a mathematical technique termed *deconvolution*. Rule-based, objective computer techniques can be applied to quantify fluorescence distribution between structures and changes in these distributions can be quantified over time. Photobleaching approaches can be applied to test how continuous an organelle is even though the underlying dimensions of the organellar membrane(s) are below the resolution of the light microscope. Modern fluorescence microscopy is a central technique for quantitative studies.

In this chapter, we outline the application of quantitative fluorescence microscopy to common problems in cell biology such as probing for organelle assembly/organization and membrane trafficking processes. We take the Golgi apparatus in mammalian cells as the prime example organelle. The Golgi apparatus is the central organelle in the secretory pathway whose primary role is to receive proteins and lipids synthesized in the endoplasmic reticulum (ER), process these in different ways often through the removal or addition of sugar residues, and package them for delivery to various destinations including the plasma membrane. In mammalian cells, the membranes of the Golgi apparatus are organized into a series of subcompartments termed *cis*, medial, and *trans* cisternae (**Fig. 1**). These membranes associate with one another in poorly defined ways to give what appears by electron microscopy (EM) to be a series of stacks organized in a ribbon-like configuration (pattern) with individual cisterna oriented in a *cis* to medial to *trans* manner. Newly synthesized proteins enter the Golgi apparatus at the *cis* side and exit at the *trans* side/*trans* Golgi network (TGN). The Golgi ribbon is found juxta-nuclearly in mammalian cells.

We present examples of quantitative image analysis based on our recent work. Much of our recent work has emphasized a quantitative approach to image analysis rather than using qualitative visualization of organelle dynamics through the tracking of green fluorescence protein (GFP)-tagged chimeric proteins. In this chapter, emphasis is given to developing principles of quantitative image analysis rather than detailed protocols. Much of actual quantitative image analysis is implemented through specific computer programs and individually written scripts or macros that function within a given program environment. The details are apt not to transfer between programs and hence we choose to be illustrative. Our goal is to demonstrate that formally intractable problems are now

Using Quantitative Fluorescence Microscopy to Probe

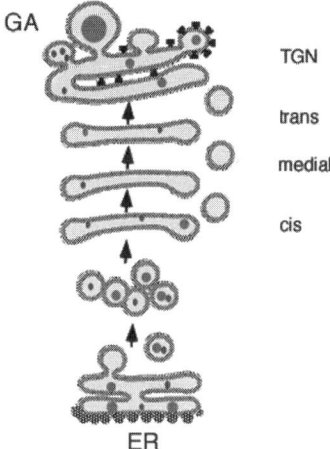

Fig. 1. Schematic depiction of the organization of the mammalian Golgi apparatus. Arrows indicate the direction of anterograde transport from the ER to the Golgi apparatus and from the *cis* to *trans* Golgi cisternae. GA, Golgi apparatus; TGN, *trans* Golgi network; ER, endoplasmic reticulum.

approachable within an individual research laboratory and provide guiding principles for general application.

2. Some Illustrated Principles of Quantitative Image Analysis
2.1. What Is Being Analyzed Is Often Submicroscopic

The resolution of a single-objective, focusing light microscope is limited by diffraction *(1)*. In practical terms, that means that even the most recent confocal microscope has an XY resolution of approx 200 nm and a Z resolution of approx 500 nm *(2)*. As illustrated in **Fig. 2**, much of the detailed structural features of an organelle such as the Golgi apparatus are below the resolution of a light microscope. Compare the distribution of the same protein, *N*-acetylgalactosyltransferase 2 (GalNAcT2) in HeLa cells by EM of immunogold-labeled cryosections (**Fig. 2A**) and by fluorescence microscopy of GFP-tagged GalNAcT2 (**Fig. 2B,C**). Note the 35-fold difference in magnification and the difference in cell area viewed in a single micrograph. At the approx 35,000 magnification necessary to readily view 10 nm immunogold by EM, only a small portion of an individual cell is visualized and in a given thin section only about 1 out of 5 to 10 cell profiles is positive for Golgi membranes. In contrast, by confocal fluorescence microscopy, the optical magnification is approx X630 to X1,300 and the entire cell area is viewed. Distributions can be rendered easily into three-dimensional projections, and Golgi-specific fluorescence can

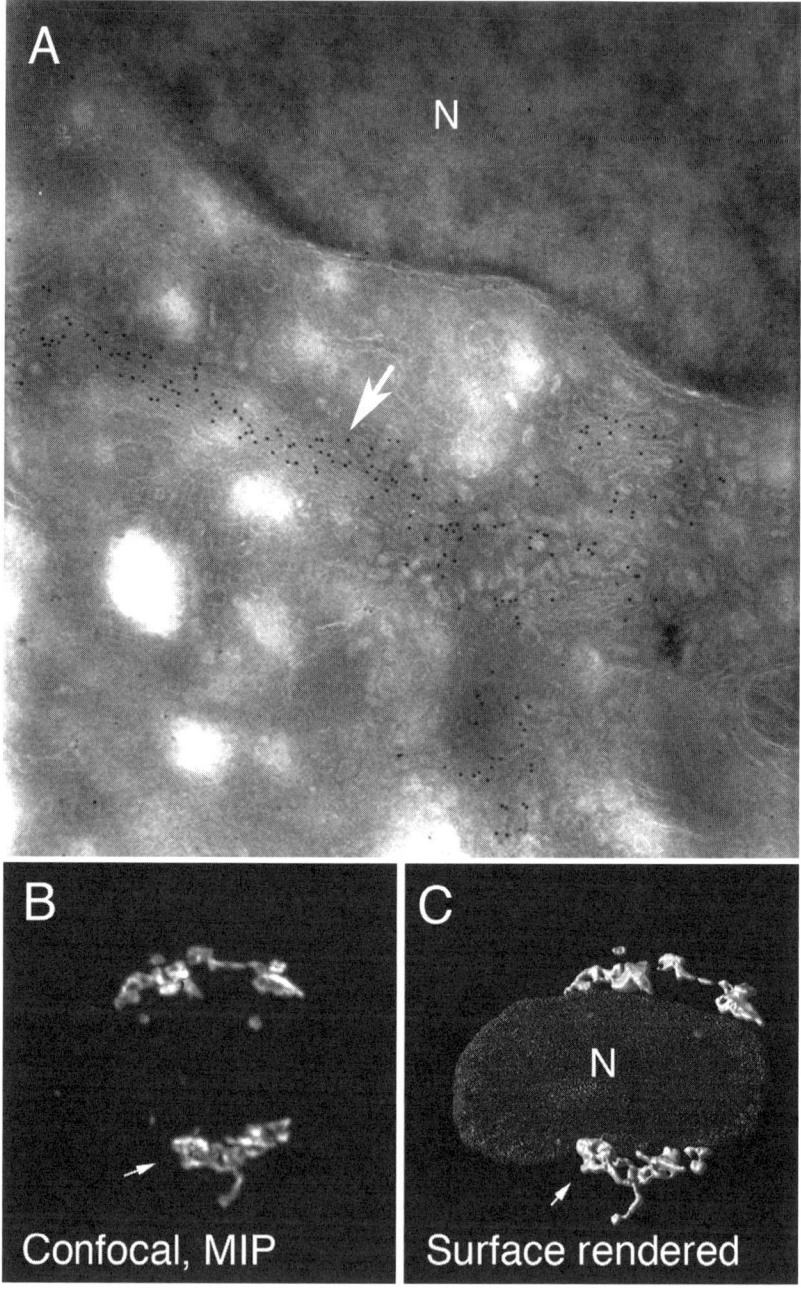

Fig. 2. (Continued)

be readily related to overall distributions of other organelles in the cell, for example, the nucleus (**Fig. 2C**). Whereas EM requires an evacuated sample compartment to allow electron penetration and cell fixation, LM can be done with fully hydrated cells. In summary, LM is rapid and simple compared to EM and can be applied to living cells. However, its resolution is limited compared with EM, thus making study of organelles more challenging.

2.2. Full Detector Resolution: Resolution Must Be Sufficient to Match the Objective

Detector devices, whether they be the photomultiplier tube of the LSCM or the charged coupled device (CCD) camera of the wide-field or spinning disk confocal microscope, potentially have the information density to fully match the resolution of X63 or X100 magnification, high numerical aperature (1.4 NA) objectives. In essence, the detector is a device for converting analog information, light waves, into digital information without detector saturation. The Nyquist criterion holds that the analog information must be oversampled by a factor of 2 to prevent loss of resolution in the conversion (for a general discussion, *see* **ref. 3**). Hence, if the microscope resolution is 200 nm in the XY plane, then the sampling interval must be at least 100 nm (i.e., XY pixel size in the digital image must be at least 100 nm). For a CCD camera, this requires operating the camera at 1×1 binning without clustering of detector elements (image pixels) together generally done in an attempt to achieve shorter exposure times or to match the image size to a low-resolution monitor. For example, exposure times are shortened by a factor of 4 when 2×2 binning is performed instead of 1×1 and the 691×525 pixel image fits nicely on a 17- or 20-inch flat panel monitor.

Fig. 2. (Continued) Comparative detail of Golgi apparatus imaging by EM and confocal fluorescence microscopy. (**A**): GalNAcT2 distribution is visualized by immunogold labeling (10 nm gold) of thawed cryosections. There is also 5 nm gold present in small amounts as label for protein disulfide isomerase in the ER. N, nucleus. In B,C Golgi distribution is visualized by the fluorescence yielded by GalNAcT2-GFP and a confocal fluorescence microscope. In B, the GalNAcT2-GFP fluorescence collected over a series of Z slices is as a maximum intensity projection (MIP) into a single plane. The Z slices were taken with the BD CARV II spinning-disk confocal accessory mounted on a Zeiss Axiovert 200 M microscope. In C, the image is surface rendered and the DAPI nucleus (N) is also shown. Arrows point to examples of Golgi apparatus. (**B**): Maximum intensity projection done with iVision for Mac, version 4.0, Biovision Technologies. (**C**): Surface rendering of deconvolved confocal image stack was done with Huygens Esssential software, version 2.9, Scientific Volume Imaging.

With 1× 1 binning, a 1,392 × 1,050 pixel image does not. For a 63x objective assuming a 1x intermediate lens at 1 × 1 binning, the pixel size is 114 × 114 nm (i.e., approaching Nyquist criterion and at 2 × 2 binning, the pixel size is 228 × 228 nm, far from Nyquist oversampling).

Using full Nyquist oversampling can truly matter in practical terms. For example, a few years ago the laboratory made the transition from video resolution cameras to current high-resolution CCD cameras. At first, monitor size was not updated and imaging habits from videocameras carried over. In effect, we were sampling with a pixel size of 362 nm. That is poor resolution compared to the capabilities of the microscope and more importantly compared with the underlying size of Golgi assembly intermediates. We were asked by a reviewer to score in a computer, rule-driven, objective manner the number of Golgi assembly intermediates at time points for a publications *(4)*. We found, in reality, that we could do this only for the image sets collected at full Nyquist oversampling. At 2 × 2 binning, object size and noise could not be distinguished *(4)*. Importantly, full Nyquist oversampling is required for any mathematically rigorous deconvolution algorithm. Deconvolution does truly improve image sharpness and signal-to-noise ratio (e.g., *see* **ref. 5***)*. However, in stating this, we freely acknowledge that mathematically rigorous deconvolution is computationally intensive. For a 200 to 300 megabyte image stack, we find that a 64-bit computer with 8 gigabytes of RAM is much more stable than a 4-gigabyte RAM machine.

In summary, maximizing the capabilities of the microscope requires using the full capabilities of the detection system. For an LSCM, that means using a X63 or X100 objective in conjunction with an appropriate zoom factor. For a wide-field microscope or spinning-disk confocal microscope, maximizing capabilities means using the full CCD array as individual sensors. The resolution is available in current state-of-the-art detectors but requires that the user be willing to take proper precautions during acquisition (e.g., oversampling) and having an appropriate monitor for the image size. Good quantitative analysis comes with using the full resolution of the microscope.

2.3. Setting Object Boundaries/Thresholds Is a Key Step in Quantitative Data Analysis and Is Simplified by Deconvolution

A key step, if not the key step, in quantitative fluorescence image analysis is recognizing the object of interest from all other objects. In general, the approach has been to identify the object of interest from its brightness (i.e., in the simplest possible case, the object of interest is the bright object and all else is low intensity background). In essence, the difference is all or nothing. In reality, the experimental situation is not so simple. We describe here two examples from our studies in membrane trafficking in the endomembrane system of mammalian cells.

Typically, in membrane trafficking problems, the desired quantity to be determined using fluorescence microscopy is how much of the specific fluorescence is within the object of interest. For example, Golgi glycosyltransferases are known to cycle between the Golgi apparatus and the ER. One might want to know how much of the enzyme is where as a function of time following a perturbation or what the steady-state distribution of the enzyme is between two compartments following the change. In HeLa cells, we have shown that the membrane area of the ER is approx 10-fold greater than that of the Golgi apparatus but, by immunogold labeling of thawed cryosections followed by EM, 90% of GalNAcT2 is associated with Golgi membranes and 10% with ER membranes *(5)*. GalNAcT2 is a marker for the entire Golgi apparatus as it is found in *cis*, medial, and *trans*-membranes *(5,6)*. Under steady-state conditions, there is therefore a 100-fold difference in the concentration of GalNAcT2 in the Golgi apparatus and ER. Biologically, the difference has importance because it can explain why ER proteins are not glycosylated by recycling Golgi enzymes; the enzyme concentration is low. But EM has its issues, as discussed previously, and fluorescent microscopy can be advantageous. We found that the issue of quantifying differences in concentration such as this could be successfully done using fluorescence microscopy. The factor that we found to be most important was treating the image as a data array rather than visual image. As shown in **Fig. 2C** (confocal microscopy) and **Fig. 3A** (wide-field microscopy), the Golgi apparatus in individual HeLa cells when visualized by eye appears as a sharp juxta-nuclearly located object of somewhat variable intensity. In a single-label experiment, the rest of the cell appears in essence black to the eye (**Fig. 2B**;

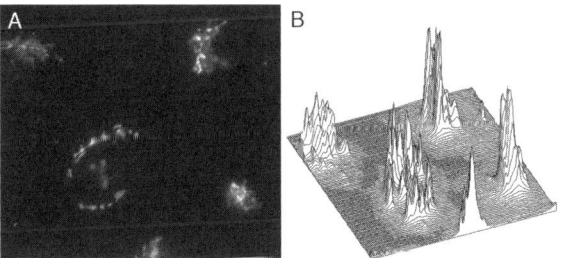

Fig. 3. The alternate appearance of the Golgi apparatus by wide field fluorescence microscopy in HeLa cells stably expressing GalNAcT2-GFP. (**A**): The data are viewed as an image. (**B**): the data are shown as an array of pixel intensity versus XY coordinate. Wide field images were collected with a Zeiss Axiovert 200 M microscope with a 100x/1.4 NA objective. The intensity plot was done with ImageJ software, version 1.37v, NIH.

Fig. 3A). However, when viewed as a data array, the situation is very different (**Fig. 3B**). Now, it is apparent that between cells there is a low-level background fluorescence signal, a higher signal intensity over the cytoplasm and, centrally located within the cells, a strong, high-intensity Golgi apparatus signal. Note that in data space there is no sharp intensity boundary between the Golgi apparatus and the cytoplasm. The visual distinctions of an image are an artifact of the interactions between the eye and the mind.

The primary issues to resolve for quantitative fluorescence analysis in this example are ones of establishing what the cytoplasmic background intensity is and how to separate this from Golgi apparatus signal using an objective threshold. In the case of cells stably transfected to express a tagged Golgi enzyme, the simplest solution to establishing cytoplasmic background is to co-culture tagged and wild-type cells and to use the cytoplasmic wild-type fluorescence as the background value. Corrected cytoplasmic fluorescence can be equated with ER fluorescence *(5)*. For the determination of intensity threshold to distinguish Golgi fluorescence from ER fluorescence, our solution has been to average the cytoplasmic fluorescence and then to score Golgi fluorescence as that above the average cytoplasmic value plus two times the standard deviation for intensity fluctuation. Corrected data then can be used to determine the Golgi and ER densities by summing fluorescence across a series of confocal Z-sections. A detailed treatment of solutions to the problems posed in this example can be found in **ref. 5**.

A second class of examples in which boundary definition was pivotal to addressing the biology arises when fluorescent signal is initially present in small structures that then accumulate in larger structures. Two examples of this situation were analyzed when we focused on tracking Shiga toxin B fragment transport from endosomes to the Golgi apparatus *(7)* and in studies of the reassembly of the Golgi from intermediates during either mitosis or Golgi reassembly following brefeldin A washout *(4)*. In these cases, an endosome or an assembly intermediate is a small structure distinctly more intense than background, which accumulates in the larger structure, the assembled Golgi apparatus. Here the key rules found were the necessity to use thresholds to distinguish bright from background as described previously and including a size criteria to distinguish small objects from large.

We note that in all cases, using the mathematical technique of deconvolution to correct for light spread, made setting thresholds for quantitative fluorescence microscopy easier. Objects were sharper and the difference between visual and quantitatively set thresholds was less. Finally, although the thresholding approach described here is the most common one for distinguishing objects from background, an alternate approach, watershed analysis, is beginning to gain notice; for example, it has been recently applied to the counting of ER exit sites and Golgi fragments *(8)*. In the watershed approach, data are grouped together as an object in analogy with water running down from a peak. The boundary then becomes the inflection line where the "water" no longer runs down the hill. The watershed approach has been implemented as a plug-in (http://www.WaterShedCounting3D.com) to the Java language, freeware program, ImageJ (http://rsb.info.nih.gov/ij/).

2.4. A Wide-Field Image Might Be Sufficient

Summing fluorescence intensities across a series of confocal Z-sections provides a logically rigorous solution to determining whole-cell distributions of fluorescence markers. However, in practice for substratum attached tissue culture cells, our experience is that a single wide-field image focused approximately mid-cell height is sufficient after deconvolution to give a realistic distribution of fluorescence between Golgi apparatus and ER *(5)*. This is because tissue culture cells are fairly flat being only a few microns thick across most of their cytoplasm. A single wide-field image focused mid-cell samples light intensities from the full cell thickness, albeit not all fully in focus. In our experience, the increased image sharpness owing to deconvolution is essential to making quantitative analysis of single wide-field images practical. In summary, deconvolved wide-field microscopy of a single plane image is not the most rigorous solution, but it is the quickest solution and requires the least equipment.

2.5. Fluorescence Co-Localization Is Again a Problem in Setting Thresholds

Quantitative co-localization of multiple fluorescent labels is a common analytical problem encountered in biological image analysis. The goal is to use a pixel-by-pixel comparison of fluorescence intensities in two or more channels to determine if the channels localize labeled proteins to the same physical location. Such analysis raises one new problem, the possibility of small pixel shifts between wavelengths. We define pixel shifts as being differences in pixel coordinate mapping owing to inability of the overall optical system to focus two different colors of light to exactly the same position. In current fluorescence microscopes (built after 2002), such differences are small to nonexistent but are substantial in older equipment. Dichroic mirrors in fluorescence filter sets and optical alignment have greatly improved as the transition from film to low-resolution CCD cameras to high-resolution detectors has been made. The possibility of pixel shifts can be easily tested by examining, for example, a sample made using a first antibody stained with a mixed population of second antibodies for two or more colors. Images from the specific channels can then be examined for pixel shift and corrected for any small pixel shifts noted. The other major problem in fluorescence co-localization is setting threshold for fluorescence intensity in each channel sufficient to distinguish between signal and background similar to what was described in **Subheading 2.3**. We present a detailed example of this in **ref. 7**.

2.6. Although Real Objects Are Three-Dimensional, Doing Quantification in Voxel Space Is Difficult

In reality, organelles and cells are three-dimensional structures. For endosomes and Golgi assembly intermediates, the underlying structure may be below the resolution of the light microscope. Although the features of the Golgi apparatus are small relative to the resolution of the light microscope, overall the Golgi apparatus in mammalian cells is a few microns in length and height. We have treated quantitative fluorescence microscopy as a project in analyzing individual images or summing intensities across a series of confocal images spaced apart in the Z dimension. Ideally, this analysis could be done with software that recognized objects in XYZ voxel space. In reality, this is a difficult problem. One interesting commercial approach has been presented recently by Scientific Volume Imaging Inc., Hilversum, the Netherlands (http://support.svi.nl/wiki/ObjectAnalyzer).

2.7. Quantitative Diffusion Measurements by FRAP Can Be Used to Test Organelle Continuity

The organization of the mammalian Golgi apparatus is remarkably plastic. Various treatments that disrupt the microtubules or deplete the cell of proteins involved in Golgi trafficking and/or organization result in changes in the appearance of the organelle. We show one such example in **Fig. 4A,B** in which the distribution of Golgi apparatus in control cells is contrasted with that in cells depleted of the protein ZW10. ZW10 has been implicated as a tether protein involved in trafficking between the Golgi apparatus and the ER *(9–11)*. In the control, the Golgi apparatus appears relatively compact, whereas in the ZW10 depleted case, it appears by fluorescence microscopy more as a cluster of what may be distinct Golgi elements. Here the Golgi apparatus is labeled with GalNAcT2-GFP. This presents the opportunity to apply fluorescence recovery after photobleaching (FRAP) as an approach to testing for Golgi continuity. FRAP, in simple terms, is an assay for protein diffusion. If the organelle is continuous, recovery of the fluorescence in a photobleached area should be rapid, however, if it is not continuous, recovery should be slower *(12)*. As shown in **Fig. 4C–H**, the FRAP assay provides effective distinction between the two

Using Quantitative Fluorescence Microscopy to Probe 189

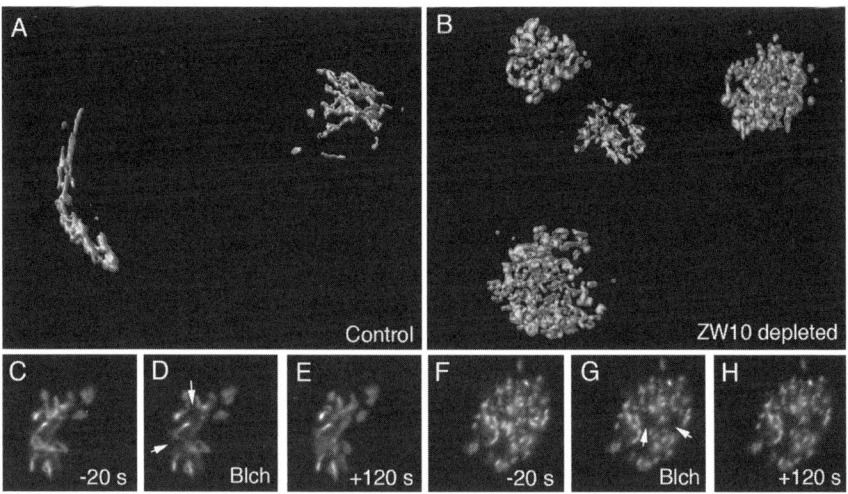

Fig. 4. Contrasting Golgi organization in control and ZW10 depleted HeLa cells as revealed by confocal microscopy and FRAP. (**A**): control Hela cells, surface-rendered Golgi apparatus, GalNAcT2-GFP. (**B**): ZW10 depleted HeLa cells, surface-rendered Golgi apparatus, GalNAcT2-GFP. (**C,D,E**): photobleaching of control Golgi apparatus, arrows in D point to bleached regions. (**F,G,H**): photobleaching of ZW10 depleted Golgi apparatus, arrows in G point to bleached regions. Recovery is rapid in the control Golgi apparatus but nonexistent in the ZW10-depleted Golgi apparatus indicating a lack of interconnections between the Golgi elements. C–H: confocal micrographs. A,B: confocal images were collected with iVision for Mac, version 4.0, BioVision Technologies and deconvolution followed by surface rendering done with Huygens Essential software, version 2.9, Scientific Volume Imaging. C–H: Photobleaching was done with a Zeiss LSM510 microscope and images processed with ImageJ, version 1.37v, NIH.

cases, a result that is consistent with the EM evidence of Hirose et al. *(10)*. Similar studies could and are being done with other organelles.

2.8. Fluorescence Line Scans Are a Good Approach to Study Distributions Within the Golgi Apparatus

A persistent question remains regarding the Golgi apparatus and proteins within it is polarity: Where is the protein distributed with respect to *cis*, medial, or *trans*-cisternae? With the widespread commercialization of confocal microscopy, LM has been repeatedly applied to solving this problem (e.g., **refs. 13** and **14**). The simplest approach is by the line scan methodologies introduced recently by Dejgaard et al. *(15)*. In this approach using multilabel markers, in our

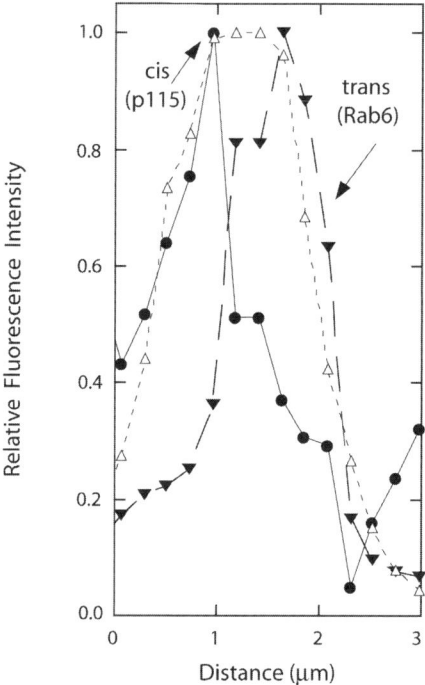

Fig. 5. The distance between *cis* and *trans* Golgi markers can be determined by LSCM line scan methodology. The distances between p115 (*cis* Golgi marker, ●), GalNAcT2-GFP (*cis* to *trans* Golgi distribution, △) and Rab6 (a *trans* Golgi marker, ▼) were determined by placing a vector perpendicular to the stained Golgi apparatus. The distances between for several vectors and individual cells can be determined to give an average distance plus and minus the error of measurement. Line scans were done with a Zeiss LSM 510 microscope and Zeiss software.

case with a *cis* and *trans* Golgi marker, a vector is placed perpendicular to the staining and the distance between peaks in marker distribution is determined. We have found this methodology to be effective (**Fig. 5**). By selecting planes with clearly defined Golgi ribbons, coupling measurements from multiple vectors within a cell and averaging over many cells, we have found the technique to be a valid means of quantifying changes in distribution within the Golgi apparatus following cell perturbation.

3. Conclusions and Perspectives for the Future

The capacity of digital fluorescence microscopy to give quantitative distributions of components at single time points and over a time series is limited only by the skill of the investigator in treating the image as a data array and by the optical resolution of the light microscope. Today, that resolution is 200 nm in the XY dimension and 500 nm in the Z dimension. However, recent research in LM points to the commercialization of fluorescence techniques such as stimulated emission depletion microscopy that can give XYZ resolution of 50 nm or less *(16)*. That resolution approaches the actual gross dimensions of many organellar features. Hence, for many situations where the resolution of the fluorescence microscope is currently an issue, this will no longer be true. We note, however, that 20 yr elapsed between the development of the confocal microscope and its widespread commercialization suggesting that the common achievement of such resolution is apt not to happen tomorrow. Fortunately, many questions regarding organelle dynamics can be addressed currently using today's technology provided careful data collection and analysis is performed. We have provided in this chapter tips to do this based on our experience.

Acknowledgments

This chapter builds on past efforts by laboratory members and was supported in part by a grant from the National Science Foundation (MCB-0549001).

References

1. Abbe, E. (1873) Beiträge zur Theorie des Mikroskops und der mikroskopishen Wahrmehmung. *Arch. Microskop. Anat.* **9**, 413–420.
2. Pawley, J. B. (2006) *Handbook of biological confocal microscopy.* Springer, New York, NY.
3. Murphy, D. B. (2001) *Fundamentals of light microscopy and electronic imaging.* Wiley-Liss, New York, NY.
4. Jiang, S., Rhee, S. W., Gleeson, P. A., and Storrie, B. (2006) Capacity of the Golgi apparatus for cargo transport prior to complete assembly. *Mol. Biol. Cell* **17**, 4105–4117.
5. Rhee, S. W., Starr, T., Forsten-Williams, K., and Storrie, B. (2005) The steady-state distribution of glycosyltransferaes between the Golgi apparatus and the endoplasmic reticulum is approximately 90:10. *Traffic* **6**, 978–990.
6. Röttger, S., White, J., Wandall, H. H., et al. (1998) Localization of three human polypeptide GalNAc-transferases in HeLa cells suggests initiation of O-linked glycosylation throughout the Golgi apparatus. *J. Cell Sci.* **111**, 45–60.
7. Starr, T., Forsten-Williams, K., and Storrie, B. (2007) Both post-Golgi and intra-Golgi cycling affect the distribution of the Golgi phosphoprotein GPP130. *Traffic* **8**, 1265–1279.
8. Gniadek, T. J., and Warren, G. (2007) WatershedCounting3D: a new method for segmenting and counting punctate structures from confocal image data. *Traffic* **8**, 339–346.

9. Andag, U., and Schmitt, H. D. (2003) Ds1p, an essential component of the Golgi-endoplasmic reticulum retrieval system in yeast, uses the same sequence motlf to interact with different subunits of the COPI vesicle coat protein. *J. Biol. Chem.* **278**, 51722–51734.
10. Hirose, H., Arasaki, K., Dohmae, N., et al. (2004) Implication of ZW10 in membrane trafficking between the endoplasmic reticulum and Golgi. *EMBO J.* **23**, 1267–1278.
11. Sun, Y., Shestakova, A., Lupashin, V., and Storrie, B. (2007). Rab6 regulates both ZW10/RINT-1 and conserved oligomeric Golgi complex-dependent Golgi trafficking and homeostasis. *Mol. Biol. Cell* **18**, 4129–4142.
12. Puthenveedu, M. A., Bachert, C., Puri, S., Lanni, F., and Linstedt, A. D. (2006) GM130 and GRASP65-dependent lateral cisternal fusion allows uniform Golgi-enzyme distribution. *Nat. Cell Biol.* **8**, 238–248.
13. Antony, C., Cibert, C., Geraud, G., et al. (1992) The GTP-binding protein rab6p is distributed from medial Golgi to the trans-Golgi network as determined by a confocal microscope approach. *J. Cell Sci.* **103**, 785–796.
14. Shima, D. T., Haldar, K., Pepperkok, R., Watson, R., and Warren, G. (1997) Partitioning of the Golgi apparatus during mitosis in living HeLa cell. *J. Cell Biol.* **137**, 1211–1228.
15. Dejgaard, S.Y., Murshid, A., Dee, K. M., and Presley, J. F. (in press) Confocal microscopy based linescan methodologies for intra-Golgi localization of proteins. *J. Histochem. Cytochem.* **55**, *709–719*.
16. Willig, K. I., Rizzoli, S. O., Westphal, V., Jahn, R., and Hell, S. W. (2006) STED microscopy revels that synaptotagmin remains clustered after synaptic vesicle exocytosis. *Nature* **440**, 935–939.

14

Measuring Secretory Membrane Traffic

A Quantitative Fluorescence Microscopy Approach

Vytaute Starkuviene, Arne Seitz, Holger Erfle, and Rainer Pepperkok

Summary

In this chapter the authors describe automated imaging methods to quantify the transport rates of transmembrane as well as soluble cargo, and to evaluate the integrity of the Golgi complex. The quantification of cargo transport rates serves as an example of fluorescence intensity-based assays, the quantification of the Golgi complex integrity—as an example of morphology-based assays. These quantitative assays could be applied for single experiments as well as for middle- and high-throughput screening approaches. Each of these assays can be used to appreciate effects caused by gene silencing by RNAi, cDNA overexpression or application of chemical compounds. For each assay the authors discuss protocols for sample preparation, parameters for automated image acquisition, strategies of image analysis, and data quantification.

Key words: Golgi complex; high-content screening microscopy; RNA interference; secretory membrane traffic.

1. Introduction

The secretory membrane traffic enables an accurate temporal and spatial distribution of newly synthesized proteins, lipids, and carbohydrates. The fidelity of the process is ensured by a concerted action of numerous regulatory processes, like protein modification and degradation, adaptive network of signaling cascades, membrane and cytoskeleton dynamics. Despite the huge amount of information about individual molecules and events accumulated by classical genetic and biochemical methods, a complete understanding of the underlying principals of the secretory pathway requires the implementation of large-scale methodologies. Systematic analyses, such as organelle proteomics *(1)* or yeast two-hybrid screening generate a comprehensive catalogue of the molecules involved, but unfortunately suffers largely from the inherent problem of not being able to acquire information on the complex temporal and spatial molecular interrelationships. Being the method free from these limitations, fluorescence microscopy is currently becoming a tool of choice to produce a large-scale functional read-out from a single cell or even subcellular structures *(2)*. In addition, this approach is applicable to living cells, thereby enabling them to enlighten the dynamic properties of the secretory pathway. Additionally, the phenotypic cellular analysis became more efficient and more focused by exploiting the whole-genome sequence information for production of full-length cDNA clones and RNA molecules to interfere with activities of defined proteins and their complexes. Next, microscopy-based approaches experienced a considerable progress in automation of various steps involved, namely sample preparation, data acquisition and evaluation, quality control, and integration into multifaceted databases. Finally, libraries of test molecules can be analyzed quantitatively by automated image analysis, what in turn provides a high degree of objectivity in data analysis and a sensitivity that allows the detection and ranking of even subtle phenotypes, which could easily be missed by manual data evaluation.

Recently, high-throughput, high-content microscopy-based functional assays have been successfully applied to address the problems of the secretory membrane traffic under the conditions of cDNA-green fluorescent protein (GFP) overexpression *(3,4)* and downregulation by RNAi *(5,6)*. As quantitative rather than qualitative fluorescence microscopy studies are the preferred choice in analyzing the complexity of the secretory pathway, a detailed description of how to perform, image, and quantitatively analysze the major events of the secretory pathway, namely cargo transfer and the integrity of the Golgi complex is presented here.

2. Materials

2.1. Functional Assays

1. Assays are performed either in NIH 3T3 cells ATCC (CRL-1658) or HeLa cells ATCC (CCL-2) growing in Dulbecco's Modified Eagle's Medium containing 10% fetal calf serum, 2 mM glutamine, 100 U/mL penicillin and 100 µg/mL streptomycin.
2. For the ectopic expression of ts-O45-G we use a recombinant adenovirus encoding ts-O45-G tagged with cyan fluorescent protein (CFP) or yellow fluorescent protein (YFP; *7*).

3. For transfection of cDNAs in HeLa cells we use FuGene6 kit (Roche), for transfection of cDNAs in 3T3 cells we use Effectene reagent kit (Qiagen), for transfection of siRNAs in HeLa cells we use Lipofectamine2000 kit (Invitrogen), for transfection of siRNAs in 3T3 cells we use Oligofectamine reagent (Qiagen). Necessary dilutions of transfection reagents, cDNAs, and siRNAs prior to applying on cells were made in serum-free Opti-MEM I medium (Gibco).
4. Protein synthesis is abolished by adding ready-made cycloheximide solution in dimethyl sulfoxide (Sigma-Aldrich).
5. Paraformaldehyde (Sigma-Aldrich), 3% solution in phosphate-buffered saline (PBS). Keep appropriate aliquots of the solution at $-20°C$.
6. Glycin, 30 mM solution in PBS. Keep the solution at $4°C$ not longer than 2 to 3 wk.
7. Triton X-100, 0.1% solution in PBS.
8. Hoechst 33342 (Sigma-Aldrich) dissolved at 1 mg/mL in PBS. Keep the solution in dark at $-20°C$.
9. Primary antibody: rabbit polyclonal anti-PC I antibody (Chemicon) diluted in PBS.
10. Secondary antibodies: anti-rabbit Alexa 568, anti-mouse Alexa 488 (Molecular Probes), Cy5- or Cy3-labeled anti-mouse and anti-rabbit antibodies (Amersham Biosciences). All antibodies are diluted in PBS prior to applying on cells.
11. Ascorbate solution: 0.25 mM ascorbate, 1 mM ascorbate-2-phosphate (Sigma-Aldrich). Keep the solution in the dark at $-20°C$.

2.2. Imaging on an Automated Wide-Field Screening Microscope

Image acquisition is done with a Scan^R system (www.olympus.com), but it can be done on any other inverted automated fluorescence microscope *(2,7)*. The major parts of Scan^R system are as such:

1. The inverse microscopy stand IX81, equipped with a Xenon light source MT20. This microscope is equipped with standard filter sets for discriminating between Hoechst 33342, CFP, GFP, YFP, Cy3, and Cy5 in a sequential imaging mode.
2. An automated table (Corvus) that allows the precise movement of the sample.
3. 12-bit charge-coupled device camera (C8484, Hamamatsu, Munich, Germany).
4. ScanR^R software controlling all components of the system.

2.3. Automated Data Evaluation

For fluorescence intensity-based assays, like ts-O45-G and PC I transport assays, the Scan^R Analysis software (www.olympus.com) was used. For evaluating morphology-based assays, such as the Golgi integrity assay, MetaMorph software (Universal Imaging Corporation) was used. However, the automated data analysis is not restricted to these programs, and other software capable of automated batch processing of images can be used.

3. Methods

The quantitative phenotypic analysis in fixed cells reaches the highest biological resolution for cellular processes that can be synchronized. As for the secretory membrane trafficking, the synchronization of cargo flow through the secretory pathway could be achieved by several methods, like temperature shifts or application of specific chemicals. These are reversible, but, unfortunately, rather general perturbations of the secretory pathway. A more specific method is to control the folding reactions in the endoplasmic reticulum (ER), but is applicable only to a few proteins so far. One of them is the ts-O45-G protein (a temperature-sensitive mutant of the vesicular stomatitis virus glycoprotein; *8*). This transmembrane protein has the feature to accumulate in the ER at 39.5°C, but moves through the secretory pathway to the plasma membrane (PM) at the permissive temperature of 32°C. Another example is the soluble protein procollagen I (PC I), whose folding requires hydroxylation of prolin residues *(9)*. By manipulating the concentration of ascorbate, which is a cofactor of the ER-localized prolyl-hydroxylase, synchronization of PC I ER exit is achieved.

3.1. ts-O45-G Secretion Assay

1. Depending on a time course of the experiment, plate HeLa or NIH 3T3 cells at a density of 5,000 to 20,000 cells per well in LabTek 8-well chamber slides (Nalge Nunc) containing 300 μL medium per well, and incubate for 24 h at 37°C (*see* **Note 1**).
2. Transfect cells with cDNAs or siRNAs using suitable transfection reagents according to manufacturers' recommendations. As a rule, 250 ng of plasmid DNA and 300 ng of siRNA are sufficient to add to each well. To achieve sufficient expression rates of products encoded by cDNAs, incubate cells for 24 h at 37°C. Optimal time of incubation in order to achieve the desired down-regulation effect by siRNAs needs to be tested individually and can vary from 12 to 72 h.
3. Overlay the cells with the recombinant adenovirus encoding ts-O45-G tagged with CFP or YFP and incubate for 1 h at 37°C (*see* **Note 2**).
4. Wash cells twice with prewarmed growth medium to remove excessive virus.
5. Transfer cells to 39.5°C and incubate for 16 to 20 h to accumulate sufficient amounts of CFP- or YFP-tagged ts-O45-G in the ER.
6. Transfer cells to 32°C in the presence of 100 μg/mL of cycloheximide to release ts-O45-G from the ER, and incubate for 1 h (*see* **Note 3**).
7. Fix cells with 3% of paraformaldehyde solution for 20 min at room temperature.
8. In order to prevent damages of the PM, quench paraformaldehyde with 30 mM solution of glycin/PBS for 5 min at room temperature.

9. Stain the fraction of ts-O45-G at the PM by a standard immunostaining procedure using a monoclonal antibody recognizing an extracellular epitope of ts-O45-G (a gift from Prof. K. Simons, Dresden, Germany) and an appropriate secondary antibody at room temperature (*see* **Note 4**).
10. Stain cell nuclei with Hoechst 33342 stain (final concentration 0.1 μg/mL) for 5 min at room temperature (*see* **Note 5**).

3.2. PC I Secretion Assay

1. Plate cells, transfect with cDNAs or siRNAs, and incubate them under the same conditions as for the ts-O45-G secretion assay (*see* **Note 6**).
2. Induce the folding of PC I by the addition of the ascorbate solution and incubate in the presence of 100 μg/mL of cycloheximide for 1 to 2 h at 37°C (*see* **Note 7**).
3. Fix cells with 3% of paraformaldehyde solution for 20 min at room temperature and permeabilize with 0.1% Triton X-100.
4. Stain for endogenous PC I by a standard immunostaining procedure with the suitable primary and secondary antibodies at room temperature (*see* **Note 8**), and stain the nuclei with Hoechst 33342.

3.3. Assay of the Golgi Complex Integrity

1. Plate, transfect, and incubate cells as for ts-O45-G and PC I secretion assay.
2. After the desired incubation times, fix and immunostain cells with the selected Golgi marker. For instance, use mouse monoclonal anti-GM130 antibody (BD Bioscience) to label *cis*-Golgi in NIH 3T3 and HeLa cells (*see* **Note 9**). Stain cell nuclei with Hoechst 3342.

3.4. Imaging on an Automated Wide-Field Screening Microscope

1. To image the ts-O-45-G secretion assay, set up three channels if working with siRNA, four channels if working under conditions of overexpression of cDNAs. Namely, use an excitation wavelength of 360–370 nm, emission wavelength of 460 nm to image nuclei stained with Hoechst 33342, excitation wavelength of 500 to 520 nm, emission wavelength of 535 to 630 nm to image ts-O45-G tagged to YFP, excitation wavelength of 620 nm, emission wavelength of 660 nm to image ts-O45-G at the PM, and excitation wavelength of 430 to 424 nm, emission wavelength of 470 to 524 nm to image cDNA-CFP overexpressed.
2. To image PC I secretion assay and Golgi integrity assays set up two channels if working with siRNA, three channels if working under conditions of overexpression of cDNAs. Namely, use an excitation wavelength of 360 to 370 nm, emission wavelength of 460 nm to image nuclei stained with Hoechst 33342, use excitation wavelength of 545 nm, emission wavelength of 620 nm to image intracellular PC I or the Golgi complex, and excitation wavelength of 430 to 424 nm, emission wavelength of 470 to 524 nm to image cDNA-CFP overexpressed.

3. Adjust image-acquisition settings, like exposure time and intensity of excitation light for each and every fluorophore separately (*see* **Note 10**). Use camera settings without binning.
4. Set software autofocus by using 2 × 2 binning. Stained cell nuclei are taken as objects for the software-orientated autofocus. A coarse autofocus (23 layers with a step width of 7.2 μm) is performed only for the first position on the first well of the Labtek chamber slide. For the other positions, a fine autofocus (13 layers with a step width of 1.8 μm) is executed (*see* **Note 11**).
5. Select the number of positions to be imaged per well. For instance, 36 positions working with a 20x/0.7 NA air UPlanApo objective (http://www.olympus.com) in 8-well LabTek chamber slide will result in imaging about 1,000 to 5,000 cells.

3.5. Automated Data Evaluation for Fluorescence-Based Assays

1. Use so-called rolling ball algorithm to correct background for all images taken as described in **Subheading 3.4**. (*see* **Note 12**).
2. Identify the cell nuclei by manually setting up the threshold value in the images of stained nuclei. For the images taken with 20x/0.7 NA air objective, thresholded objects smaller than 800 pixels and larger than 5,000 pixels are excluded from further analysis. That ensures that interfering particles, like dust, also cells growing too close to each other or touching each other are excluded. Also cell nuclei touching the image border are not considered for further analysis.
3. The thresholded images are used to create a binary mask by setting the background to "0" and the nucleus to a pixel value of "1."
4. Dilate the binary mask so that it encompasses the maximum area of each cell.
5. The dilated binary mask is multiplied with the other nonthresholded fluorescence channels to calculate the average intensity of each fluorophore within the mask. That way average fluorescence intensities representing the total amount of ts-O45-G, ts-O45-G on the PM or endogeneous PC I are calculated.
6. Use the "gating" function to discard cells, which interfere with the further analysis (*see* **Note 13**), cells with irregular nuclear shape, out-of-focus images, and to separate cells transfected with cDNA from nontransfected ones.
7. Calculate the rate of the ts-O45-G transport by dividing the average fluorescence intensity of ts-O45-G at the PM by the average fluorescence intensity of total ts-O45-G recorded within the same dilated mask. This needs to be done for every cell separately (*see* **Note 4**).
8. Calculate the rate of the PC I transport by evaluating the average fluorescence intensity of the PC I-specific fluorescence within the dilated mask for each cell separately.

3.6. Automated Data Evaluation for Morphology-Based Assays

The "MetaMorph" software allows one to apply automatically consecutive image-processing steps in a so-called "journal." This "journal" is applied to a set

Measuring Secretory Membrane Traffic 199

of two or three different images (colors; *see* **Subheading 3.4**.) and the journal is looped for all images in a directory automatically.

1. Use stained nuclei image as the first image in the "journal." Not only by thresholding, but also by low-pass filtration, binarization and a small particle filter of this image the localizations and geometries of all nuclei are identified. All particles smaller than a particular value are excluded from the following calculations.
2. Use the image of so-called "transfection" channel (if working with cDNAs) or an empty image (if working with siRNAs) as the second image in the "journal." After correcting the background by subtracting a user-defined fixed grey value, thresholding, low-pass filtration, and binarization, the output image is multiplied with the before-processed nuclei image. The result of these steps is an image with only the nuclei of the transfected cells if working with cDNAs or all cell nuclei if working with siRNAs.
3. Dilate nuclei so that it encompasses the maximum area of each cell. By using the "create regions around objects" function, define regions of interest (ROI) and assign to each of them a number (*see* **Note 14**). If working with cDNAs, from all nuclei defined in **step 1** deduce the nuclei of the transfected cells (**step 2**) resulting in an image with the nuclei of the nontransfected cells. Also define ROIs and assign to each of them a number.
4. Use the Golgi complex image as the third image in the "journal." Subtract the background and threshold the image so that only Golgi-specific structures are taken into consideration. Transfer the previously created ROIs to this image, and calculate the number of Golgi complex fragments and the average fluorescence intensity of each fragment in each ROI. That is done separately for transfected and nontransfected cells.
5. The generated values are automatically exported to an Excel spreadsheet, where transfected and nontransfected cells are differentially named and processed separately (*see* **Note 15**).
6. Use the function "loop" in the "journal" to ensure automatic batch processing of all images in an analogous way (*see* **Note 16**).

Once the calculations for all assays are done, derive the average and standard deviaton values of the selected features for an entire cell population. If working with cDNAs overexpressed, compare the transfected to the nontransfected cells, if working with siRNAs, compare the cells transfected with a control "scrambled siRNA" to the cells transfected with the "test siRNA." Depending on the experimental outline and numbers of cells imaged, diverse methods can be used to decide whether a test molecule is an effector. For instance, a pooled standard error of the difference between the means of two populations can be calculated, and only these test molecules for which the difference between the means are two times higher than the pooled standard error are considered to be effectors *(4)*. Alternatively, molecules with average values higher than two standard deviations of the mean of the controls can be taken as effectors *(6)*. If working with cDNAs overexpressed, a phenotypic effect can be correlated with the expression level of a GFP-tagged cDNA product in a linear function *(4)*.

4. Notes

1. Here and in the following paragraphs protocols are given to work in LabTek 8-well chamber slides, which enable to pursue middle-scale experiments; they are superior in terms of optical properties, reproducibility, and are nontoxic to cell cultures. By using different formats, the protocols need to be scaled appropriately. Particularities to experiment with cell arrays are described in *(6)*. The assays are not restricted to HeLa or NIH 3T3 cells: any type of cell culture, which can be infected by the recombinant adenovirus is suitable.
2. The titre of the recombinant adenovirus and the duration of incubation depend on the cell type. For instance, HeLa cells are efficiently infected in 45 min, longer incubation times might cause cell death. To the contrary, NIH 3T3 cells require a longer incubation time than 1 h to achieve that essentially all cells are infected.
3. Saturation of the arrival of ts-O45-G at the PM might vary in different cell types and needs to be tested experimentally.
4. This step is needed to detect the fraction of ts-O45-G, which arrives at the PM and to compare it to the total amount of ts-O45-G produced in the cell *(2–4)*. This approach enables to account for different levels of ectopically expressed ts-O45-G and, consequently, to compare the behavior of ts-O45-G from cell to cell and from experiment to experiment.
5. Stained cell nuclei are used for autofocus and automated image-evaluation routines. A low concentration of Hoechst 33342 and short incubation times ensure that the signal is relatively weak and does not interfere with the emission wavelengths of the other fluorophores.
6. Only cells producing collagen, for instance NIH 3T3 fibroblasts, are recommended for this assay. The assay also can be pursued in primary fibroblasts.
7. Incubation times depend on the cell type and need to be tested. Using the mixture of ascorbate and ascorbate-2-phosphate ensures the presence of constant ascorbate concentration over the whole course of experiment.
8. The anti-PC I antibody recommended here is well suitable for NIH 3T3 cells. By using other type of cells, other antibodies might be needed.
9. Virtually any type of cells and numerous antibodies can be used for this assay. Importantly, an antibody should be specific only for the Golgi complex and does not stain punctuate structures, like transport carriers, and so on.
10. Images must not contain saturated pixels. To prevent a sample from photobleaching it is preferred to minimize the excitation light intensity.
11. 2×2 Binning during the autofocus procedure minimizes the required exposure time and light intensity. The software-based autofocus function works well only when sufficient numbers of cells are plated.
12. For this type of image, the size of the rolling ball was set to 200. Generally, the ball radius should be at least as large as the diameter of the largest object in the image that is not part of the background.

13. For the calculation of the secretion rates it is necessary to discard cells, which have too high or too low amounts of ts-O45-G expressed. Both conditions might lead to a false arrival of ts-O45-G at the PM or (a) owing to the overload of cell with too much protein expressed or (b) owng to overcoming trafficking block when underexpressed. As PC I is an endogenous protein, no particular selection is needed.
14. Numbering of ROIs is necessary to correlate the measurement and visual inspection of the images.
15. Excel must be opened while working with the MetaMorph software (when data export to Excel is used in the "journal") in order to ensure data tranfer and saving. For data calculation and analysis write a macro in Excel. For instance, sorting of nontransfected cell from the transfected cells can be done by the "autofilter" function.
16. Despite of the capability for sophisticated multiparameter analysis, Metamorph is rather suitable for middle-scale experiments. For instance, a time-consuming feature of Metamorph is that the images prior to image processing need to be uploaded and sometimes even renamed.

References

1. Gilchrist, A., Au, C.E., Hiding, J et al. (2006) Quantitative proteomics analysis of the secretory pathway. *Cell* **127**, 1265–1281.
2. Pepperkok, R., and Ellenberg, J. (2006) High-throughput fluorescence microscopy for systems biology.*Nat. Rev. Mol. Cell Biol.* **7**, 690–696.
3. Liebel, U., Starkuviene, V., Erfle, H. et al. (2003) A microscope-based screening platform for large-scale functional protein analysis in intact cells. *FEBS Letters* **554**, 394–398.
4. Starkuviene, V., Liebel, U., Simpson, J.C., et al. (2004) High-content screening microscopy identifies novel proteins with a putative role in secretory membrane traffic. *Genome Res.* **14**, 1948–1956.
5. Pelkmans, L., Fava, E., Grabner, H., et al. (2005) Genome-wide analysis of human kinases in clathrin- and caveolae/raft-mediated endocytosis. *Nature* **436**, 128–133.
6. Simpson, J.C., Cetin, C., Erfle, H., et al. (2007) An RNAi screening platform to identify secretion machinery in mammalian cells. *J. Biotechnol.* **129**, 352–365.
7. Starkuviene, V., and Pepperkok, R. (2007) The potential of high-content high throughput microscopy in drug discovery. *British J Pharm,* **152**, 62–71.
8. Zilberstein, A., Snider, M.D., Porter, M. and Lodish, H.F. (1980) Mutants of vesicular stomatitis virus blocked at different stages in maturation of the viral glycoprotein. *Cell* **21**, 417–421.
9. Kao, W.W., Berg, R.A., and Prockop, D.J. (1975) Ascorbate increases the synthesis of procollagen hydroxyproline by cultured fibroblasts from chick embryo tendons without activation of prolyl hydroxylase. *Biochim. Biophys. Acta.* **411**, 202–215.

15

A Correlative Light and Electron Microscopy Method Based on Laser Micropatterning and Etching

Julien Colombelli, Carolina Tängemo, Uta Haselman, Claude Antony, Ernst H.K. Stelzer, Rainer Pepperkok, and Emmanuel G. Reynaud

Summary

Correlative microscopy is a hybrid method that allows the localization of events observed under visible, ultraviolet, or infrared light, at molecular and submolecular levels, combining two microscopy techniques. However, the main limitation of correlative microscopy is to develop a labeling technique that can be easily used first in light and then in electron microscopy. Laser etching is a well-established method to create precisely designed shapes or volumes in various materials including glass. We have applied this technique to develop a new correlative light and electron microscopy method and to apply it in our study of the Golgi apparatus. The location of the cell of interest is laser-inscribed into the glass allowing a simple follow-up in light and fluorescence microscopy. Furthermore, the glass surface is laser-etched and upon fixation and flat embedding, the inverse ridge can be localized as well as the cell of interest, which is then processed for electron microscopy.

Key words: Correlative; electron microscopy; Golgi apparatus; laser etching; laser micropatterning; light fluorescence microscopy; plasma-induced ablation; pulsed laser; TEM.

1. Introduction

Correlative microscopy refers to experiments in which the same sample is examined by two or more imaging techniques (e.g., fluorescence and electron microscopy [EM]). Light microscopy (LM) and EM each have certain advantages and limitations for the investigation of organelle structure. LM provides a high temporal resolution of fluorescent features in a living cell such as vesicle trafficking (*1*). However, the spatial resolution of LM is limited to the range of the wavelength. This limitation constrains the understanding of the organelle organization for which information is often required below the 100 nm level. EM gives high-resolution spatial analysis of structural features but provides only static images and is not applicable to dynamic investigations.

Correlative microscopy provides an opportunity to combine the advantages of both techniques and establish functional connections between dynamic imaging and high resolution. However, the success of correlative microscopy requires a labeling technique that is simply followed in a cell in LM as well as in EM. Many methods have been introduced for the correlative observation under both light and electron microscopes such as the combine use of gold- and fluorochrome-labeled antibodies (*2*), FluoroNanogold labeling (*3*), gold locator grid (*4*), photo-etched locator coverslip (*5*), acid-etched coverslip (*6*), or simply by engraving the nearby glass with a diamond pen (*7*).

The method presented here offers a simple technique based on laser-inscribing and laser-etching using a laser nanosurgery set up.

2. Materials

1. Glass bottom dishes (Matek).
2. CO_2-independent medium (Gibco BRL).
3. Bsc1 cell line stably transfected with a GalT-green fluorescent protein (GFP) under the control of a cytomegalovirus promoter (CMVp).
4. Cell-location inscription in glass and glass-surface etching along defined patterns were performed with a pulsed ultraviolet laser (wavelength 355 nm, pulse duration 470 ps, JDSUniphase, Milpitas, *see* **Note 1**) coupled into a conventional inverted microscope (Axiovert 200 M, Carl Zeiss) with scanning galvanometric mirrors (GSILumonics, Billerica).
5. Fluorescence imaging with an ORCA camera (Hamamatsu).
6. Heating stage insert P (Carl Zeiss), or alternatively incubation chamber for inverted microscope.
7. Forceps.
8. Phosphate-buffered saline (PBS) without calcium and magnesium: 0,2 g KCl, 0,2 g KH_2PO_4, 1,15 g Na_2HPO_4, 8 g NaCl per liter for 1X final solution.
9. Glutaraldehyde (EMS,), 25% solution in distilled water.
10. Cacodylate buffer: 0,2 M cacodylic acid (Serva).
11. Fixative solution: 2.5% glutaraldehyde, 50 mM cacodylate buffer, 2% sucrose, 0.05 M KCl, 0. 4 mM $MgCl_2$, 0.04 mM $CaCl_2$.
12. Postfixation solution: 2% OsO_4, 50 mM cacodylate buffer.
13. Leica-Reichert Ultramicrotome (Leica Microsystems).
14. A diamond knife with a 35° angle for plastic sectioning (2.1 mm, Diatome).
15. Copper-palladium slot grids (2 × 1 mm) (Plano) freshly coated with Formvar (1% polyvinyl formal in chloroform).

16. Epon resin (Serva).
17. Uranyl acetate (UA; Fluka-AG) 2% in 70% methanol.
18. Lead citrate solution (Reynolds solution).
19. UA, 0.5% solution in water.
20. Propylene oxide.
21. Ethanol.
22. CM120 Biotwin transmission electron microscope (TEM) (FEI Company) operating at 100 KV and equipped with a Keenview bottom-mounted camera (Soft Imaging System).

3. Methods

To perform correlative fluorescence and EM, we use the same laser-coupled based microscope to achieve two different effects based on plasma-induced ablation *(8)*: glass inscription and glass etching. The combination of a pulsed laser and a high numerical aperture objective lens is required for sufficient laser intensity (*see* **Note 1**). Optical properties of bulk glass can be modified locally to reveal specific locations in space by transmission microscopy (e.g., phase contrast). As described before *(9)*, the effect of plasma formation in glass by a single laser pulse can extend from a few hundreds of nanometers to several microns. Here we inscribe user's defined patterns within the bulk glass of regular microscopy coverslips several microns below the sample of interest in order to mark the cell location. Subsequently, the sample is fixed according to a regular EM fixation protocol. Before cell embedding, the sample is placed back in the laser-coupled microscope, the cell of interest is located and the glass surface is etched along new user's defined patterns with the same laser to generate grooves. Embedded sample results in inverse ridges, which are visible in a stereo microscopy for further location during EM sectioning.

3.1. Cell Preparation

Cells are seeded on Matek glass dishes 24 to 48 h prior to the laser-inscribing step (*see* later). Cells can be either transfected (transitionally or stably), microinjected or treated with a particular drug. In the presented example, Bsc1 cells were stably transfected with a GFP-Golgi construct (Clontech) which contains the targeting motif of Gal T under the control of a CMVp and seeded in order to be subconfluent at the start of the experiment. Prior to the experiment, cells are incubated in CO_2-independent media allowing an easier transfer of the cells to the different microscopes used during the protocol.

3.2. Cell Location by In Situ Glass Laser Inscription

Cell location prior to fluorescent imaging was inscribed within the glass volume of the coverslip 10 to 20 μm below the cell. The laser was scanned to inscribe patterns square or rectangular shapes surrounding the correspond-

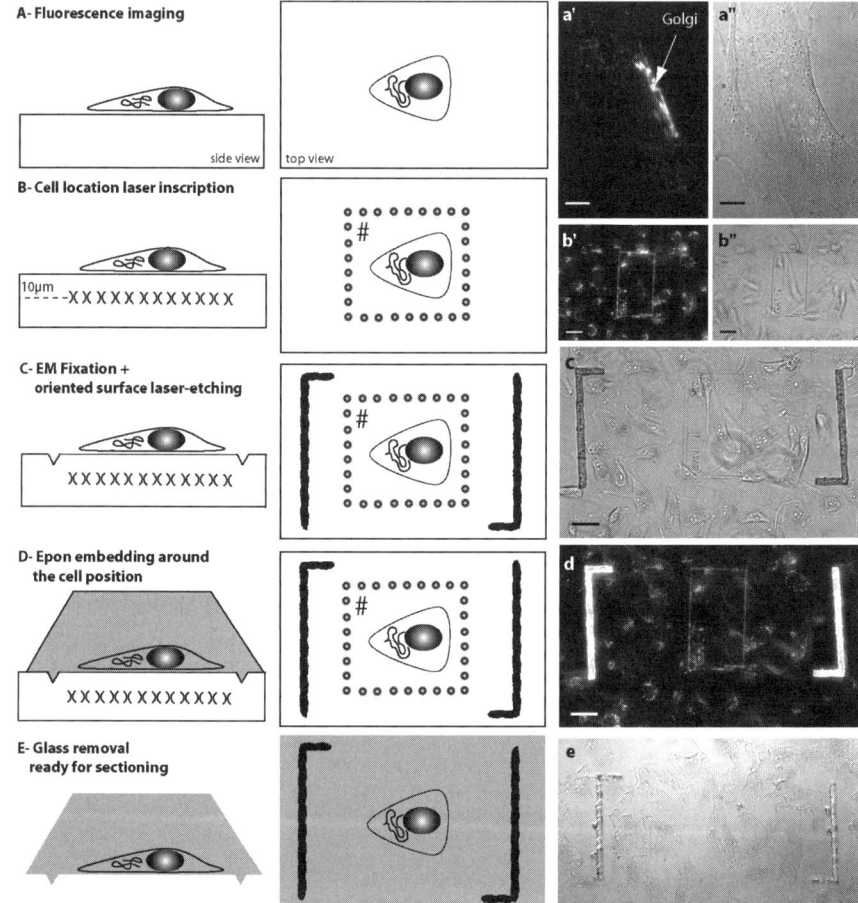

Fig. 1. Overview of the technique. (**A**) The cell of interest is located on the coverslip and follows up by light microscopy (fluorescence in **a'**, transmission in **a"**, scale bar: 10 μm). (**B**) The cell position is labeled in the glass coverslip to allow long-time follow up of the cell as well as transfer from one microscope to another (darkfield contrast in **b'**, phase contrast in **b"**, scale bar: 50 μm). (**C**) After fixation for electron microscopy, the cell can be further localized by etching the surface of the coverslip. Scale bar: 50 μm. (**D**) The Epon is poured onto the cell and covers the etched glass groove. Scale bar: 50 μm. (**E**) The ridges on the surface of the Epon block allow the localization of the cell.

A Correlative Light and Electron Microscopy Method Based

ing cell outline, as shown in **Fig. 1**. Laser pulses were scanned either manually with the microscope stage or automatically with the laser scanners. Laser pulses were spaced as much as possible to reduce the energy deposited in the glass, but in order to still allow the pattern to be visible in phase contrast microscopy (*see* **Note 2**).

1. Mount sample on inverted fluorescent microscope, find cell of interest and focus onto it.
2. Defocus 10 to 20 μm down axially (*see* **Note 3** about chromatic aberrations).
3a. If automatic (*see* **Note 4**), define a pattern along which the laser will induce glass inscription. For imaging purposes, the patterns should not overlap with the cell outline laterally in x and y. Start laser-scanning procedure at high speed up to 1 kHz repetition rate, if automatic.
3b. If manual (*see* **Note 4**), start laser emission at low repetition rate and scan manually the microscope stage in fine-tune positioning to inscribe around the cell outline. Typically, a repetition rate of 10 Hz allows to manually space laser pulses inscriptions sufficiently so as to be seen in transmission afterward. The energy levels to use in correspondence to the laser source may differ (*see* **Note 4**). With low numerical aperture transmission illumination set up, the out-of-focus cell outline can be viewed while scanning the stage so as to avoid overlapping the laser trace to the region of interest.
4. Repeat **steps 1** to **3** if several cells have to be located, considering proper spacing for optimal EM embedding preparation (*see* **Note 5 and 6**).
5. Sample is ready for fluorescent imaging and fixation.

3.3. Fluorescent Imaging

Cells were imaged using an inverted Zeiss Axiovert 200 M microscope (Carl Zeiss) with 20x, 40x, and 63x objectives to image the laser-inscribed patterns, the selected cell and the fluorescent Golgi apparatus, respectively (**Fig. 2A,B,D**). However, after laser-inscribing cells can be transferred to any kind of light microscope as the inscribed pattern within the glass can be easily found.

3.4. Fixation Prior to Etching and Electron Microscopy

At the end of the LM observation, the samples are processed for flat embedding fixation.

1. Before fixation, wash the adherent cells two to three times with PBS, prewarmed (37°C).

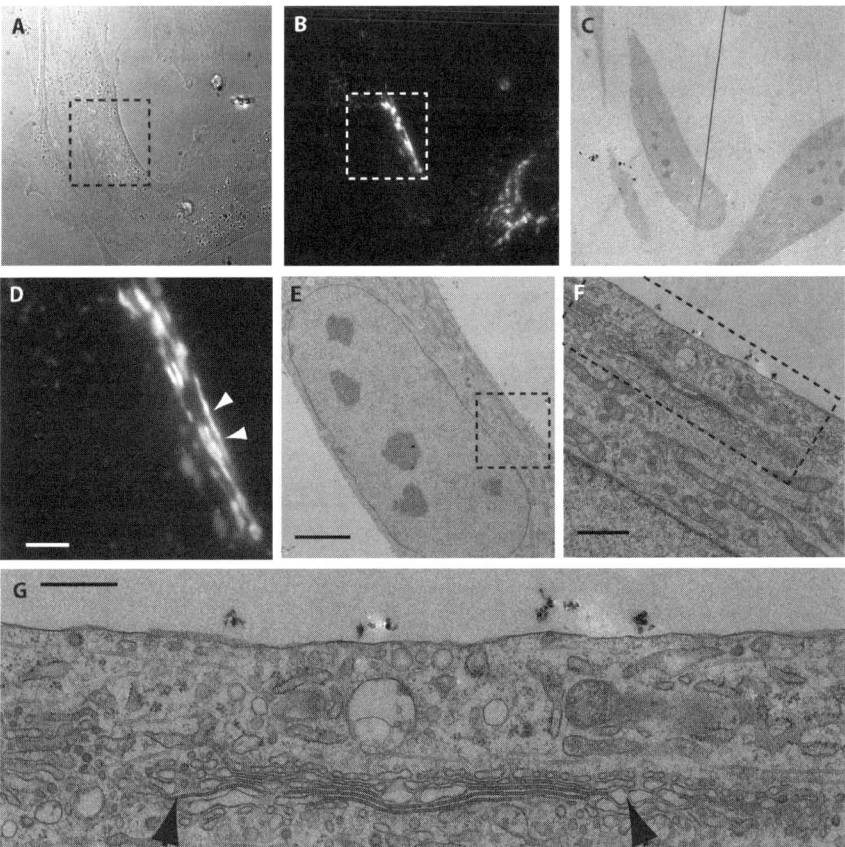

Fig. 2. Correlation of light and electron microscopy images of the Golgi apparatus. (A) Differential interference contrast of the cell followed up in Fig. 1. (B) Fluorescence imaging of (A) showing the Golgi apparatus. (C) Low-magnification overview of one Epon section showing the same cell as in (A). (D) Magnified view in fluorescence of (B) showing the Golgi apparatus, highlighting one Golgi stack (white arrows). Scale bar: 5 μm. (E) Electron microscopy image of the same area than in (D). Scale bar: 5 μm. (F) Higher magnification of the boxed area in (E) showing the highlighted cisternae from (D). Scale bar: 1 μm. (G) Higher magnification of the boxed area in (F) showing the highlighted cisterna in (D), black arrows. Scale bar: 0.5 μm.

2. Coverslips are first quickly rinsed in the fixative solution and then immersed in fresh fixative solution at room temperature for 30 min. Be careful not to dry the samples.
3. Fixative is washed away with 50 mM cacodylate buffer (5 × 2 min), and the samples are kept on ice during incubation with post-fixation solution for 40 min.
4. Rinse extensively the sample with water before laser etching.
5. Sample can be either etched at this point or later, *see* later.

6. Rinse with H_2O.
7. Incubate the samples in 0.5% UA in H_2O for 30 min.
8. Rinse with H_2O 30 min. Keep samples cold.
9. Sample is ready for etching.

3.5. Laser Etching

Two L-lines were etched into the glass close to the surface of the coverslip. Grooves are etched one on each side of the cell of interest. Laser-induced plasma formation at the glass-culture medium boundary results in glass removal and projection (8). The laser was scanned across user-defined shapes (e.g., lines or splines) composed of close laser pulses (with a linear density of two pulses per micron) at a repetition rate of 500 Hz. The pulse energy sufficient to remove minimal amounts glass was measured. With 300 nJ per pulse, material removal occurs with accuracy below 1 μm. With 400 to 600 nJ per pulse, the removed area extends over 5 μm laterally, with 2 to 5 depth in axial direction. Glass removal is repeated several times laterally to extend over a surface at least 20 × 100 μm. This allows a clear identification on the Epon block with a low magnification microscope (binocular, stereo microscope etc...). At these energy levels, plasma formation is unavoidably accompanied by cavitation, i.e. the formation of transient hypersonic bubbles. Since those bubbles can potentially harm the surrounding cells (fixed cell detachment) at a 20 μm distance from the etched area, etching was performed at least 100 μm away from the cell of interest.

1. Find the cell of interest with a 10- or 20-fold phase-contrast lens (Epiplan 20x/ numerical aperture [NA] Ph2 in this study). Laser inscription in glass is better visible with darkfield contrast by using a Ph3 ring contrast with a Ph2 lens.
2. Change objective to high NA objective (C/Apo 63x/1.2 W or similar high NA lens) and focus on the cell of interest.
3. Displace the stage 100 μm away from the cell and perform glass etching along an oriented pattern. The thickness of the pattern can vary according to the laser energy used. However, to avoid high mechanical effects (*see* **Note 5**) that could damage neighboring cells of interest, it is better to use energy values close to the threshold of plasma formation and repeat glass etching axially so as to enlarge the etched area.
4. Displace the stage on the opposite side of the cell and perform glass etching similarly along an oriented pattern.
5. Switch to 20x lens again and record pictures for correlation during the sectioning protocol. To record better pictures, wash glass particles away from the field of view (*see* **Note 5**).
6. Repeat **steps 1** to **5** according to the number of cells located originally in fluorescence (*see* **Note 6**).

3.6. Staining and Plastic Embedding

1. Samples are washed extensively with water to remove glass splitter
2. *En bloc* UA staining is performed in 0.5% UA in water for 30 min in order to increase the background contrast.
3. Dehydrate samples in graded ethanol from 40°, 50°, 70°, 95°, to 100° (5 min for each steps, except for 100°, twice, 10 min; *see* **Note 7**). At the end of the dehydratation process, the coverslips are then dipped for 30 s in small beakers containing ethanol 100° (twice, 30 s).
4. Dip coverslips in propylene oxide just before flat embedding.
5. Coverslips are then quickly turned upside down onto the top of a beam capsule filled with pure Epon resin so that the cell layer is facing the Epon (flat embedding).
6. Capsules with their coverslips on top are placed upside down onto a plastic support to avoid bubbles sticking on the coverslip side, and put in an oven set at 60°C for polymerization of the resin over night.
7. The next day the coverslips are detached by dipping the capsules with their coverslip attached to it alternatively in hot water ($\sim$100°C) and liquid nitrogen.
8. Finally, Epon blocks are then further polymerized at 60°C for 24 h.

3.7. Trimming of the Blocks and Sectioning

1. The surface of the polymerized blocks is observed using a dissecting microscope and the ridges on the Epon resin are localized.
2. Once the marks are identified, the block is trimmed with a razor blade in order to shape a precise block face by progressively removing of the Epon around the marked area so as to leave just an area precisely framing the area defined by the marks and making a trapezoid shape, the height of which is matching its width (*see* **Note 8**).
3. Serial ultrathin sections (50 to 54 μm thickness) are prepared using a Leica-Reichert Ultramicrotome (Leica Microsystems) and contrasted with methanolic UA (2%) and lead citrate (Reynolds solution). The cells of interest are found in the TEM scope by comparing the outline of the cells in the EM scope with the pattern of the cells recorded by LM in phase-contrast mode.
4. Pictures are taken on CM120 Biotwin TEM microscope (FEI) operating at 100 KV and equipped with a Keenview bottom-mounted camera (Soft Imaging System) (**Fig. 2C,E,F,G**)

4. Notes

1. Laser source. To perform controlled plasma formation and, therefore, glass nanopatterning or etching, short-pulsed lasers providing a high peak power must be used in combination with a high NA lens. As far as the literature can show, there is not a single combination to achieve efficient glass removal or labeling. However, as plasma formation is a nonlinear effect, solid-state lasers providing

a Gaussian beam profile are more efficient as the diffraction limit is best reached in practice, hence providing the highest peak intensity (in Watts/cm2 [W/cm2]) for the lowest energy per pulse (in Joule [J]). To avoid the great expansion of cavitation bubbles and thermal effects, laser pulses below the nanosecond are usually preferred, although a few nanoseconds can be also suitable. More details are found in recent reviews (8), however threshold for plasma formation in glass and water varies around 100 to 200 GW/cm2. Depending on the laser pulse duration and repetition rate, this corresponds to a few nJ up to hundreds of nJ of energy. As glass etching occurs well above this threshold, it is safe to choose a laser source that provides at least one to two orders of magnitude higher.

2. Laser inscription. During laser inscription, laser pulses were spaced as much as possible to reduce the energy deposited in the glass, but in order to still allow the pattern to be visible in phase-contrast microscopy. With 300 nJ pulses, the effect of plasma formation results in a change in glass optical properties that become visible in phase-contrast microscopy. Energy levels between 400 and 600 nJ per pulse are sufficient to inscribe with an extent of several microns with each pulse and, therefore, make them visible even a low magnification (5x, 10x, 20x).

3. Chromatic aberrations. Depending on the NA of the objective lens, chromatic aberrations can occur at 355 nm wavelength. Defocusing below the cell location to inscribe the cell position should include those aberrations in order to make sure that the laser position is well controlled and not too close to the glass medium interface axially (i.e., the cell). Water immersion objectives used in this protocol show an axial shift not superior to 2 μm at 355 nm wavelength.

4. Laser cutter systems. There is a great difference between commercially available systems and self-implemented systems in research labs that, to date consists mainly of the use of scanning mirrors across the sample. Commercial systems or other systems reduced to the minimum number of components displace the beam through the microscope stage, either manually or automatically. With galvanometric mirrors, scanning can be very fast as compared to microscopy stages and ablation can virtually occur simultaneously at any location across the whole field of view. For glass inscription, speed is here not an issue, however manual stage scanning is definitely less performing and requires good ability of the user as well as thorough software control of lasers pulses in order to label only the region of interest.

5. Collateral damage. At the energy levels required to etch glass, plasma formation was shown to be unavoidably accompanied by cavitation (i.e. the formation of transient hypersonic bubbles). As those bubbles were observed to potentially harm the surrounding cells, that is to simply detach fixed cells, at a 40 μm distance from the etched area, etching was therefore performed more than 100 μm away from the cell of interest.

6. Cell numbers. A maximum of four cells can be marked and isolated easily on the same slide.

7. Dehydration. Note that the coverslips get loose in ethanol 100° and need to be detached from their Matek glass dish.
8. Preparation of block face trimming. It may be necessary to re-trim the block face so as to further reduce its area if more convenient. Use a saw to separate the different areas of interest on the Epon block before initially trimming each sample.
9. Stage heating insert. Set temperature displayed on the device often does not correspond to the actual medium temperature. On the stage insert P we have used, it was measured that the culture medium stays at 37°C when the stage is set to 42.7°C. Setting 37°C on the device results in a measured 34°C. Use of Lab grease between the chamber and its lid can be effective in limiting the heat dissipation.
10. Glass projections. During the glass etching prior to cell embedding, small particles of glass are projected in the medium. They float around in the area of the sample and can cause two problems. First, they are highly refringent in water medium and therefore much contrasted by phase contrast microscopy. Pipetting in and out the medium allows to easily removing practically all particles from the field of view in order to take relevant pictures of the sample of interest. Moreover, and most importantly, those glass particles could be embedded in Epon before sectioning and may later damage the diamond knife. Therefore, they should be carefully washed out by extensively rinsing the coverslips prior to embedding. Also, as glass particles can be phagocytosed by living cells and would lead to a higher probability of sectioning glass particles, glass etching must be performed after fixation.

Acknowledgments

Alfons Riedinger for software implementation of the laser surgery platform. This research was partially supported by the VDI-TZ and the German ministry for research and development (BMBF) by grant FKZ 13N8287.

References

1. Simpson, J.C., Nilsson, T., and Pepperkok, R. (2006) Biogenesis of tubular ER-to-Golgi transport intermediates. *Mol. Biol. Cell.* **17**, 723–737.
2. Sun, X.J., Tolbert, L.P., and Hildebrand, J.G. (1995) Using laser scanning confocal microscopy as a guide for electron microscopic study: a simple method for correlation of light and electron microscopy. *J. Histochem. Cytochem.* **43**, 329–335.
3. Takizawa, T. and Robinson, J.M. (2003) Ultrathin cryosections: an important tool for immunofluorescence and correlative microscopy. *J. Histochem. Cytochem.* **51**, 707–714.
4. Svitkina, T.M. and Borisy, G.G. (1998) Correlative light and electron microscopy of the cytoskeleton of cultured cells. *Methods Enzymol.* **298**, 570–592.

5. Kislauskis, E.H., Zhu, X., and Singer, R.H. (1997) β-Actin messenger RNA localization and protein synthesis augment cell motility. *J. Cell Biol.* **24**, 1263–1270.
6. Small, J.V. (1985). Simple procedures for the transfer of grid images onto glass cover slips for the rapid relocation of cultured cells. *J. Microscopy* **137**, 171–175.
7. Chen, W.T. (1981) Surface changes during retraction-induced spreading of fibroblasts. *J. Cell. Sci.* 1981 **49**, 1–13.
8. Vogel, A and Venugopalan, V. (2003) Mechanisms of pulsed laser ablation of biological tissues. *Chem. Rev.* **103**, 577–644.
9. Colombelli, J., Grill, S.W., and Stelzer, E.H.K. (2004) Ultraviolet diffraction limited nanosurgery of biological tissues. *Rev. Sci. Instr.* **75**, 472–478.

16

Approaches to Investigate the Role of Signaling in ER-to-Golgi Transport

Laura J. Sharpe, Ximing Du, and Andrew J. Brown

Summary

There have been many indications that kinases play a role in signaling the transport of proteins through the secretory pathway. Specifically, the serine/threonine kinase Akt affects the transport of cholesterol regulatory components from the endoplasmic reticulum (ER) to the Golgi. However, elucidating the target of Akt in this process has proven to be challenging. Here we describe an approach devised to investigate the Akt target(s) based on examining a potential candidate.

Key words: Akt; fluorescence microscopy; immunoprecipitation; SCAP; SREBP

1. Introduction

Cell functioning relies on the secretory pathway, where proteins are synthesized in the endoplasmic reticulum (ER), and transported to the Golgi apparatus, from which they are sorted and transported to their functional locations within the cell *(1,2)*. Kinases are involved in the regulation of protein functions, and there is a possibility that this regulation extends to protein transport through the early secretory pathway, as has been shown for the late secretory pathway *(3,4)*. H89, an inhibitor of serine/threonine kinases, has been shown to block ER to Golgi transport *(5,6)*. This, therefore, provides evidence that the secretory pathway is regulated in some way by kinases and phosphorylation. Data from the authors' laboratory indicate that Akt plays a crucial role in the ER to Golgi transport of cholesterol regulatory components *(7)*. Akt inhibition by LY294002

(LY) or a dominant-negative form of Akt (DN-Akt) prevents the transport of these components from the ER to the Golgi, and stimulation of Akt using a growth factor leads to increased transport *(7)*.

In mammalian cells, cholesterol homeostasis is controlled by the sterol regulatory element-binding protein (SREBP)-2 transcription factor *(8)*. Upon cholesterol depletion, the SREBP-2 precursor is transported from the ER to the Golgi by SREBP cleavage activating protein (SCAP) via COPII vesicles. In the Golgi, SREBP-2 is cleaved by Site-1 Protease (S1P) and Site-2 Protease (S2P), leading to the release of mature SREBP-2, which is translocated to the nucleus where it up-regulates the transcription of target genes involved in cholesterol synthesis and uptake. On the other hand, when cellular cholesterol levels are high, the transport of SREBP/SCAP from the ER to the Golgi is blocked because of a cholesterol-induced binding of SCAP to another ER protein, INSIG-1 or -2 *(8)*. In this way, the SREBP-2 processing is switched off to facilitate a return to normal cholesterol levels. Therefore, ER to Golgi transport of SREBP-2 is critical for the regulation of cholesterol metabolism (**Fig. 1**).

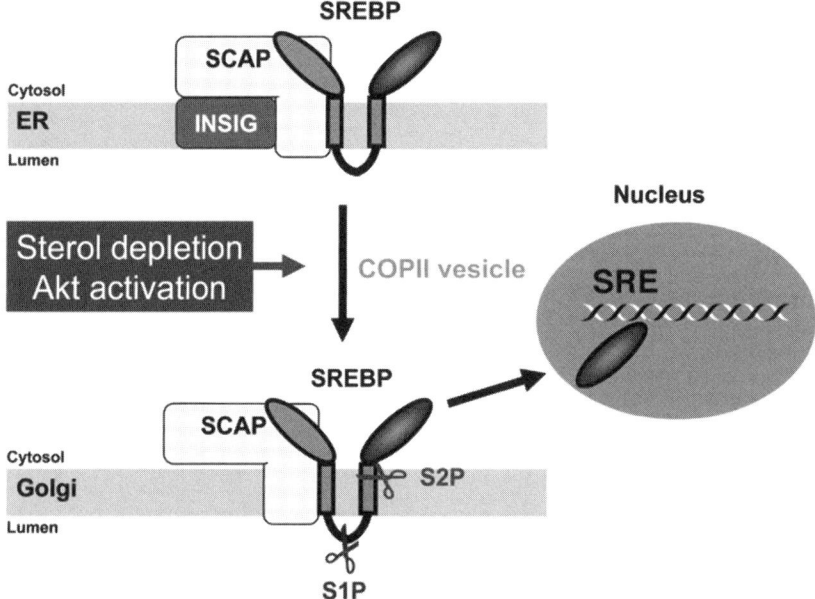

Fig. 1. Endoplasmic reticulum (ER)-to-Golgi transport of SREBP-2 is crucial for cholesterol homeostasis in mammalian cells. Sterol depletion or Akt activation leads to the transport of the SREBP/SCAP complex in a COPII vesicle from the ER to the Golgi. In the Golgi, SREBP-2 is subsequently cleaved by Site-1-Protease (S1P) and Site-2-Protease (S2P), leading to the release of a free transcription factor that is translocated to the nucleus, where it binds to the sterol regulatory element (SRE) and in turn activates cholesterogenic target genes.

Apart from sterol status, we have shown that the activity of Akt kinase also controls the transport of SREBP/SCAP from the ER to the Golgi *(7)* (**Fig. 1**). Here, we employ fluorescence microscopy, bioinformatics, and protein immunoprecipitation (IP) as approaches to identify the putative Akt target(s) involved in ER to Golgi transport of the SREBP/SCAP complex.

2. Materials
2.1. Preparation of Cells for Microscopy

1. Plasmid encoding DsRed-monomer-Golgi, a fluorescent protein marker of the Golgi (Clontech; *see* **Note 1**).
2. FuGene 6 transfection reagent (Roche; *see* **Note 2**).
3. Dulbecco's Modified Eagle's Medium: Ham's Nutrient Mixture F12 (DMEM/F12).
4. $1 \times$ Phosphate-buffered saline (PBS).
5. Fixing solution: 3% (v/v) formaldehyde in PBS.
6. Mounting medium containing anti-fade reagent is preferable for fluorescence microscopy (e.g., from Biomeda).
7. Adobe Photoshop 7.0, or a similar software program to enable image overlays.

2.2. Identification of Potential Akt Phosphorylation Sites

1. NCBI's Reference Sequence Project (http://www.ncbi.nlm.nih.gov/).
2. Scansite (http://scansite.mit.edu).

2.3. Examination of Conservation Throughout Evolution

1. BLAST (http://www.ncbi.nlm.nih.gov/).
2. ClustalW (http://www.ebi.ac.uk/clustalw/index.html) (*see* **Note 3**).

2.4. Immunoprecipitation of Protein

1. Protein G-Sepharose beads.
2. RIPA buffer: 0.1% (w/v) sodium dodecyl sulphate (SDS), 0.1% (w/v) NP-40, 1% (w/v) Na-deoxycholate, 150 mM NaCl, 20 mM Tris-HCl, pH 7.4, 5 mM ethylenediamine-tetraacetic acid (EDTA), 1 mM Na_3VO_4.
3. Protease inhibitors (PI) and phosphatase inhibitors (PPI; e.g., cocktails of these are available from Sigma).
4. Green fluorescent protein (GFP) rabbit polyclonal antibody (Abcam; *see* **Note 4**).
5. Normal rabbit immunoglobulin (Ig)G antibody (Santa Cruz).

2.5. Protein Precipitation

1. Loading buffer: Four volumes of 1% SDS lysis buffer (10 mM Tris-HCl, pH 7.6, 100 mM NaCl, 1% [w/v] SDS) and one volume of 5× loading buffer (250 mM Tris-HCl, pH 6.8, 10% [w/v] SDS, 25% [v/v] glycerol, 0.2% [w/v] bromophenol blue, 5% [v/v] β-mercaptoethanol).

2.6. SDS-PAGE

1. Running buffer: 25 mM Tris, 192 mM glycine, 0.1% (w/v) SDS.
2. Transfer buffer: 18 mM Tris-HCl, 150 mM glycine, 20% (v/v) methanol.
3. Nitrocellulose membranes (*see* **Note 5**).

2.7. Western Blotting

1. PBS Tween-20 (PBST): 1 L 1× PBS, 1 mL Tween-20.
2. Blocking solution: 5% (w/v) skim milk powder in PBST.
3. Antibody diluting solution: 5% (w/v) bovine serum albumin (BSA) in PBST.
4. Primary antibodies: Anti-GFP (Abcam) and Anti-phospho-Akt substrates (Cell Signaling Technology).
5. Secondary antibody: Anti-rabbit conjugated to horseradish peroxidase (HRP).
6. Immobilon™ Western Chemiluminescent HRP Substrate detection system (ECL, Millipore; *see* **Note 6**).
7. Film suitable for chemiluminescence detection (e.g., Hyperfilm from GE Healthcare).
8. Developer and fixer solutions.

3. Methods

To confirm the link between Akt and the ER to Golgi transport of a cholesterol-regulatory component, SCAP, a Golgi marker plasmid, is transfected into a cell-line stably expressing SCAP fused to GFP. In the search for potential targets of phosphorylation by Akt, a bioinformatics-based approach is first employed to identify plausible candidates, and then the conservation throughout evolution of the identified site is examined. An IP is then performed in which GFP antibody is used to pull down GFP–SCAP, this is separated by SDS-polyacrylamide gel electrophoresis (PAGE) and then analyzed by Western blotting with an antibody specific to phosphorylated substrates of Akt.

3.1. Preparation of Cells for Microscopy

1. Set up Chinese Hamster Ovary (CHO) cells stably expressing GFP–SCAP in 6-well plates at a density of 2×10^5 cells per well in DMEM/F12 containing penicillin (100 U/mL), streptomycin (100 μg/mL), and L-glutamine (2 mM),

Approaches to Investigate the Role of Signaling 219

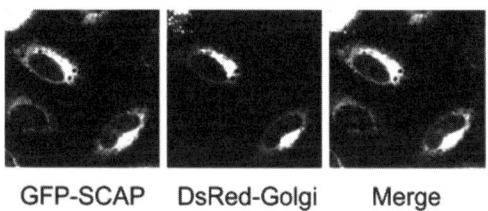

GFP-SCAP   DsRed-Golgi   Merge

Fig. 2. Compactin induces movement of green fluorescent protein (GFP)-SCAP from the endoplasmic reticulum (ER) to the Golgi. Treatment with compactin inhibits cholesterol synthesis and induces movement of SCAP from the ER to the Golgi. Co-localization of the GFP-SCAP and the Golgi marker is observed as a yellow area in the merged image. Co-incubation with LY294002 to inhibit Akt prevents this Golgi localization, which is reflected by a marked decrease in co-localization of the GFP-SCAP with the Golgi marker (not shown).

supplemented with 5% lipoprotein-deficient serum prepared from newborn calf serum (see **Note 7**).
2. After 24 h, transfect cells with cDNA encoding pDsRed-monomer-Golgi (1 μg per well for a 6-well plate) using FuGene 6 transfection reagent. Mix 1 μg of plasmid with 3 μL of FuGene 6 transfection reagent in 100 μL of serum-free medium (DMEM/F12) and incubate at room temperature for 15 min. Add 100 μL of the cDNA mixture dropwise to each well (see **Note 8**).
3. After 24 h, rinse cells once with PBS, then supply fresh medium supplemented with appropriate treatment agents for the desired length of time.
4. After treatment, rinse cells with PBS.
5. Fix cells with 1 mL of 3% (v/v) formaldehyde/PBS for 15 min at room temperature.
6. Discard formaldehyde and rinse cells twice with PBS.
7. Invert the cover slips on 20 μL of mounting medium on a clean glass slide, and allow to dry (~15 min).
8. Seal the edges of the cover slips with nail polish to prevent the cells drying out (see **Note 9**).
9. For confocal fluorescence microscopy, DsRed-monomer excitation and emission maxima are 557 nm and 585 nm, respectively, and GFP excitation and emission maxima are 488 nm and 508 nm, respectively. The images obtained can be merged using Adobe Photoshop 7.0. See **Fig. 2** for an example of results obtained using CHO cells stably expressing GFP–SCAP *(9)*, and transfected with DsRed-monomer-Golgi.

3.2. Identification of Potential Akt Phosphorylation Sites

1. Obtain human amino acid sequences of suspected Akt targets using the Protein function at http://www.ncbi.nlm.nih.gov/.

Table 1
Scansite Results for Human Cholesterol Regulatory Proteins

	Stringency			Percentile	Sequence	a.a. position
	High	Medium	Low			
SCAP	–	√		0.220	RKRMVSY	228–234
SREBP-2	–	√		0.985	MERSWSV	771–777
INSIG-1	–	–	√	3.070	RSAAMSG	63–69
SREBP-1a	–	–	√	4.201	RLALCTL	517–523
SREBP-1c	–	–	√	4.201	RLALCTL	487–493
INSIG-2	–	–	√	4.649	IQRNVTL	51–57

A lower percentile score indicates a higher stringency site. So of these proteins, SCAP is the most likely candidate for phosphorylation by Akt. SREBP-2 is mainly involved in cholesterol-related genes, SREBP-1c in fatty acid-related genes, and SREBP-1a plays roles in both.

2. Submit these sequences to Scansite (http://scansite.mit.edu), using the "Scan a Protein by Input Sequence" function and selecting the Akt kinase motif. *See* **Table 1** for an example using cholesterol-regulatory proteins.
3. Select the sequence with the highest stringency site (lowest percentile score) for further investigation.

3.3. Examination of Conservation Throughout Evolution

1. Obtain amino acid sequences for the protein of interest from species other than humans by using the Protein function at http://www.ncbi.nlm.nih.gov/ or by using the Basic Local Alignment Search Tool (BLAST) at the same Web site.
2. Align these sequences using ClustalW (http://www.ebi.ac.uk/clustalw/index.html) and examine conservation in the region of interest between species. *See* **Table 2** for an example of an alignment performed for SCAP. (*See* **Note 10**.)

3.4. Immunoprecipitation of Protein

1. Set up CHO cells stably expressing GFP–SCAP in 10-cm dishes at a density of 1×10^6 cells per dish in DMEM/F12 containing penicillin (100 U/mL), streptomycin (100 μg/mL), and L-glutamine (2 mM), supplemented with 5% lipoprotein-deficient serum prepared from newborn calf serum (*see* **Note 7**).
2. At approx 80% confluence, treat the cells with appropriate agents.
3. Transfer the cells to ice, and wash twice with cold PBS.
4. Aspirate the PBS, and then add 1 mL of RIPA buffer containing 10 μL of PI and 2 μL of PPI.

Table 2
An Alignment of SCAP Sequences

Akt phosphorylation consensus sequence	RX—RXXSH
Species	Sequence
Homo sapiens	RK—RMVSY
Macaca mulatta	RK—RMVSY
Bos taurus	RK—RLVSY
Canis familiaris	RK—RMVSY
Mus musculus	RK—RMVSY
Rattus norvegicus	RK—RMVSY
Danio rerio	KK—RVVTY
Drosophila melanogaster	LRARSRIIQY
Strongylocentrotus purpuratus	IHSVSPQISF
Caenorhabditis elegans	QTNRKRKIEF
Schizosaccharomyces pombe	EQ———

Conservation of the sequence is high amongst mammals, but absent in lower eukaryotes. In the consensus sequence, S is the phosphorylated serine, H is a large hydrophobic amino acid, and X is any amino acid.

5. Scrape the cells in RIPA buffer and transfer to microfuge tubes.
6. To generate homogenized cell lysates, pass the cells through a 22-gauge needle 10 times.
7. Transfer 1 mL of protein G-Sepharose stock to a 1.5 mL microfuge tube (make sure the slurry is in suspension) and centrifuge for 30 s at 1,000 g.
8. Rinse the protein G-Sepharose beads with 0.5 mL of ice-cold RIPA buffer by rotating for 2 min at room temperature and centrifuging for 30 s at 1,000 g. Repeat five times (*see* **Note 11**).
9. Resuspend protein G-Sepharose beads in one volume of RIPA buffer to make a 50% slurry and then store at 4°C until ready to use (*see* **Note 12**).
10. Preclear the cell lysates from **step 6** by adding 40 µL of protein G-Sepharose and rotating at 4°C for 1 h (*see* **Note 13**).
11. Centrifuge the cell lysates at 20,000 g for 10 min at 4°C, and transfer the resultant supernatant to new microfuge tubes.
12. Add 1 µL of GFP antibody to the precleared samples. For a negative control, add 1 µg of same species IgG antibody instead of the GFP antibody. Incubate at 4°C with gentle rotation for approx 16 h.
13. Centrifuge the samples for 1 min at 1,000 g. Subject the supernatant to protein precipitation (**Subheading 3.5.**), and resuspend the pellet (protein G-Sepharose beads bound to the antibody and the protein of interest) in 1 mL of RIPA buffer containing 10 µL of PI and 2 µL of PPI. Transfer to new microfuge tubes.

14. Wash the pellet with RIPA buffer three times by rotating for 10 min at 4°C, centrifuging for 1 min at 1,000 g at 4°C, discarding the supernatant and repeating.
15. After the final wash, centrifuge the samples at 20,000 g for 10 min at 4°C and discard the supernatant.
16. Add 100 μL of loading buffer to the protein G-Sepharose beads. Denature and elute the proteins from the protein G-Sepharose beads by boiling at 95°C for 5 min with occasional gentle vortexing.
17. Centrifuge the mixtures at 20,000 g for 5 min at room temperature and analyze the supernatants by SDS-PAGE (**Subheading 3.6.**).

3.5. Protein Precipitation

1. Add 4 mL of methanol to 1 mL of the supernatant from the IP (**Subheading 3.4.**, step 13) and vortex.
2. Add 1 mL of chloroform and vortex.
3. Add 3 mL of water and vortex.
4. Centrifuge at 15,000 g for 2 min and remove the top layer.
5. Add 4 mL of methanol and vortex.
6. Centrifuge at 15,000 g for 2 min and remove all supernatant.
7. Completely dry the pellet and resolubilize in 200 μL of loading buffer (*see* **Note 14**).

3.6. SDS-PAGE

1. Boil the pellet and supernatant samples from the IP for 5 min at 95°C.
2. Analyse 55 μL of the pellet and 20 μL of the supernatant by 10% (w/v) SDS-PAGE. Suggested running conditions are 200 V for approx 1 h, or longer as required (*see* **Note 15**).
3. Transfer to a nitrocellulose membrane. Suggested transfer conditions are 200 V for 1 h.

3.7. Western Blotting

1. Block the membrane in blocking solution with rocking for 1 h.
2. Rinse briefly with PBST.
3. Incubate the membrane in primary antibody (either phospho-Akt substrates antibody or GFP antibody) with rocking for 1 h (*see* **Note 16**).
4. Wash three times for 10 min in PBST with agitation.
5. Incubate in secondary antibody with rocking for 30 min.
6. Wash three times for 10 min in PBST with agitation.
7. Visualize antibodies using the ECL detection system—mix equal volumes of the two reagents and leave at room temperature for approx 10 min, then incubate the membrane in this solution for 5 min.

Approaches to Investigate the Role of Signaling 223

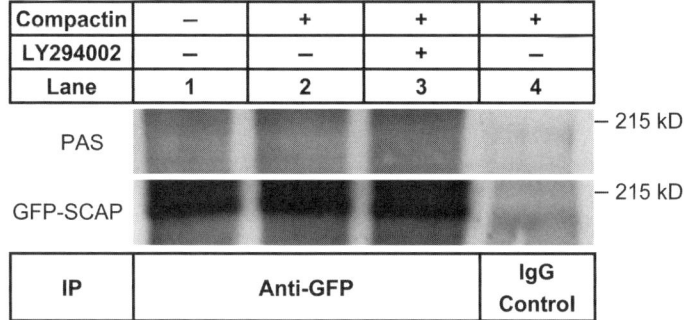

Fig. 3. Akt does not phosphorylate SCAP in Chinese Hamster Ovary (CHO) cells. Under these conditions, the cholesterol synthesis inhibitor, compactin, activates Akt, which is blocked by LY294002, an inhibitor of phosphatidylinositol 3 kinase (PI3K), upstream of Akt. The anti-phospho-Akt Substrates (Anti-PAS) blot shows that no phospho-Akt substrates-detected band is present after immunoprecipitating with the anti- green fluorescent protein (GFP) antibody. The anti-GFP blot shows a band corresponding to the size of GFP-SCAP (~175 kD) indicating that the immunoprecipitation has worked effectively, and that it is specific to the anti-GFP antibody, as no corresponding band is detected in the IgG control condition (*see* **Note 18**).

8. Expose the membrane to film for 30 s, develop the film, and then adjust exposure times accordingly (*see* **Note 17**). *See* **Fig. 3** for an example of an IP using GFP antibody in GFP–SCAP-expressing cells. No bands were observed using the phospho-Akt-substrates antibody, indicating that SCAP is not a target for Akt.

4. Notes

1. A similar result can be obtained using immunofluorescence to stain with appropriate Golgi marker antibodies, but the use of a fluorescent Golgi marker plasmid has been found to be much simpler to use.
2. Other transfection reagents may be used, but our preference is for FuGene 6 because of its low toxicity to cells.
3. BoxShade 3.2 (http://www.ch.embnet.org/software/BOX_form.html) can be used for presentation of alignment results.
4. We have found that antibodies directed against epitope tags such as GFP perform far better in IPs than those directed against specific proteins.
5. Polyvinylidene difluoride membranes may also be used, but they must be soaked in 100% methanol before use.
6. Using this ECL detection system, most primary antibodies are effective at a dilution of 1:10,000, and the secondary antibody at a dilution of 1:20,000. If using an alternative detection system, dilutions can be adjusted as necessary.

7. Cells and growth medium should be appropriate for the protein being investigated and the cell system being employed.
8. To maximize transfection efficiency, change the medium in the wells to antibiotic-free medium prior to adding the cDNA mixture.
9. Protect the slides from direct light as this decreases fluorescence, and allow the nail polish to dry on the bench for approx 30 min. The slides can then be stored at 4°C for at least 2 wk prior to microscopy if necessary.
10. The structure of the protein, if available (e.g., from the Structure Database at http://www.ncbi.nlm.nih.gov/), can also be examined in relation to the potential accessibility to the site for the kinase. In the case of SCAP, where the topology is known *(10)*, the sequence of interest is located within a portion of the protein in the lumen of the ER, so access to the site is restricted, decreasing the likelihood of Akt phosphorylating SCAP.
11. Thorough washing of the protein G-Sepharose beads is necessary as the commercial stock is preserved in ethanol.
12. This slurry stock can be stored at 4°C for at least 2 wk. Before being added to cell lysates, the slurry stock should be washed three times with RIPA buffer.
13. The purpose of this step is to remove any proteins from the cell lysate that are pulled down indiscriminantly by the protein G-Sepharose beads.
14. A portion of this precipitated protein from the supernatant is analyzed to determine the efficiency of the IP, as there should be no protein detected in the supernatant when using the same antibody for Western blotting as that used for the IP.
15. Monitor the migration of the molecular-weight markers to ensure that the size corresponding to the protein of interest is at least halfway into the gel, to increase resolution in this region.
16. Blocking and primary antibody incubations can be performed either overnight at 4°C, or for 1 h at room temperature, depending on which is more convenient.
17. The heavy-chain and light-chain IgGs from the antibody used for the IP will be detected very strongly if using antibodies from the same species for both the IP and the Western blotting. These will be visible at approx 50 and approx 25 kD, so proteins that are of a similar size to this will not be visible owing to overshadowing by the detection of the IgG components on the film.
18. The reciprocal experiment could also be performed using the phospho-Akt substrates antibody for the IP and the GFP antibody for detection by Western blotting. In this way, a positive control could also be included in the form of Western blotting for a known Akt substrate following the phospho-Akt substrates IP.

References

1. van Vliet, C., Thomas, E. C., Merino-Trigo, A., Teasdale, R. D., and Gleeson, P. A. (2003) Intracellular sorting and transport of proteins. *Prog. Biophys. Mol. Biol.* **83**, 1–45.
2. Gurkan, C., Stagg, S. M., LaPointe, P., and Balch, W. E. (2006) The COPII cage: unifying principles of vesicle coat assembly. *Nat Rev Mol Cell Biol.* **7**, 727–738.

3. Schu, P. V., Takegawa, K., Fry, M. J., *et al.* (1993) Phosphatidylinositol 3-Kinase Encoded by Yeast VPS34 Gene Essential for Protein Sorting. *Science* **260**, 88–91.
4. Wilson, M. L., and Guild, S. B. (2001) Effects of wortmannin upon the late stages of the secretory pathway of AtT-20 cells. *Eur. J. Pharmacol.* **413**, 55–62.
5. Aridor, M., and Balch, W. E. (2000) Kinase signaling initiates coat complex II (COPII) recruitment and export from the mammalian endoplasmic reticulum. *J. Biol. Chem.* **275**, 35673–35676.
6. Lee, T. H., and Linstedt, A. D. (2000) Potential role for protein kinases in regulation of bidirectional endoplasmic reticulum-to-Golgi transport revealed by protein kinase inhibitor H89. *Mol. Biol. Cell* **11**, 2577–2590.
7. Du, X. M., Kristiana, I., Wong, J., and Brown, A. J. (2006) Involvement of Akt in ER-to-Golgi transport of SCAP/SREBP: A link between a key cell proliferative pathway and membrane synthesis. *Mol. Biol. Cell* **17**, 2735–2745.
8. Goldstein, J. L., DeBose-Boyd, R. A., and Brown, M. S. (2006) Protein sensors for membrane sterols. *Cell* **124**, 35–46.
9. Nohturfft, A., Yabe, D., Goldstein, J. L., Brown, M. S., and Espenshade, P. J. (2000) Regulated step in cholesterol feedback localized to budding of SCAP from ER membranes. *Cell* **102**, 315–323.
10. Nohturfft, A., Brown, M. S., and Goldstein, J. L. (1998) Topology of SREBP cleavage-activating protein, a polytopic membrane protein with a sterol-sensing domain. *J. Biol. Chem.* **273**, 17243–17250.

17

Recruitment of Coat Proteins to Peptidoliposomes

Gregor Suri, Martin Spiess, and Pascal Crottet

Summary

Intracellular transport between compartments within the cell is generally mediated by membrane vesicles. Their formation is initiated by activation of small GTPases that then recruit cytosolic proteins to the membrane surface to form a coat, interact with cargo proteins, and deform the lipid bilayer. Liposomes proved to be a useful tool to study the molecular mechanisms of these processes *in vitro*. To analyze the involvement of membrane proteins, the cytosolically exposed sequences may be coupled chemically to reactive lipids in the membrane. Here we describe the use of such peptidoliposomes presenting lipid-coupled cytosolic tails of cargo proteins for the *in vitro* analysis of the membrane recruitment of AP-1 adaptors in the process of forming AP-1/clathrin coats. AP-1 recruitment is mediated by the GTPase Arf1, requires specific lipids, and cargo signals. Interaction with cargo induces AP-1 oligomerization already in the absence of clathrin.

Key words: Arf1; clathrin adaptor protein; coat protein; liposome; membrane traffic; peptidoliposome; protein sorting.

1. Introduction

Liposomes are widely used to study molecular processes at membrane surfaces *in vitro*. They have been particularly useful in the area of membrane trafficking to reconstitute the assembly of cytosolic coat proteins at the lipid bilayer and the formation of coated vesicles from purified components. Lipid and protein compositions, the order of addition of individual components and the conditions (temperature, nucleotides, etc.) can be easily manipulated and help to define the minimal machinery of coat assembly and their molecular mechanisms.

From: *Methods in Molecular Biology, vol. 457: Membrane Trafficking,*
Edited by: A. Vancura, DOI: 10.1007/978-1-59745-261-8_17, © Humana Press, Totowa, NJ

The three best characterized coats are coat protein (COP) I mediating intra-Golgi and Golgi to endoplasmic reticulum (ER) transport, COPII for vesicles derived from the ER, and clathrin with various associated adaptor proteins (APs) for pathways between the plasma membrane, endosomes, and the *trans*-Golgi network *(1)*. In all systems, coat recruitment is initiated by a small GTPase that is activated at the membrane by nucleotide exchange from GDP to GTP. The generation of COPI vesicles *in vitro* requires the heteroheptameric coatomer complex and ADP-ribosylation factor 1 (Arf1; *2,3*). COPII consists of two components: Sec23/24 is first targeted by Sar1·GTP to the membrane to recruit the second layer of Sec13/31 *(4)*. Clathrin coats are similarly composed of two layers, heterotetrameric adaptor complexes and clathrin *(5)*.

In all systems, specific lipid compositions are required for coat recruitment, acidic phospholipids for COPI *(3)*, and phosphoinositides for COPII *(4)* and clathrin adaptors *(6)*. In addition, cargo proteins influence coat formation. For example, presentation of cytoplasmic cargo sequences on the membrane surface is important for efficient recruitment of COPI and AP-1 clathrin adaptors *(2,6)* and enhances the stability of membrane-bound Sec23/24 of COPII *(7)*. Different approaches have been used to study the role of cargo proteins. Cargo SNAREs, such as Bet1p and Sec22p, were purified and reconstituted into liposomal membranes by detergent dialysis *(7)*. Alternatively, cytoplasmic sequences of cargo proteins were coupled to liposomes via modified lipids. This can be accomplished by preparing fusion proteins with glutathione-*S*-transferase (GST) that bind tightly to a glutathione-phosphatidylethanolamine conjugate incorporated into the liposomal membrane *(8)* or by covalently coupling synthetic peptides to a maleimide lipid via the sulfhydryl group of a cysteine side chain *(2,6,9)*.

Here we describe the methods to study the *in vitro* recruitment of AP-1 adaptors to peptidoliposome by a floatation assay and to analyze the oligomeric state of recruited AP-1 by sedimentation into a density gradient. We provide protocols to produce adaptor complexes from calf brain clathrin-coated vesicles (CCVs), myristoylated Arf1 from a bacterial expression system, and peptidoliposomes using maleimide lipids, as well as protocols of the floatation and the density gradient sedimentation assays.

2. Materials

Dithiothreitol (DTT) is added just before use from a 1 M frozen stock. All the buffers for fast protein liquid chromatography (FPLC) are filtered through 0.2-µm nitrocellulose filters (Millipore) and degassed.

2.1. Isolation of Cytosol and CCVs

1. Two calf brains fresh from the slaughterhouse (*see* **Note 1**).
2. CB-6 Three-speed extra-large capacity blender from Waring Laboratory or equivalent.

3. Loose-fitting dounce homogenizer (Bellco Biotechnology) small (7 mL) and medium (40 mL) size.
4. Phosphate-buffered saline (PBS): 12.5 mM sodium phosphate buffer, pH 7.6, 125 mM NaCl.
5. Phenylmethylsulfonyl fluoride (PMSF) at 0.5 M in dimethyl sulfoxide.
6. Buffer A: 0.1 M MES-NaOH, pH 6.6, 0.5 mM MgCl$_2$, 1 mM EGTA, 0.2 mM DTT.
7. Buffer B: 0.1 M MES-NaOH, pH 7.0, 0.5 mM MgCl$_2$, 1 mM EGTA, 0.2 mM DTT, 12.5% (w/v) Ficoll 400, 12.5% (w/v) sucrose (*see* **Note 2**).

2.2. Preparation of Mixed Adaptors

1. Fast protein liquid chromatography (FPLC) system (e.g., äkta FPLC system from GE Healthcare).
2. Superose 6 prep grade is packed into an HR 16/60 column (1.6 × 60 cm; both from GE Healthcare).
3. Monoclonal antibodies for immunoblotting (all used at 1:1,000 dilution): 100/3 against γ-adaptin of AP-1 (~90 kD), 100/2 against α-adaptin of AP-2 (doublet at ~100 kD; both from Sigma), and anti-clathrin heavy chain (from BD Biosciences).
4. Stripping buffer (3X): 1.5 M Tris-HCl, pH 7.0, 6 mM ethylenediamine-tetraacetic acid (EDTA), 0.6 mM DTT. This buffer can be stored at –20°C in 15-mL Falcon tubes. Falcon tubes do not resist shock-freezing.
5. Superose running buffer: 0.5 M Tris-HCl, pH 7.0, 2 mM EDTA, 0.2 mM DTT.
6. Protease inhibitor cocktail (500X): 5 mg/mL benzamidine, 1 mg/mL pepstatin A, 1 mg/mL leupeptin, 1 mg/mL antipain, 1 mg/mL chymostatin (all from Sigma); dissolved in 40% dimethyl sulfoxide (DMSO)/60% ethanol and stored at –20°C.

2.3. Preparation of Myristoylated Arf1

1. Plasmids pET-mArf1* encoding bovine Arf1 with codons 3 to 7 replaced by those from yeast Arf2p *(10)* and pBB131 encoding yeast myristoyl-CoA:protein *N*-myristoyltransferase (NMT; *11)* were provided by Drs. Stuart Kornfeld and Jeffrey Gordon (Washington University, St. Louis, MO), respectively.
2. Competent *Escherichia coli* BL21(DE3) from Stratagene.
3. Monoclonal antibody for immunoblotting 1D9 against Arf (used at 1:5,000 dilution; from Alexis).
4. Diethylaminoethyl (DEAE) Sephacel (from GE Healthcare) in an Econo-Column (2.5 × 10 cm) from Bio-Rad.
5. Superdex 75 column (HighLoad 26/60 prep grade; 2.6 × 60 cm; from GE Healthcare).
6. Amicon Ultra–15 centrifugal filter devices with a 10 kD cut-off from Millipore.
7. Luria broth (LB): 10 g/L bactotryptone (Applichem), 5 g/L yeast extract (Applichem), 10 g/L NaCl.
8. Ampicillin (1,000X): 100 mg/mL in water; store frozen.
9. Kanamycin (500X): 50 mg/mL in water; store frozen.

10. Isopropyl-1-thio-β-D-galactopyranoside (IPTG) is dissolved at 1 M in water and stored frozen.
11. Brij 58 is prepared at 10% (w/v) in water and autoclaved.
12. Myristic acid is dissolved at 0.5 M in ethanol and stored frozen.
13. DEAE buffer: 50 mM Tris-HCl, pH 8.0, 1 mM MgCl$_2$, 10 µM GDP, 0.02% (w/v) NaN$_3$, 1 mM DTT.
14. Superdex running buffer: 50 mM Tris-HCl, pH 7.5, 1 mM MgCl$_2$ 10 µM GDP, 0.02% (w/v) NaN$_3$, 10% (w/v) sucrose, 1 mM DTT.

2.4. Preparation of Peptidoliposomes

1. Synthetic peptides LY (amino acid sequence CRKRSHAGYQTI) and LA (CRKRSHAGAQTI) correspond to the cytoplasmic domain of Lamp1 (lysosome-associated membrane protein-1) and a transport-deficient mutant with the critical tyrosine mutated to alanine, respectively. Peptides were purchased at more than 70% purity from, e.g., NeoMPS. The N-terminal cysteines were added to the sequences for coupling.
2. Glass tubes resistant to liquid nitrogen, e.g., 12-cm test tubes NS 14.5/23 for vacuum from Glas Keller.
3. Nucleopore polycarbonate membrane with 400-nm pore size and prefilters (drain disc 10 mm PE) from Whatman.
4. Mini-Extruder from Avanti Polar Lipids. For description *see* http://www.avantilipids.com/extruder.html.
5. Egg L-α-phosphatidylcholine (PC) from Avanti Polar Lipids.
6. Soybean L-α-phosphatidylcholine, also called azolectin (Sigma). This is a mixture of phospholipids, containing only 20% PC, however.
7. MMCC-DOPE (*N*-[{4-maleimidylmethyl}cyclohexane-1-carbonyl]-1,2-dioleoyl-sn-glycero-3-phosphoethanolamine) from Avanti Polar Lipids is dissolved in chloroform/methanol 2:1 (v/v). Aliquots of 125 nmoles are dried in silanized 1.5-mL reaction tubes and stored at –20°C.
8. Chloroform/methanol 2:1 (v/v).
9. 1,1,-Dichloromethane.
10. Liposome buffer: 10 mM HEPES-NaOH, pH 6.5, 100 mM NaCl, 0.5 mM EDTA.

2.5. Floatation Assay

1. Assay buffer: 10 mM HEPES-NaOH, pH 7.0, 150 mM NaCl, 10 mM KCl, 2 mM MgCl$_2$, 0.2 mM DTT.
2. Sucrose solutions: 20% and 60% (w/v) in assay buffer.
3. 5'-Guanylylimidodiphosphate (GMP-PNP; Fluka) is prepared at 10 mM and GTP (Sigma) at 100 mM in water and stored at –80°C.
4. TCA: 100% (w/v) 2,2,2-trichloroacetic acid in water.
5. Acetone (analysis grade).

2.6. Sedimentation Assay

1. Gradient Master from BioComp Instruments (Fredericton) or alternative gradient maker.
2. Sucrose solutions: 10% and 25% (w/v) in assay buffer containing 0.2% (w/v) Triton X-100.
3. Triton X-100 at 20% (w/v) in water.

2.7. General Materials

1. Standard equipment and materials for sodium dodecyl sulfate-polyacrylamide gel electrophoresis (12.5% acrylamide) and immunoblot analysis.
2. Bradford Protein Assay (Bio-Rad) or equivalent.

3. Methods

For centrifugations, an example of rotors and speeds are provided in parentheses. Unless expressly stated, all procedures are performed at 4°C.

3.1. Isolation of Cytosol and CCVs

Intracellular membranes released from homogenized calf brains are recovered in a low-speed supernatant and separated from the cytosol by high-speed centrifugation. In a subsequent medium-speed centrifugation with Ficoll and sucrose, CCVs are enriched in the supernatant. After dilution, they are collected by high-speed centrifugation. The procedure is based on **ref. 12**.

1. Two calf brains fresh from the slaughterhouse are transported on ice to the cold-room (*see* **Note 3**).
2. Fat, brain stem, meningae, and blood clots are removed with paper towels.
3. The cleaned brains are placed in a beaker filled with 1 L PBS to measure their volume (~ 200 mL per brain).
4. The brains are placed in the Waring blender together with half their volume of buffer A (*see* **Note 4**). PMSF (0.5 mM) is added and the blender is turned on three times for 8 s on medium speed.
5. The homogenate is centrifuged at 8,000g for 30 min (Sorvall GS3 rotor, 7,000 rpm). Keep the pellets for **step 7**.
6. The supernatant is carefully collected (*see* **Note 5**) and the membranes are pelleted by centrifugation at 180,000g for 80 min (Kontron TFT 45.94 rotor, 40,000 rpm). The high-speed supernatant (i.e., the cytosol) is collected (*see* **Note 6**), aliquoted, shock-frozen in liquid nitrogen, and stored at –80°C. Protein concentration is typically 15 to 30 mg/mL. The high-speed pellets are kept for **step 8**.
7. The low-speed pellets of **step 5** are resuspended in an equal volume of buffer A and—after addition of 0.5 mM PMSF—re-homogenized in the blender as before

and centrifuged again at 8,000g for 30 min. The supernatant is collected and all membranes pelleted at 180,000g for 80 min. This second high-speed supernatant is discarded.

8. The high-speed pellets from **steps 6** and **7** are resuspended in 1 mL buffer A per tube with the help of a spatula, collected, and homogenized with five to six strokes in a dounce homogenizer.
9. The suspension (~8 mL) is mixed with an equal volume of buffer B, dounce homogenized again, and centrifuged at 60,000g for 40 min (Kontron TFT 45.94 rotor, 24,000 rpm; *see* **Note 7**).
10. The supernatants are collected, diluted with three volumes of buffer A, and centrifuged at 180,000g for 80 min (Kontron TFT 45.94 rotor, 40,000 rpm) to pellet the CCVs.
11. The pellets are resuspended in an equal volume of buffer A, homogenized in a small dounce homogenizer, and centrifuged in 1.5-mL tubes in a microfuge for 12 min to remove aggregated material.
12. The supernatant containing the CCVs is collected, frozen as 750-μL aliquots in liquid nitrogen, and stored at –80°C. Typically four aliquots are obtained from two brains.

3.2. Preparation of Mixed Adaptors

Clathrin coats are released from CCVs with high concentrations of Tris, and adaptors enriched by gel filtration. The procedure is based on **ref. 13**.

1. Two 750-μL aliquots of CCVs are thawed, mixed with an equal volume of stripping buffer containing 0.5 mM PMSF and 1X proteinase inhibitor cocktail in a small dounce homogenizer, and kept on ice overnight.
2. The mixture is dounce homogenized again and centrifuged at 100,000g for 35 min (Beckman TLA 100.3 rotor, 70,000 rpm) to pellet the membranes. If the supernatant is still turbid, it is centrifuged again.
3. Two milliliters of the clear supernatant are fractionated on the Superose 6 column pre-equilibrated in Superose running buffer at a flow rate of 0.5 mL per minute and collected in 1-mL fractions.
4. Samples of 25 μL of every second fraction from 41 to 72 are analyzed by SDS-gel electrophoresis and immunoblotting for the presence of AP-1 and clathrin. AP-1 is typically found in fractions 55 to 63 (*see* **Note 8**).
5. AP-1-containing fractions are pooled, supplemented with 1X protease inhibitor cocktail and 0.5 mM PMSF, and may be stored at 4°C for several weeks. Typically 10 mL at 100 μg/mL protein are obtained.

3.3. Preparation of Myristoylated Arf1

Arf1 and NMT are expressed in bacteria and myristic acid is exogenously provided to produce efficiently myristoylated Arf1. The protein is purified from freeze-thawed cells by ion exchange chromatography and gel filtration. The procedure is based on **ref. 14**.

1. Competent BL21(DE3) bacteria are mixed with approx 0.1 μg each of pET-mArf1* and pBB131 encoding Arf1 and NMT, respectively, incubated on ice for 1 h, and heat-shocked at 42°C for 2 min.
2. The bacterial suspension is incubated with 1 mL LB at 37°C for 1 h before addition of 4 mL LB with 100 μg/mL each of ampicillin and kanamycin and incubation overnight with shaking to select for transformants carrying both plasmids.
3. The overnight culture is used to inoculate 2.5 L LB with 100 μg/mL each of ampicillin and kanamycin. The culture is grown until OD_{600} reaches 0.6 to 0.8.
4. Arf1 and NMT expression is induced by adding 1 mM isopropyl-β-D-thiogalactopyranoside (IPTG). Simultaneously, 0.5% (w/v) Brij 58 (to increase the solubility of myristic acid) and 500 μM myristic acid are added, and the culture is incubated 4 h at 30°C (a temperature that favors myristoylation).
5. The cells are harvested by centrifugation at 5,000g for 15 min (Sorvall GS3 rotor, 5,400 rpm).
6. The pellets are resuspended with PBS, transferred into a 50-mL Falcon tube, and pelleted again for freezing at –80°C.
7. To the frozen pellet (~5 mL), 25 μL 0.5 M PMSF and 50 μL 500X protease inhibitor cocktail are added, and the bacteria are lysed by three cycles of freezing in dry ice/methanol and thawing in luke warm water.
8. Bacteria are resuspended in 25 mL DEAE buffer and incubated on ice for 1 h before centrifugation at 20,000g for 15 min (Sorvall SS34 rotor, 13,000 rpm).
9. The supernatant is collected and loaded directly onto the 25-mL DEAE Sephacel column equilibrated and run with DEAE buffer at a flow rate of 1 mL per minute.
10. The flow-through contains Arf1 and is collected until OD_{280} is below 1.0.
11. These Arf1 containing fractions are concentrated by ultrafiltration (Amicon Ultra filter) to approx 2 mL.
12. The sample is fractionated at a flow rate of 1 mL per minute on the Superdex 75 column equilibrated with Superdex running buffer. Fractions of 3 mL are collected.
13. Fractions 51 to 80 are tested for the presence of Arf1 and contaminating proteins by analyzing 25-μL samples on parallel 15% SDS-gels by immunoblotting and Coomassie staining, respectively.
14. Arf1 containing fractions (typically fractions 60–70) are pooled (*see* **Note 9**) and concentrated to 1 mg/mL by ultrafiltration.
15. The purified protein is aliquoted, shock-frozen in liquid nitrogen, and stored at –80°C. Approximately 0.5 mg Arf1 is typically purified from a 2.5 L culture.

3.4. Preparation of Peptidoliposomes

Liposomes are prepared by extrusion through a defined pore-size filter. A maleimide-containing lipid is included to allow chemical coupling of cysteine-terminal peptides. The procedure is based on **refs. 6** and **15**.

1. Five μmoles of the desired lipids, for example 3.8 mg egg PC, soybean lipids, or a 1:1 mixture of the two, are dissolved in 1 mL chloroform/methanol in a test tube that can stand liquid nitrogen.
2. 125 nmoles (2.5 mol %) MMCC-DOPE is dissolved in 100 μL chloroform/methanol and added to the lipids (see **Note 10**).
3. The organic solvent is evaporated under a stream of nitrogen
4. The lipids are redissolved in 2 mL of 1,1-dichloromethane and dried again under nitrogen.
5. Meanwhile, the extruder is assembled and washed with liposome buffer, and 0.28 μmoles (400 μg) of the peptides to be coupled (approximately fourfold excess over the reactive lipid, assuming half of it is exposed to the outside of the liposome) are weighed out and dissolved in 100 μL liposome buffer.
6. The dried lipids are suspended in 1 mL liposome buffer by five cycles of vortexing, shock-freezing in liquid nitrogen, and thawing under warm tap water (*see* **Note 11**).
7. Lipids are passed 11 times through the extruder unit (*see* **Note 12**).
8. The liposomes are immediately mixed with 100 μL 4 mg/mL peptide and incubated for 1 h at room temperature (*see* **Note 13**).
9. The final peptidoliposomes are stored at 4°C with 0.02% (w/v) NaN_3 for up to 2 wk (*see* **Notes 14** and **15**).

3.5. Floatation Assay

Peptidoliposomes are incubated with coat proteins, Arf1, and nucleotides to allow for Arf1 activation and coat recruitment, and then floated on a sucrose step gradient. Proteins associated with the liposomes or remaining in the loading zone are collected by trichloroacetic acid (TCA) precipitation and detected by immunoblot analysis. The procedure is based on **refs. 6** and **9**.

1. Cytosol is centrifuged at 170,000*g* for 30 min (Beckman TLA100.3 rotor, 85,000 rpm) to remove aggregates.
2. Peptidoliposomes (100 μL containing ~0.5 μmol lipid) are mixed with 5 μg Arf1, 0.2 m*M* GMP-PNP or 2 m*M* GTP, either 10 μg mixed adaptors or 1 mg cytosol, and reaction buffer to 200 μL. The mixture is incubated for 30 min at 37°C to allow nucleotide exchange on Arf1 and protein recruitment.
3. The reaction is mixed in a 4-mL ultracentrifuge tube with 0.4 mL 60% sucrose solution (to a final concentration of 40% sucrose). The mixture is carefully overlayed with 1 mL of 20% sucrose solution. After removal of any foam at the top with a pipet, 20% sucrose solution is added to a total of 3.82 mL.
4. The tubes are balanced and 180 μL assay buffer is overlayed to facilitate liposome collection afterwards.
5. Samples are centrifuged at 300,000*g* for 1 h (Kontron TST60.4 rotor, at 55,000 rpm).
6. Four 1-mL fractions are collected from the top.

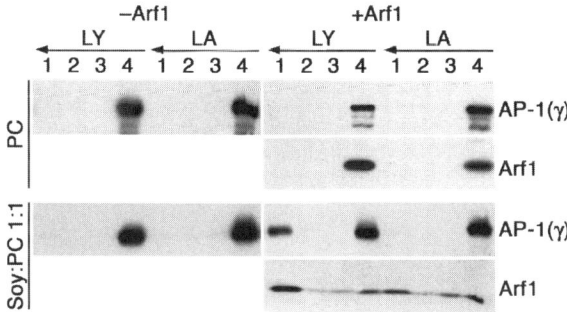

Fig. 1 AP-1 recruitment to peptidoliposomes is Arf1-, lipid-, and signal-dependent. Peptidoliposomes made of 100% PC or a 1:1 PC/soybean lipid mixture and presenting LY or LA peptides were incubated with mixed clathrin adaptors and with or without Arf1 and GMP-PNP. After flotation on a sucrose step gradient, four fractions were collected from the top and analyzed by immunoblotting for γ-adaptin (a subunit of AP-1) or Arf1. Recruitment of AP-1 required Arf1, lipids other than PC, and the presentation of the Lamp1 sequence with an intact tyrosine motif (LY). While Arf1 is activated on PC membranes (not shown), it also requires non-PC lipids for binding to liposomes. Arrows indicate the direction of floatation from the bottom to the top fraction of the gradient. (Reprinted in modifed form from **ref. 6** with permission of The American Society for Cell Biology.)

7. TCA precipitation: the fractions are mixed with 160 μL TCA in 1.5-mL tubes and centrifuged in a microfuge at maximal speed for 15 min.
8. To the pellets, 850 μL ice-cold acetone is added, and the tubes are centrifuged again for 5 min at 4°C.
9. The pellets are air-dried for 15 to 20 min at room temperature and dissoved in 60 μL SDS-sample buffer by pipeting up and down and by heating at 95°C for 10 min (*see* **Note 16**).
10. Samples are separated by gel electrophoresis on a 12.5% SDS-gel and analyzed by immunoblotting for AP-1 (γ-adaptin) and Arf1. An example result is shown in **Fig. 1** (and using cytosol in **Fig. 3**, left lanes).

3.6. Sedimentation Assay

To analyze the oligomerization state of liposome-associated protein, membranes in the floated fraction are solubilized and the protein centrifuged into a sucrose density gradient. Protein in the gradient fraction is detected by immunoblot analysis. The procedure is based on **ref. 9**.

1. The top 350 μL of a floatation gradient (**Subheading 3.5**, **step 5**), which contain most of the floated liposomes, are collected and mixed with 350 μL assay buffer to dilute the sucrose.
2. The lipid membranes are solubilized by addition of 0.5% Triton X-100.

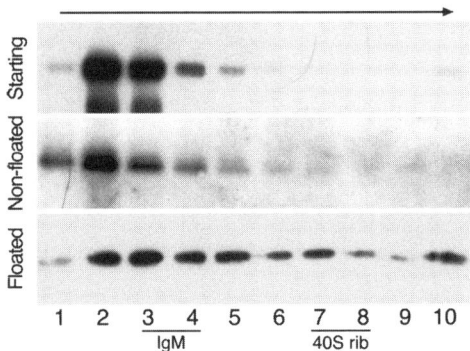

Fig. 2 AP-1 recruited to peptidoliposomes forms high-molecular weight complexes. Peptidoliposomes made of soybean lipids and presenting LY peptides were incubated with clathrin adaptors, Arf1, and GMP-PNP. After floatation on a sucrose step gradient, the floated liposomes were solubilized with Triton X-100 and centrifuged into a 10 to 25% sucrose velocity gradient (horizontal arrow). Ten fractions were collected and analyzed by immunoblotting for γ-adaptin. For comparison, the nonfloated fraction and the original adaptors were analyzed in parallel. The positions of the sedimentation markers IgM (19S) and 40S ribosomes are indicated. Individual AP-1 adaptors (~300 kD) have a sedimentation coefficient of 7.7S. AP-1 recruited to sorting peptides show oligomerization to high-molecular-weight complexes even in the absence of clathrin. (Reprinted in modified form from **ref. 9** with permission of The American Society for Cell Biology.)

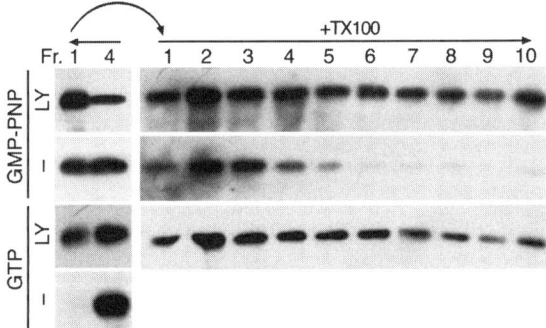

Fig. 3 AP-1 recruitment from cytosol. Bovine brain cytosol was supplemented with Arf1, 0.2 mM GMP-PNP or 2 mM GTP, and liposomes with or without LY peptides as indicated and incubated for 30 min at 37°C. After a first step gradient floatation, the floated fraction 1 was solubilized with Triton X-100 and sedimented into a sucrose gradient as in **Fig. 2**. The fractions were analyzed by immunoblotting for γ-adaptin. Unlike AP-1 from purified adaptors, AP-1 from cytosol can be recruited to liposomes with Arf1/GMP-PNP even in the absence of sorting peptides. However, LY peptides are required for AP-1 to oligomerize. If GTP is used, stable recruitment is observed only with sorting peptides. These results indicate the presence of a cytosolic factor recruiting

3. The sample is loaded onto a 4.3-mL 10 to 25% sucrose gradient with 0.2% Triton X-100 and centrifuged at 100,000g for 5 h (Kontron TST 55.5 rotor, at 30,000 rpm).
4. Ten 0.5-mL fractions are collected from the top and 125 µL of 5X SDS-sample buffer are added for analysis by SDS-gel electrophoresis and immunoblotting for AP-1 (γ-adaptin). An example result is shown in **Figs. 2** and **3**.

4. Notes

1. Alternatively, 10 pig brains may be used.
2. Ficoll 400 is hard to dissolve. Warm up the solution and add Ficoll 400 step by step with continued stirring.
3. To process two brains, it is convenient in the following procedure to use two TFT45.94 rotors and two ultracentrifuges in parallel. For upscaling to six brains (2 × 3 brains), one needs two GS3 and three TFT45.94 rotors and appropriate centrifuges.
4. The buffer volume is kept low to retain high cytosol concentration.
5. It is really important to only take the red supernatant. Pour the supernatant over the white pellet into a beaker. It is better to take less supernatant than to have contamination with material from the pellet.
6. It is important not to contaminate the cytosol with the pellet. Carefully remove the cytosol with a glass pipet, leaving behind the last 1 cm of supernatant above the pellet.
7. To fill tubes to their minimal filling level or to balance them, a 1:1 mixture of buffers A and B may be added.
8. AP-2, AP180, and some clathrin are the major contaminants.
9. The later fractions of the Arf1 peak are usually purer than the earlier ones.
10. In addition, (N-[7-nitrobenz-2-oxa-1,3-diazol-4-yl]-1,2-dihexadecanoyl-sn-glycero-3-phosphoethanolamine [NBD-PE]; from molecular probes), a fluorescent phospholipid analog, may be added at 1 mol% to quantify liposome recovery in the subsequent experiments by fluorimetry (λ_{ex} 450 nm, λ_{em} 530 nm).
11. As soon as the liposome buffer is added to the dried lipids it is important to work quickly because of hydrolysis of the maleimide group.
12. Be very careful during the extrusion not to break the membranes. If you are stuck, back up a little before pushing on. After extrusion, disassemble the extruder and make sure the filter is intact. If the filter is damaged, repeat the extrusion.
13. To determine coupling efficiency, *see* **ref. 16**. The coupling reaction at pH 6.5 is essentially complete within 10 min at room temperature *(16)*.

Fig. 3. (Continued) AP-1 to membranes in a cargo-independent, GTPase-sensitive manner. They further show that membrane recruitment is not sufficient to induce AP-1 oligomerization, but that binding to cargo signals is necessary. (Reprinted in modifed form from **ref. 9** with permission of The American Society for Cell Biology.)

14. Some peptides induce liposome aggregation with time, which may cause unspecific trapping of protein in the floatation assay. These liposomes have to be used more quickly or even on the same day only.
15. For most uses, the free, excess peptides do not have to be removed (e.g., by gel filtration or dialysis), because the concentration in solution is much lower than that at of coupled peptides at the membrane surface.
16. Especially when using cytosol, pellets from the bottom fraction may be hard to dissolve. Also, if samples turn yellow, the pH can be adjusted by pipeting ammonia vapour onto them.

Acknowledgments

Our work was supported by grant 3100A0-109424 from the Swiss National Science Foundation.

References

1. Kirchhausen, T. (2000) Three ways to make a vesicle. *Nat. Rev. Mol. Cell Biol.* **1**, 187–198.
2. Bremser, M., Nickel, W., Schweikert, M., et al. (1999) Coupling of coat assembly and vesicle budding to packaging of putative cargo receptors. *Cell* **96**, 495–506.
3. Spang, A., Matsuoka, K., Hamamoto, S., Schekman, R., and Orci, L. (1998) Coatomer, Arf1p, and nucleotide are required to bud coat protein complex I-coated vesicles from large synthetic liposomes. *Proc. Natl. Acad. Sci. U S A.* **95**, 11199–11204.
4. Matsuoka, K., Orci, L., Amherdt, M., Bednarek, S. Y., et al. (1998) COPII-coated vesicle formation reconstituted with purified coat proteins and chemically defined liposomes. *Cell* **93**, 263–275.
5. Robinson, M. S. and Bonifacino, J. S. (2001) Adaptor-related proteins. *Curr. Opin. Cell Biol.* **13**, 444–453.
6. Crottet, P., Meyer, D. M., Rohrer, J., and Spiess, M. (2002) ARF1.GTP, tyrosine-based signals, and phosphatidylinositol 4,5-bisphosphate constitute a minimal machinery to recruit the AP-1 clathrin adaptor to membranes. *Mol. Biol. Cell* **13**, 3672–3682.
7. Sato, K., and Nakano, A. (2005) Dissection of COPII subunit-cargo assembly and disassembly kinetics during Sar1p-GTP hydrolysis. *Nat. Struct. Mol. Biol.* **12**, 167–174.
8. Matsuoka, K., Morimitsu, Y., Uchida, K., and Schekman, R. (1998) Coat assembly directs v-SNARE concentration into synthetic COPII vesicles. *Mol. Cell* **2**, 703–708.
9. Meyer, D. M., Crottet, P., Maco, B., Degtyar, E., Cassel, D., and Spiess, M. (2005) Oligomerization and dissociation of AP-1 adaptors are regulated by cargo signals and by ArfGAP1-induced GTP hydrolysis. *Mol. Biol. Cell* **16**, 4745–4754.

10. Liang, J. O., Sung, T. C., Morris, A. J., Frohman, M. A., and Kornfeld, S. (1997) Different domains of mammalian ADP-ribosylation factor 1 mediate interaction with selected target proteins. *J. Biol. Chem.* **272**, 33001–33008.
11. Duronio, R. J., Jackson-Machelski, E., Heuckeroth, R. O., et al. (1990) Protein N-myristoylation in Escherichia coli: reconstitution of a eukaryotic protein modification in bacteria. *Proc. Natl. Acad. Sci. U S A.* **87**, 1506–1510.
12. Campbell, C., Squicciarini, J., Shia, M., Pilch, P. F., and Fine, R. E. (1984) Identification of a protein kinase as an intrinsic component of rat liver coated vesicles. *Biochemistry* **23**, 4420–4426.
13. Keen, J. H. (1987) Clathrin assembly proteins: affinity purification and a model for coat assembly. *J. Cell Biol.* **105**, 1989–1998.
14. Liang, J. O. and Kornfeld, S. (1997) Comparative activity of ADP-ribosylation factor family members in the early steps of coated vesicle formation on rat liver Golgi membranes. *J. Biol. Chem.* **272**, 4141–4148.
15. Mayer, L. D., Hope, M. J., and Cullis, P. R. (1986) Vesicles of variable sizes produced by a rapid extrusion procedure. *Biochim. Biophys. Acta* **858**, 161–168.
16. Schelte, P., Boeckler, C., Frisch, B., and Schuber, F. (2000) Differential reactivity of maleimide and bromoacetyl functions with thiols: application to the preparation of liposomal diepitope constructs. *Bioconjug. Chem.* **11**, 118–123.

18

SNARE-Mediated Fusion of Liposomes

Jérôme Vicogne and Jeffrey E. Pessin

Summary

Lipid-mixing assay is now commonly used to study protein, temperature and ion-dependent membrane fusion events. This assay has been crucial to demonstrate the ability of neuronal and non-neuronal soluble NSF attachment receptor (SNARE) to promote spontaneous fusion of liposomes. This lipid-mixing assay is based on the fluorescence resonance energy transfer (FRET) capability between a donor fluorescent lipid and a quenching lipid. When fusion between donor fluorescent liposomes and nonfluorescent acceptor liposome occurred, FRET decreases. This assay allows a real-time reading of SNARE-mediated liposome fusion.

Key words: FRET; fusion; lipids; liposomes; SNARE.

1. Introduction

Assays of membrane fusion report either the mixing of membrane lipids or the mixing of the contents of the fused vesicles *(1,2)*. Lipid-mixing assay has been mostly used to demonstrate the ability of divalent metal ions (e.g.: calcium) or proteins (SNAREs) to promote spontaneous membrane fusion *in vivo* and *in vitro (3–5)*.

The lipid-mixing assay described here has been introduced by Struck, Hoekstra, and Pagano and is based on nitrobenzoxadiazol (NBD)–rhodamine energy transfer *(6)*. In this method (**Fig. 1**), membranes labeled with a combination of fluorescence energy transfer donor (NBD) and acceptor lipid (rhodamine) probes are mixed with unlabeled membranes. Fluorescence resonance energy transfer (FRET) decreases when the average spatial separation of the probes

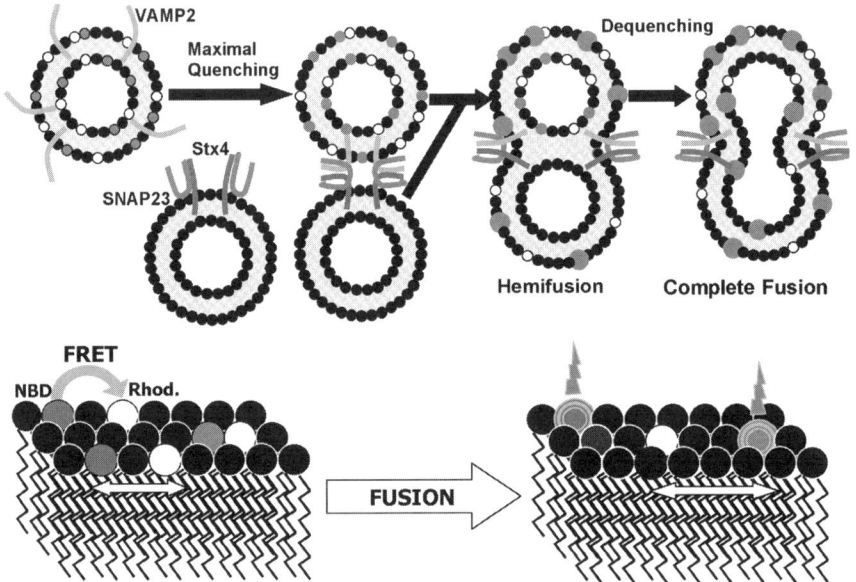

Fig. 1. Principles of lipid mixing assay. In the reconstituted v-SNARE vesicle nitrobenzoxadiazol (NBD) fluorescence (gray) is quenched by rhodamine (white). When a fusion event occurs between donor and acceptor vesicles, the distance between NBD and rhodamine increases, reducing the quenching (FRET) and increasing the NBD signal.

is increased upon fusion of labeled membranes with unlabeled membranes. The reverse detection scheme, in which FRET increases upon fusion of membranes that have been separately labeled with donor and acceptor probes, has also proven to be a useful lipid-mixing assay *(7)*.

This assay is now commonly used to determine the fusogenic properties of viral proteins and more recently, those of soluble NSF attachment receptors (SNARE). SNAREs belong to a large family of proteins involved in the fusion of cellular compartments in eukaryotes *(5,8,9)*. There are two families of SNAREs depending on their structures and cellular localization. Vesicle (v)-SNAREs (or R-SNAREs) possess a single coil-coiled domain and a transmembrane domain. They are localized in "transit compartment" (e.g., secretory vesicles). The second family, target (t)-SNAREs (or Q-SNAREs) is dimeric. One partner belongs to the family of syntaxin isoforms. Syntaxins possess a single SNARE domain (coiled-coil), a transmembrane domain, and an N-terminal extension involved in the binding of regulator factors (e.g., Sec1/Munc family). The second partner, a SNAP isoform, possesses two coiled-coil domains and is usually linked to the plasma membrane by palmitoylation. They are localized on the acceptor/target compartment *(10–12)*. Fusogenic properties of neuronal SNAREs (VAMP2–

Stx1/SNAP25) have been extensively studied because they induce the extremely fast release of synaptic vesicles. The exact mechanism of fusion is still under investigation and requires the participation of many partners. However, SNAREs alone, neuronal and non-neuronal, have been demonstrated to be able to promote an efficient rate of fusion when they are reconstituted in liposomes *(13–15)*.

The assay methodology described here is a general protocol to analyze the fusogenic properties of SNAREs or any similar transmembrane proteins using the lipid-mixing assay and is based on the VAMP2–Stx4/SNAP23 SNARE isoforms.

2. Materials

2.1. SNAREs Purification

1. Stx4/SNAP23 expression plasmid: mouse Stx4/SNAP23 coding sequences are sequentially ligated in pET duet (Novagen). This results in the generation of a His-6-Stx4 plus a SNAP23 bacterial coexpression plasmid. His-6-Stx4/SNAP23-pETduet is transformed in *Rosetta Escherichia coli* (Novagen). This strain improves Stx4 expression and reduces the amount of SNAP23 proteolysis (*see* **Note 1**).
2. VAMP2 expression plasmid: human VAMP2 coding sequence is cloned in Topo-D pET200 that gives a His-6-VAMP2 tag protein. VAMP2 expression vector is transformed in Bl21 Star (DE3) *E. coli* (Invitrogen).
3. Luria broth (LB): 10 g/L bactotryptone, 5 g/L yeast extract, 10 g/L NaCl.
4. Circlegrow medium (Bio101).
5. Ampicillin or carbenicillen (1,000X): 100 mg/mL in water; store frozen.
6. Kanamycin (500X): 50 mg/mL in water; store frozen.
7. Cell-washing buffer: 20 mM Tris-HCl, pH 7.4, 150 mM KCl.
8. Breaking buffer: 10% glycerol, 50 mM Hepes-KOH, pH 7.4, 400 mM KCl, 2 mM 2-mercaptoethanol, 4% Triton X-100, four tablets of Complete Protease Inhibitor (Roche) per 50 mL.
9. Wash buffer: 10% glycerol, 25 mM Hepes-KOH, pH 7.4, 400 mM KCl, 2 mM 2-mercaptoethanol, 1% Triton X-100.
10. Exchange buffer: 10% glycerol, 25 mM Hepes-KOH, pH 7.4, 100 mM KCl, 2 mM 2-mercaptoethanol, 0.8% or 1% octyl-β-D-glucopyranoside (OG).
11. 1 M isopropyl-thio-β-galactopyranoside (IPTG).
12. OG.
13. Ni-NTA agarose beads (Quiagen).
14. 3 M midazole, pH 7.5, titrated with acetic acid.
15. 10 mL affinity chromatography columns (Pierce).

2.2. Proteoliposomes Production and Fusion

1. Reconstitution buffer: 10% glycerol, 25 mM Hepes-KOH, pH 7.4, 100 mM KCl, 2 mM 2-mercaptoethanol.

2. 1-Palmitoyl, 2-oleoyl phosphatidylcholine (POPC; 16:0–18:1 acyl side chains), 25 mg/mL in chloroform (Avanti Lipids).
3. 1,2-Dioleoyl phosphatidylserine (DOPS), 10 mg/mL in chloroform (Avanti Lipids).
4. 1-Palmitoyl, 2-oleoyl phosphatidic acid (POPA), 10 mg/mL I chloroform (Avanti Lipids).
5. 1,2-dioleoyl phosphoinositol 4,5 *bis*-phosphate, ammonium salt (PIP2), 0.1 mg/mL in chloroform (Avanti Lipids).
6. *N*-(7-nitro-2,1,3-benzoxadiazole-4-yl)-1,2-dipalmitoyl phosphatidylethanolamine NBD-DPPE), 1 mg/mL in chloroform (Avanti Lipids).
7. Rhodamine-DPPE (*N*-[lissamine rhodamine B sulfonyl]1,2-dipalmitoyl- phosphatidylethanolamine), 1 mg/mL in chloroform (Avanti Lipids).
8. Chloroform (high-performance liquid chromatography grade).
9. Histodenz powder (Sigma).
10. Nitrogen or argon tank.
11. Biobeads SM2 adsorbent (Biorad).
12. Sodium dithionite ($Na_2O_4S_2$) powder protected under argon.
13. Organic solvent-safe pipets (*see* **Note 2**).
14. Ultraclean borosilicate glass tubes (12 × 75 mm).
15. Dialysis cassette (100–500 μL capacity; 3.5 k cut-off; Pierce).
16. SW55Ti rotor and adaptors (Split adaptor 356860, Beckman).
17. 5 × 41 mm ultraclear centrifuge tubes (Beckman).
18. Thermo-regulated plate reader fluorometer (96-wells) with integrated automatic injectors and shaker (i.e., Fluoroscan Ascent, Thermo Electron).
19. Black 96-well plastic plates.

3. Methods

3.1. SNAREs Purification

1. A 100 mL overnight preculture of His-6-VAMP2 cells in standard LB medium containing 30 μg/mL kanamycin is used to start a 2-L culture in Circlegrow medium containing 30 μg/mL kanamycin.
2. The cells are induced with 0.8 mM IPTG after reaching an OD_{600} of 1.0. Four hours later, the cells are collected by centrifugation (6,000*g*, 15 min at 4°C), and washed once in cell-washing buffer (20 mM Tris-HCl, pH 7.4, 150 mM KCl). At this stage, the bacterial pellet can be frozen and store at −20°C.
3. The pellet is resuspended in 40 mL of breaking buffer and bacteria are thoroughly sonicated three times for 1 min on ice. The lysates are next centrifuged for 15 min at 18,000 *g* in a JA-20 rotor (Beckman) at 4°C.
4. The resulting supernatant is next centrifuged for 1 h at 278,000*g* in a Ti70 rotor at 4°C (Beckman).
5. Each 10 mL of the supernatant is then incubated with 1 mL of Ni-nitrilotriacetic acid agarose beads (Qiagen) equilibrated in washing buffer in a 10-mL affinity chromatography column for 2 h at 4°C under rotation.

6. The beads are washed "in batch" four times with 10 mL of washing buffer. The column is next open and after beads are settled, the Triton X-100 is exchanged with OG by adding 4 × 10 mL of exchange buffer containing 50 mM imidazole.
7. Proteins are eluted from beads with 500 μL of 100 to 500 mM imidazole in exchange buffer. Fractions eluted from 200 to 400 mM imidazole are combined.
8. For the coexpression of His-6-Stx4 and SNAP23, cells are grown in 100 mg/mL ampicillin and 25 mg/mL chloramphenicol in LB. Cells are harvested, and t-SNAREs purified as described for His-6-VAMP2.
9. Typical protein yields are approx 2 mg/L bacterial culture of Stx4/SNAP23 complex and approx 5 mg/L of His-6-VAMP2. The purity of the preparations is then assessed by sodium dodecyl sulfate-polyacrylamide gel electrophoresis (SDS-PAGE) and Coomassie blue staining and is typically greater than 90%.

3.2. Proteoliposomes Production

1. For a standard vesicle preparation, 100 μL of a 3 mM premixed lipids in chloroform (i.e., 300 nmol total lipids) POPC:DOPS in a 85%:15% molar ratio are dried in a 7.5 mm glass test tube by a gentle stream of nitrogen (*see* **Note 3**).
2. For fluorescent vesicle preparation (defined as donor vesicle), the premixed lipid solution in chloroform contains POPC:DOPS:NBD-DPPE:rhodamine-DPPE in an 82:15:1.5:1.5 molar ratio.
3. Any remaining traces of chloroform are then removed under vacuum for 30 min in the dark to protect NBD and rhodamine from photobleaching.
4. v-SNARE-containing vesicles are prepared by dissolving the lipid film in 100 μL of a solution containing VAMP2-His6 (∼5 mg/mL) in 1% (w/v) OG as eluted from Ni-NTA agarose.
5. t-SNARE-containing vesicles are prepared using 100 μL of solution containing the t-SNARE complex (∼2 mg/mL of total protein) and 0.8% (w/v) OG.
6. Protein-free donor vesicles are prepared by dissolving the lipid film with 100 μL of reconstitution buffer containing 0.8% (w/v) OG.
7. Other lipids can be added by adjusting the other molar lipid ratios.
8. In all cases, the lipids film is dissolved by gentle agitation on a shaker for 15 min at room temperature in the dark.
9. Vesicles are formed by rapid dilution: while vortexing vigorously, add 200 μL (2× sample volume) of reconstitution buffer (at room temperature), thereby diluting the detergent OG below its critical micellar concentration promoting vesicle formation.
10. Detergent is removed by dialysis (in 3.5 kDa cutoff cassette) against 2 L of room temperature reconstitution buffer containing 2 g of Biobeads SM2 per Liter of dialysate. The beads will trap the detergent during dialysis. The dialysis container is kept at 4°C over night in the dark with gentle stirring.

Higher amount of vesicles can be prepared using the same protocol by proportional adjustment of the above reagents.

3.3. Proteoliposomes Purification

Vesicles are recovered and concentrated by flotation in a Histodenz (Sigma) step gradient:

1. Each 300 μL dialysate is mixed with 300 μL of 80% (w/v) Histodenz dissolved in reconstitution buffer giving 600 μL of 40% Histodenz dialysate. This dialysate is divided equally into two 5 × 41-mm ultraclear centrifuge tubes (Beckman). The samples are overlaid with 350 μL 30% (w/v) Histodenz followed by 50 μL of reconstitution buffer lacking glycerol (*see* B).
2. The samples are then centrifuged in a SW55Ti rotor (Beckman) with the appropriate adaptors at 246,000 g for 4 h at 4°C. The vesicles are harvested from the 0 to 30% Histodenz interface in 80 μL per tube and then combined (*see* **Note 5**).
3. Briefly vortex the vesicle to homogenize the buffer and check for the presence of vesicle aggregates.
4. Vesicles can be stored under nitrogen or argon in the dark at 4°C up to 4 days without significant loss of fusion efficiency.
5. SNAREs yield incorporation into vesicles is directly dependent on the lipid composition and original SNAREs concentration.
6. In the case of a POPC:DOPS (85:15) reconstitution, typically 750 VAMP2 copies (corresponding to a 1:20 VAMP2:lipid molar ratio) and 75 copies (1:300) of the Stx4/SNAP23 complex are incorporated per vesicle (*see* **Note 6**).
7. For the incorporation of other SNARE isoforms, the number of SNAREs incorporated must be determined by estimating the protein:lipid ratio. This requires first the determination of the average vesicle size. Predicted vesicle diameter in standard condition is approx 60 ± 15 nm. However, this must be experimentally determined (e.g., dynamic light scattering and/or electron microscopy) because many factors can affect vesicle homogeneity and size.
8. Protein recovery/concentration can be easily determined by standard protein assay. Lipids recovery can be calculated by fluorescent dye recovery (NBD–rhodamine) or ^{3}H lipid tracer recovery *(14)*.

3.4. Fusion Assay

Fusion reactions mixtures are typically designed at 1:1 v-SNARE:t-SNARE molar ratio to allow for multiple rounds of fusion to occur. In standard experiments, that implies the addition of a 10 times volume excess of t-SNARE vesicles compare to v-SNARE in the reaction volume.

In black 96-well plates or strips on ice, add per well:

1. 50 μL Reconstitution buffer.
2. 5 μL VAMP2 proteoliposomes.
3. 45 μL Stx4/SNAP23 proteliposomes.
4. Mix gently to avoid formation of bubbles (*see* **Note 7**).
5. Proceed to plate reading or incubation.

SNARE-Mediated Fusion of Liposomes

For overnight incubation, wells must be seal with tape to avoid evaporation and keep in the dark.

Other reagents (chemicals, proteins, toxins, etc.) can be added, but must be devoid of any trace of detergent. The balance is made with reconstitution buffer up to 100 μL.

3.5. Fusion Detection and Analysis

Membrane fusion is monitored in black 96-well plates in a fluorometer preincubated at 37°C with the excitation filter set at 460 nm (half-bandwidth 25 nm) and the emission filter set at 538 nm (half-bandwidth 25 nm).

1. Fluorescence is measured every 2 min for 2 to 5 h.
2. At the end of the reaction, 10 μL of a 2.5% (w/v) Triton X-100 solution is added to each well.
3. The plate is shaken in the fluorometer at a setting of 1000 oscillations for 30 s.
4. Fluorescence is monitored every 2 min for 10 min to determine maximum of NBD dequenching.
5. Results can be expressed in raw data (absolute fluorescence values) or can be normalized for each well as "percentage of maximum fluorescence" (**Fig. 2**).
6. Conversion to percentage of maximum fluorescence of NBD–DPPE by addition of Triton X-100 at the end of reaction was performed as follow:

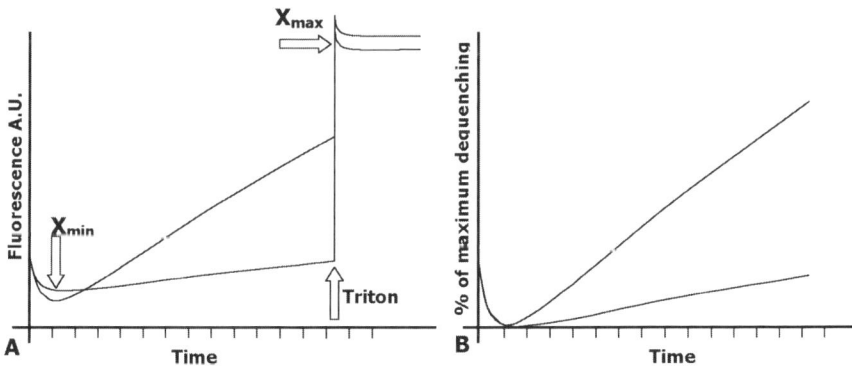

Fig. 2. Theoretical curves of membrane lipid fusion by nitrobenzoxadiazol (NBD) fluorescence dequenching. (**A**) Following the mixing of reconstituted vesicles containing fusogenic proteins, NBD dequenching can be observed as function of time. X_{min} correspond to the lowest value of NBD signal. At the end of the reaction Triton-X100 is added, solubilizing the vesicles and allowing a maximum NBD dequenching (X_{max}) (**B**) After conversion to percentage of maximum dequenching, normalized curves can be compared for each time value.

7. For each well, the lowest NBD fluorescence signal is set to 0%, and the maximal signals reached after detergent addition is set to 100% fluorescence (14).

$$(X_t - X_{min})/(X_{max} - X_{min}) \times 100 = P_{t(\% max)}$$

with:

X_t, the NBD fluorescence value as function of the time
X_{min}, the lowest NBD fluorescence signal
X_{max}, the maximal signal reached after detergent addition
$P_{t(\%max)}$, Percentage of maximum fluorescence as function of the time.

It is also possible to convert the data in round of fusion but this method requires the establishment of a calibration curve (*see* **Note 8**).

3.6. Hemifusion Determination Assay

To assess the proportion of hemifused vesicles during a reaction, fluorescent donor vesicles must be preincubated in 45 μL of reconstitution buffer containing 1 mM sodium dithionite for 10 min at 37°C and then transferred to 4°C before mixing with acceptor vesicles.

In this condition, the outer membrane leaflet of NBD is efficiently reduced to nonfluorescent *N*-(7-amino-2,1,3-benzoxadiazole-4-yl). This should decrease the effective NBD–DPPE content to approx 50%.

The total lipid-mixing and inner leaflet-mixing fluorescence signals are measured simultaneously. The percentage of hemifused vesicles can be calculated as a function of time and is defined as:

$$2(P_T - P_I)/[2(P_T - P_I) + P_I] \times 100$$

where P_T is the percentage of maximum for total lipid mixing and P_I is the percentage of maximum for inner leaflet mixing (i.e., after sodium dithionite treatment; *16*).

Hemifusion assay can reveal unexpected aspect of the SNARE-mediated fusion not necessary detectable with the "complete fusion assay" described previously (*see* Note 9)

3.7. Western Blot Analysis

1. At the end of the reaction 1 to 5 μL of the reaction mixture can be analyzed by Western blot to assess the proportion of each SNAREs.
2. Use of an anti-His-6-antibody has the advantage to simultaneously detect VAMP2 and Stx4 and, because they have the same epitope, to give a direct determination of the v-SNARE:t-SNARE ratio.

3. Coomassie blue or other protein-staining methods can be used but the presence of Triton X-100 and lipids can alter the quality of the electrophoresis. Additionally, various protein stains do not necessary label all proteins with equal efficiency.

4. Notes

1. Any standard expression vectors and epitope tags are suitable for bacterial expression, however, the dual-expression vector "pET duet" appears to be the easiest method to obtain a 1:1 molar ratio for Stx and SNAP. Additionally, His-6-tag allows a fast and efficient purification without adding a cumbersome protein extension that could impair SNAREs functionality.
2. Only glass and Teflon tubes can be used in presence of chloroform. Plastic caps or rings will dissolve into the lipid/chloroform mixture modifying its properties. However, plastic tips and tubes can be used after dried lipids are resuspended in any aqueous buffer.
3. Concentration of lipids in organic solvent is always approximate, even from commercial source and must be assessed regularly, particularly when tubes are often opened. Concentration can be easily measured using a micro-scale and weighing lipids after solvent evaporation.
4. Sucrose gradient can be used but Histodenz or Nicodenz gradient are usually easier to handle. Sucrose stock solutions can be frozen but repetitive freezing and thawed cycles can create the formation of crystals.
5. One or several pink rings of fluorescent vesicle are usually visible after centrifugation, whereas the unlabeled vesicles are barely observable as a milky ring against a dark background. The lowest ring must be recovered because it contains the proteoliposomes with the highest SNARE concentration.
6. Many parameters affect the protein incorporation in liposomes, particularly when using this "detergent dilution" method. Intrinsic charges of lipids and proteins affect the final incorporation. Mainly, the addition of charges lipids (like phosphatidylethanolamine, phosphatidic acid [PA], phosphatidylinositol, etc.) will dramatically affect the SNARE incorporation and recovery *(13)*.
7. When adding the reagent, start with the "reconstitution buffer" and gently mix the other reagent by pumping up and down the pipet tips. Buffer of the purified vesicles still contains Histodenz and has a significant higher density. Therefore, gentle but efficient mixing is crucial to obtain a homogenous reaction volume.
8. At the beginning of fusion the reaction, a sudden drop in the fluorescence intensity is usually observed. This drop is mostly caused by the temperature shift from 4 to 37°C. If no fusion reaction occurs, the signal stabilizes after 10 to 15 min of incubation (**Fig. 2**).
9. Efficient SNARE-dependent fusion has been shown in many settings, *in vivo*, to require the generation of both PIP2 and PA. However, the direct implication of these key lipids in fusion reaction mechanism has only recently begun to be investigated. We have used the assay system described here to assess the specific role of PIP2 and PA in the donor and acceptor membranes during *in vitro*

fusion *(13)*. By selectively modifying the different lipid compositions of donor and acceptor liposomes, we demonstrated that the presence of PA in Stx4/SNAP23 vesicles markedly enhances the fusion rate, whereas its presence in VAMP2 vesicles is inhibitory. In contrast, addition of PIP2 to Stx4/SNAP23 vesicles inhibits the fusion reaction, and in VAMP2 vesicles is stimulatory. Using the hemifusion assay, we also demonstrated that the optimal distribution of phospholipids triggers the progression from the hemifused state to full fusion. These findings reveal an unanticipated dependence of SNARE complex-mediated fusion on asymmetrically distributed acidic phospholipids and provide mechanistic insights into the roles of phospholipase D and PIP kinases in the late stages of regulated exocytosis.

References

1. Duzgunes, N. and Wilschut, J. (1993) Fusion assays monitoring intermixing of aqueous contents. *Methods Enzymol.* **220**, 3–14.
2. Hoekstra, D. and Duzgunes, N. (1993) Lipid mixing assays to determine fusion in liposome systems. *Methods Enzymol.* **220**, 15–32.
3. Duzgunes, N., Allen, T. M., Fedor, J., and Papahadjopoulos, D. (1987) Lipid mixing during membrane aggregation and fusion: why fusion assays disagree. *Biochemistry.* **26**, 8435–8442.
4. Blumenthal, R., Gallo, S. A., Viard, M., Raviv, Y., and Puri, A. (2002) Fluorescent lipid probes in the study of viral membrane fusion. *Chem. Phys. Lipids.* **116**, 39–55.
5. Ungermann, C. and Langosch, D. (2005) Functions of SNAREs in intracellular membrane fusion and lipid bilayer mixing. *J. Cell Sci.* **118**, 3819–3828.
6. Struck, D. K., Hoekstra, D., and Pagano, R. E. (1981) Use of resonance energy transfer to monitor membrane fusion. *Biochemistry.* **20**, 4093–4099.
7. Uster, P. S. (1993) In situ resonance energy transfer microscopy: monitoring membrane fusion in living cells. *Methods Enzymol.* **221**, 239–246.
8. Schuette, C. G., Hatsuzawa, K., Margittai, M., et al. (2004) Determinants of liposome fusion mediated by synaptic SNARE proteins. *Proc. Natl. Acad. Sci. U S A.* **101**, 2858–2863.
9. Sollner, T. H. (2003) Regulated exocytosis and SNARE function. *Mol. Membr. Biol.* **20**, 209–220.
10. Antonin, W., Fasshauer, D., Becker, S., Jahn, R., and Schneider, T. R. (2002) Crystal structure of the endosomal SNARE complex reveals common structural principles of all SNAREs. *Nat. Struct. Biol.* **9**, 107–111.
11. Bonifacino, J. S. and Glick, B. S. (2004) The mechanisms of vesicle budding and fusion. *Cell.* **116**, 153–166.
12. Fasshauer, D. (2003) Structural insights into the SNARE mechanism. *Biochim. Biophys. Acta.* **1641**, 87–97.
13. Vicogne, J., Vollenweider, D., Smith, J. R., Huang, P., Frohman, M. A., and Pessin, J. E. (2006) Asymmetric phospholipid distribution drives in vitro reconstituted SNARE-dependent membrane fusion. *Proc. Natl. Acad. Sci U S A.* **103**, 14761–14766.

14. Weber, T., Zemelman, B. V., McNew, J. A., et al. (1998) SNAREpins: minimal machinery for membrane fusion. *Cell.* **92**, 759–772.
15. Parlati, F., Weber, T., McNew, J. A., Westermann, B., Sollner, T. H., and Rothman, J. E. (1999) Rapid and efficient fusion of phospholipid vesicles by the alpha-helical core of a SNARE complex in the absence of an N-terminal regulatory domain. *Proc. Natl. Acad. Sci. U S A.* **96**, 12565–12570.
16. Xu, Y., Zhang, F., Su, Z., McNew, J. A., and Shin, Y. K. (2005) Hemifusion in SNARE-mediated membrane fusion. *Nat. Struct. Mol. Biol.* **12**, 417–422.

19

Use of Polarized PC12 Cells to Monitor Protein Localization in the Early Biosynthetic Pathway

Ragna Sannerud, Michaël Marie, Bodil Berger Hansen, and Jaakko Saraste

Summary

A prerequisite for understanding the cellular functions of an unknown protein is the establishment of its subcellular localization. As increasing numbers of novel proteins of the biosynthetic pathway are currently being identified, accessible new methods are required to facilitate their localization. Differentiating rat pheochromocytoma (PC12) cells reorganize their biosynthetic membrane compartments as they develop neurite-like processes. The authors recently showed that polarization of these cells involves the expansion of the intermediate compartment (IC) between the rough endoplasmic reticulum (RER) and the Golgi apparatus. Tubules emerging from the vacuolar parts of the IC move to the developing neurites accumulating in their growth cones, whereas the vacuoles, like RER and Golgi, remain in the cell body. Thus, polarized PC12 cells enhance the resolution for immunofluorescence microscopic mapping of protein localization in the early biosynthetic pathway. The authors also describe here a rapid cell fractionation protocol employing velocity sedimentation in iodixanol gradients that allows one-step separation of the pre-Golgi vacuoles, tubules, and RER.

Key words: Biosynthetic pathway; cell fractionation; endoplasmic reticulum; immunofluorescence microscopy; PC12 cells; pre-Golgi tubules; velocity sedimentation.

1. Introduction

Biochemical and genetic studies carried out with mammalian and yeast cells initially paved the way for the discovery of proteins that operate in the biosynthetic-secretory pathway *(1,2)*. However, although many of these proteins have been biochemically well characterized, knowledge regarding their precise intracellular localization is still in many cases limited. More recently, the employment of modern methods of genomics and proteomics have resulted in the identification of a multitude of novel proteins that function in the secretory pathway *(3–5)*. In parallel, microscopy-based technologies have been developed to facilitate high-content screening of the cellular localization of these unknown proteins *(6)*.

Owing to its technical simplicity and accessibility as compared to the more demanding electron microscopic methods, immunofluorescence microscopy has become the most commonly used technique to study the subcellular localization of proteins *(7)*. However, an inherent limitation of this technique is its low resolution. In particular, light microscopic assignment of proteins to the compartments of the early biosynthetic pathway, including the rough and smooth endoplasmic reticulum (ER), the pre-Golgi intermediate compartment (IC), and the Golgi apparatus, is problematic because the boundaries of these dynamic compartments are still poorly defined. Therefore, transport inhibitors, such as low temperature *(8)*, brefeldin A *(9)*, or microtubule depolymerizing compounds *(10)* have been frequently used to facilitate protein localization at the ER–Golgi boundary.

Polarized epithelial cells and neurons provide valuable models for the analysis of the intracellular compartments that operate in intracellular membrane traffic *(11–14)*. Cultured rat pheochromocytoma (PC12) cells—a neuroendocrine cell line derived from adrenal cancer tissue —express both the constitutive and regulated secretory pathways and thus have been commonly used to study post-Golgi trafficking *(15–17)*. Upon treatment with nerve growth factor (NGF), these cells become highly polarized and form neurite-like extensions *(18)*. The authors recently observed that the NGF-induced differentiation of PC12 cells is accompanied by a dramatic reorganization of the early biosynthetic pathway involving selective transfer of IC-derived pre-Golgi tubules to the developing neurites and their growth cones *(19)*. Because the redistribution of these membranes can be readily followed by light microscopy (LM: **Fig. 1**), the polarized PC12 cells provide a useful assay to map the localization of cellular components within the early biosynthetic compartments.

This chapter describes the immunofluorescence microscopic technique that the authors used to localize different proteins in the polarized PC12 cells. Moreover, these cells are easy to propagate in large numbers and can be readily applied for cell fractionation analysis *(17,19)*. To study the compositional properties of the IC by biochemical methods, the authors have adapted a previously described velocity sedimentation protocol *(20)*, presented in detail

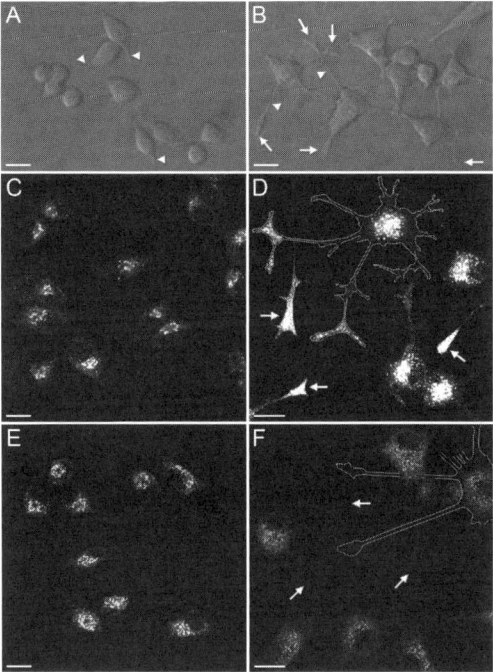

Fig. 1. PC12 cells were bound on poly-L-lysine-coated glass cover slips and incubated overnight in complete growth medium at 37°C, followed by 48 h incubation in differentiation medium without (A, C, and E; control cells) or with 10 ng/mL NGF (B, D, and F; NGF-treated cells). The cells were fixed and processed for immunofluorescence microscopy using antibodies against p58 (C and E) and Rab1 (D and F), as described in Methods. Note the strong Rab1 signal in the extensions and growth cones (arrows) of the differentiated cells (D), whereas p58 is not detectable in the cell processes (F). The contours of two diffentiating cells in D and F are outlined. A and B : Differential interference contrast (DIC) images showing the morphological and size differences of the control and NGF-treated cells. The images were acquired with a Leica TCS NT confocal microscope. Bars: 10 μm.

here, for the rapid separation of the vacuolar and tubular domains of the IC.

2. Materials

Prepare all solutions using double-distilled milli-Q water (ddH$_2$O).

2.1. Growth and Differentiation of PC12 Cells

1. PC12 cells: Our laboratory uses a subclone of PC12 cells, which in the right conditions readily adhere to cell culture substrate. No pretreatment of the plastic surfaces with collagen- or poly-L-lysine is required. However poly-L-lysine is used to attach the cells to glass cover slips (see **Subheading 3.1., step 4**).
2. Growth medium: RPMI 1640 medium supplemented with 10% heat-inactivated horse serum, 5% heat-inactivated fetal bovine serum (FBS), L-glutamine (2 mM), penicillin (25 U/mL), and streptomycin (25 μg/mL). (see **Note 1**).
3. Differentiation medium: Contains 1/10 of the normal amount of serum. Complete growth medium diluted with 1:10 with RPMI 1640 and supplemented with L-glutamine and the antibiotics *as step 2*.
4. Nerve growth factor: (NGF, 2.5S, Cat. no. 13257-019, Invitrogen). The lyophilized protein is reconstituted to the concentration of 20 μg/mL with growth medium and stored in 25 μL aliquots at –20°C. For differentiation of PC12 cells, this NGF stock is freshly thawed, diluted 1:2,000 (final concentration 10 ng/mL) in the differentiation medium, which is added to the cells.
5. Stock solution (0.5 mg/mL) of poly-L-lysine (1:1 mixture of mol. wt 30,000–70,000 and mol. wt. 70,000–150,000) in sterile ddH$_2$O. Stored in aliquots at –20°C and diluted 1:10 with sterile ddH$_2$O prior to use.
6. Sterile (autoclaved) ddH$_2$O.
7. 25 and 75 cm^2 Cell culture flasks and 6-well culture plates.
8. Round glass cover slips (18 mm in diameter).

2.2. Immunofluorescence Microscopy

1. Phosphate-buffered saline (PBS, 10x stock): 27 mM KCl, 1.37 M NaCl, 100 mM Na$_2$HPO$_4$, 18 mM KH$_2$PO$_4$ (adjust to pH 7.4), autoclave and store at room temperature. Prepare working solution by diluting 1:10 with ddH$_2$O.
2. 0.2 M Phosphate buffer (PB): For 100 mL of buffer, mix 28 mL of 0.2 M Na$_2$HPO$_4$ and 72 mL of 0.2 M NaH$_2$PO$_4$. If necessary, adjust to pH 7.2 with either one of the components.
3. Fixative: 3% paraformaldehyde (PFA) in 0.1 M PB, pH 7.2. Make fresh for each experiment. Prepare first 6% (w/v) solution of PFA in ddH$_2$O (3 g in 50 mL) in a glass container and add five drops of 1 M NaOH. Heat the solution slowly up to approx 60°C (use a magnetic stirring bar and a mixing hot plate in a fume hood) to dissolve the PFA. Cool the solution on ice and add an equal volume of 0.2 M PB. Filter through a double filter paper in a funnel. Keep the fixative in a tightly capped container at 4°C until use.
4. Quench solution: 50 mM NH$_4$Cl in PBS.
5. Washing buffer (WB): PBS containing 0.2% bovine serum albumin (BSA, fraction V) and 0.02% Na-Azide.
6. Washing buffer with saponin (WB-saponin): WB supplemented with 0.2% saponin.

7. Blocking buffer (BB): WB-saponin supplemented with 5% goat serum (when the secondary antibodies are prepared in goat) or e.g., FBS. Filter through a 0.2-μm filter attached to a 10-mL syringe. Keep at 4°C until use.
8. Primary antibodies: Polyclonal or monoclonal antibodies are stored at –20°C either frozen in small aliquots or unfozen in 50% glycerol. Appropriate dilutions of the antibodies are prepared in BB.
9. Secondary antibodies: In most cases we use commercial goat anti-rabbit or anti-mouse F(ab)$_2$-fragments conjugated to fluorescein isothiocyanate (FITC) or Texas Red. Secondary antibodies are also diluted in BB.
10. Mounting medium: Vectashield with or without DAPI (Vector Laboratories, Inc) or an equivalent mounting medium.
11. Objective slides washed with 70% ethanol and dried with clean tissue paper.

2.3. Cell Fractionation

All solutions must be ice cold.

2.3.1. Homogenization of PC12 Cells

1. Washing solution 1: Hank's balanced salt solution (HBSS).
2. Washing solution 2: PBS.
3. Washing solution 3: 140 mM NaCl, 30 mM KCl, 10 mM ethylenediaminetetraacetic acid (EDTA), 25 mM Tris-HCl, pH 7.4.
4. Homogenization solution (HS): 130 mM KCl, 25 mM NaCl, 1 mM ethyleneglycoltetraacetic acid (EGTA), 25 mM Tris-HCl, pH 7.4. Add one tablet of protease inhibitors (PI; Complete Mini^{TM}, EDTA-free; Boehringer, Mannheim, Germany) to 10 mL of HS.
5. Homogenizer: A ball-bearing cell-cracker (EMBL, Heidelberg) containing a precision-made stainless-steel ball. For PC12 cells we use a ball of 8.010μmm in diameter giving a clearance of 10 μm. Two 1-mL plastic syringes are attached to the outlets of the homogenizer. The original construction of this homogenizer is described in **ref. 21**.
6. An inverted microscope equipped with phase-contrast optics.
7. 15-mL Capped plastic tubes with a conical bottom.
8. A bench-top centrifuge accommodating the above tubes.

2.3.2. Velocity Gradient Centrifugation

1. Iodixanol 60% stock solution in water (OptiPrep^{TM}, Axis-Schield).
2. Diluent: 130 KCl, 25 mM NaCl, 6 mM EGTA, 150 mM Tris-HCl, pH 7.4 (*see* OptiPrep^{TM}; Application Sheet S27 found at the Axis-Schield home pages).
3. Iodixanol 50% (v/v) working solution: Mix one volume of iodixanol stock solution with five volumes of the diluent.
4. HS (*see* **Subheading 2.3.1., step 4**).

5. 1,000 x CLAP stock solution: A PI cocktail containing 10 mg/mL each of chymotrypsin, leupeptin, antipain, and pepstatin in dimethyl sulfoxide (DMSO). Keep frozen in 20 µL aliquots at –20°C.
6. SW41 UltraClear™ ultracentrifuge tubes (Beckman).
7. Beckman LE-80 K ultracentrifuge with a SW41 rotors, or another ultracentrifuge with a swinging bucket rotor giving identical centrifugation conditions.

3. Methods
3.1. Growth and Differentiation of PC12 Cells

1. PC12 cells in grown in a humidified incubator at 37°C in 5% CO_2 atmosphere. Use a sterile luminar flow hood and prewarm the growth medium to 37°C. The cells must be passaged before they have reached 80 to 90% confluency. *For routine passage* the cells are cultivated in 25 cm^2 culture flasks in 5 mL of growth medium and passaged at 1:3 and 1:6 dilutions twice a week.
2. To detach the cell monolayers hit the flask several times against the palm of your hand. Pass the cell suspension up and down in a 5-mL pipet (pipet forcefully against a corner of the flask) to obtain a homogenous suspension without cell clumps. Check the efficiency of release and resuspension using the inverted microscope. Transfer appropriate volumes of the suspension to new 25 cm^2 flasks containing appropriate amounts of fresh medium. Gently mix the cell suspensions by vertical movements (to get an even monolayer) and place the flasks in the CO_2 incubator.
3. *For immunofluorescence microscopy* the glass cover slips are washed with several changes of 70% ethanol, air-dried (or rapidly flamed), and placed in 6-well culture plates (one cover slip per well).
4. The cover slips are treated with 100 µL of poly-L-lysine (50 µg/mL) for 5 min in the laminar flow hood, washed five times (remove also poly-L-lysine from under the cover slips) with sterile ddH_2O and finally left in sterile PBS (1 mL per well).
5. The cells from one 25 cm^2 culture flask are detached and resuspended as described *in step 2* (*see* **Subheading 3.1.**) and pelleted for 3 min at 300*g* in a bench-top centrifuge. They are washed once with 10 mL of sterile PBS (by resuspension and pelleting) and finally resuspended in 2 mL of sterile PBS.
6. PBS is removed from the wells (**step 4**) by suction and 50 µL aliquots of the cell suspension (**step 5**) are placed on each cover slip. The density of bound cells is checked in the inverted microscope and more cells can be added, if required. The cells are placed for 5 min in the CO_2 incubator, whereafter 2 mL of growth medium is added into each well. After growth of the cells overnight, the medium is replaced by differentiation medium without or with 10 ng/mL NGF (in case of control and NGF-treated cells, respectively) and the cells are grown in the CO_2 incubator for the desired periods of time (up to 3 d).
7. *For cell fractionation experiments* 5 mL of resuspended cells from one 25 cm^2 flask are transfered to one 75-cm^2 flask containing 10 mL of fresh growth medium

(dilution 1:3). The cells are used after 1 to 2 d (*see* **Note 2**). Four full (80–90% confluent) 75-cm^2 flasks give enough sample material for two velocity gradients. NGF-treatment of cells in the 75-cm^2 flasks is carried out as described for cells on cover slips. Minimum volume of NGF-containing medium to cover the cells in 75-cm^2 flasks is approx 7 mL. When growing differentiated PC12 cells in larger amounts, the starting density has to be reduced as compared to control cells because NGF stimulation results in considerable increase in cell size (**Fig. 1**).

3.2. Immunofluorescence Staining

All steps are carried out at room temperature maintaining the cells on cover slips in the 6-well plates.

1. The growth medium is removed from each well by suction and 1 mL of the PFA fixative, prewarmed to 37°C, is immediately added to each well. Fixation is for 30 min at room temperature.
2. Remove the fixative and wash the cells three times with WB.
3. Block unreacted aldehyde groups by incubating the cells for 15 min in 1 mL of quench solution in a shaker.
4. Permeabilize the cells and block unspecific antibody binding sites by incubation for 30 min in BB (0.5 mL per well).
5. Remove BB by suction and incubate the cells with the primary antibody appropriately diluted in BB. Make a dry circular area around each cover slip with a Pasteur pipet connected to a suction device and add approx 25 to 40 µL of the antibody on the cover slip. Incubate for 1 to 2 hr at room temperature. For overnight incubation (to increase the signal) place the 6-well plates in a humidified chamber (a closed plastic box with PBS-impregnated paper in the bottom).
6. Remove the primary antibody and wash the cells with WB-saponin (three quick changes), followed by washing for 2 h on a shaker (change the solution twice during this long washing period) (*see* **Note 3**).
7. Remove WB-saponin by suction and incubate the cells with the secondary antibody appropriately diluted in BB. *See* **step 4** for application of the antibody on the cover slips. Incubate for 2 h in the dark (cover the 6-well plate with aluminium foil).
8. Remove the secondary antibody and wash the cells as described in **step 5** (*see* **Note 4**).
9. Wash the cover slips three times with PBS and mount them (cell-side down) on small drops (~20 µL) of mounting medium placed on clean objective slides. Carefully remove excess mounting medium from the edges of the cover slips by suction. Prior to microscopy carefully wash the top of the cover slips with ddH$_2$O to remove salt precipitates. Avoid getting water under the cover slips.

3.3. Separation of IC Subdomains

All steps are carried out on ice using ice-cold solutions. Centrifugations are carried out at 4°C.

3.3.1. Homogenization of Cells

This method describes the preparation of samples for two velocity sedimentation gradients from four 75-cm^2 culture flasks (80–90% confluent). Increase the amount of cells as required keeping in mind that the SW41 rotor contains six buckets.

1. Remove the growth medium from each flask and wash the cell monolayers quickly but carefully with 10 mL of HBSS. Avoid release of cells (see **Note 5**).
2. Add 10 mL of fresh HBSS into the flasks and release the cells by hitting the flasks. Resuspend the cells by pipeting (see **Subheading 3.1., step 2**) and transfer the four cell suspensions in 15-mL plastic tubes.
3. Pellet the cells by centrifugation for 5 min at 300g in a bench-top centrifuge.
4. Resuspend the four pellets each in 10 mL of PBS and centrifuge as in **step 3**.
5. Resuspend the four pellets each in 2.5 mL of washing solution 3, combine them into one 15 mL tube and centrifuge as in **step 3** (see **Note 6**).
6. Finally, resuspend the single cell pellet (~100 µL of packed cells) in 0.75 mL of HS (with PIs; Complete Mini™). Resuspend extensively by pipeting 20 times through a 1-mL micropipet tip.
7. Assemble the homogenizer (with the 8.010-mm ball inside) during the above centrifugation steps. Attach tightly two 1 mL syringes half-filled with HS to the outlets of the homogenizer and keep it on ice. Remove air bubbles from the chamber of the homogenizer by passing HS between the two attached syringes. Discard excess solution by emptying the syringes, leaving the cell cracker filled with HS (chamber volume is ~0.4 mL).
8. For homogenization of the cells, draw the cell suspension into one of the syringes, then attach it tightly back into its outlet. Pass the cells 20 times back and forth through the chamber of the homogenizer (i.e., between the two syringes) by pressing sequentially and firmly the pistons of the two syringes.
9. Place a drop of the homogenate on an objective glass and monitor cell breakage by phase-contrast optics in the inverted microscope (use an objective that gives high magnification). The released nuclei appear dark and round (with clearly visible nucleoli) as compared with unbroken cells, which are brighter and smooth-surfaced ("tennis ball-like" appearance). Release of cytoplasmic components can be evaluated because the small and dark mitochondria are visible. If cell breakage less than 70 to 80% pass the homogenate through the cell cracker 5 to 10 more times and monitor cell breakage again (see **Note 7**).
10. Collect the total homogenate, including the approx 0.4 mL inside the chamber, by pressing air into the homogenizer using one of the syringes.

Use of Polarized PC12 Cells to Monitor Protein Localization

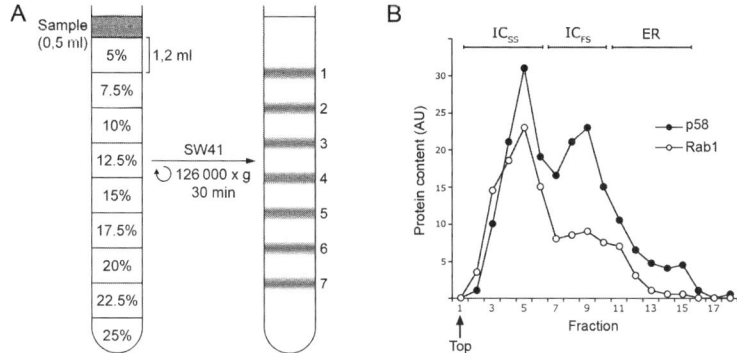

Fig. 2. (**A**) Schematic representation of the iodixanol velocity sedimentation gradient before (left) and after (right) centrifugation. After centrifugation seven bands are visible in the gradient, roughly positioned at the interfaces of the original iodixanol layers. Eighteen fractions of 0.5 mL each were collected from the top, diluted, and the membranes were concentrated by pelleting. (**B**) To study the relative p58 and Rab1 contents of the fractions, samples of equal volume were run in SDS-PAGE and analyzed by immunoblotting using specific antibodies. Note that the Rab1- and p58-containing membranes in the upper part of the gradient are separated into two distinct, slowly (IC_{SS}) and fast sedimenting (IC_{FS}) components, corresponding to tubular (Rab1-enriched) and vacuolar (p58-enriched) IC elements, respectively. About 70% of Rab1 associates with IC_{SS} whereas p58 is more equally distributed between the two components. The IC_{SS} peak of p58 corresponds to small transport vesicles (**ref. 23**). RER elements sediment towards the bottom of the gradients (not shown, but *see* **refs. 18** and **20**), separating from IC_{SS} and IC_{FS}. Minor pools of p58 (~18%) and Rab1 (~8%) appear to associate with RER. The very bottom (~2 mL) of the gradient does not contain immunoreactive material.

3.3.2. Preparation of Low-Speed Supernatants

1. Centrifuge the homogenate (total volume ~1.2 mL) for 10 min at 600g in a benchtop centrifuge to pellet the nuclei and cell debris.
2. Remove carefully the post-nuclear supernatant (PNS), transfer it to another 15-mL tube and centrifuge for 10 min at 3000g to pellet heavy mitochondria and lysosomes. Transfer the 3000g supernatant (volume ~1 mL) into a 1.5-mL tube and keep it on ice. This is the sample material for velocity sedimentation.

3.3.3. Velocity Sedimentation Gradients

1. Prepare the iodixanol step gradients just before use to avoid linearization of the gradients (**Fig. 2A**).
2. Prepare 25, 22.5, 20, 17.5, 15, 12.5, 10, 7.5, and 5% iodixanol solutions by mixing appropriate volumes of the iodixanol working solution (50%) and HS (without

PIs) and keep them on ice. Making 5 mL of each solution is enough for the preparation of four gradients. Using a 2-mL plastic or glass pipet prepare the step gradients by carefully ("finger control") pipeting 1.2 mL of each of the solutions into SW41 UltraClear™ ultracentrifuge tubes (**Fig. 2A**). To avoid mixing of the layers keep the tube at a 45-degree angle and pipet the solutions dropwise along the inside wall of the tubes.
3. Pipet carefully 0.5 mL of the 3,000g supernatant on top of the gradient. If necessary, balance the gradients (0.1 g accuracy) by adding sample (if any remains left) or drops of HS supplemented with PIs.
4. Centrifuge the gradient for 30 min at 126,000g for 30 (maximum acceleration, brake set at slow) in a Beckman LE-80 K ultracentrifuge using the SW41 rotor.
5. After centrifugation, seven bands are visible in the gradient (**Fig. 2A**).
6. Collect 18 fractions (0.5 mL) from top of the gradients using a 1-mL micropipet.
7. Dilute the fractions with HS to give a final iodixanol concentration of less than 5%.
8. Concentrate the membranes by centrifugation for 90 min at 100,000g in a Beckman SW41 rotor.
9. Resuspend the membrane pellets carefully in 50 μL of HS supplemented with protease inhibitors (both Complete Mini™ and 1xCLAP) and freeze in aliquots at –80°C (*see* **Note 8**).

4. Notes

1. The pH of the growth medium is important for optimal binding of the cells to plastic surfaces. In wrong pH conditions the cells tend to preferentially adhere to one another, forming clumps and ending up growing in suspension. After mixing of the components, the pH of the growth medium is acidic (yellowish) and should be adjusted with 10 to 15 drops of 1 M NaOH giving it pinkish color (pH ~7.0). Also, the source and batch of the horse serum should be tested for optimal cell attachment. Note that some sera may induce spontaneous neurite formation.
2. To obtain optimal breakage of cells during homogenization, it is preferable to use freshly passaged cells. For example, using approx 1-d-old cells passaged in the previous morning is suitable as a standard. With time, the cells deposit increasing amounts of extracellular matrix and adhere more strongly to their substrate. Therefore, they become more resistant to mechanical detachment, as well as breakage during homogenization.
3. Many antibodies, particularly monoclonal, but even polyclonal, fail to recognize their corresponding, native antigens in PFA-fixed cells. The antigens in question may be part of large protein complexes and thus remain shielded. We have successfully used guanidine-HCl, which by disintegrating such complexes seems to render many initially nonreactive antibodies reactive in immunocytochemistry (*see* **ref. 22**). Simply, an additional step is included in the staining protocol: After 15 min incubation in BB (**step 3**), BB is removed by suction and the cells are treated with 6 M guanidine-HCl, 50 mM Tris-HCl, pH 7.5. Add 50 μL of this solution on the cover slip, incubate for 3 min and wash excessively (five times)

with WB-saponin to remove all guanidine-HCl. Continue permeabilization and blocking in BB for another 15 μmin, before adding the primary antibodies.
4. If double staining with two primary (e.g., rabbit and mouse) antibodies is carried out, repeat **steps 4** to **7**. In this case, for example, FITC-coupled goat anti-rabbit F(ab)$_2$-fragments and Texas Red-coupled goat anti-mouse F(ab)$_2$-fragments can be used as the secondary antibodies for the two successive staining steps. Alternatively, if any crossreactivities between them can be completely ruled out, the two primary and the two secondary antibodies can be combined in successive incubations, shortening the staining time by half.
5. If release of cells is a problem, detach the cells already in their growth medium and carry out all the washing steps by pelleting and resuspension of the cells.
6. Extensive washing of the cells is required to remove all medium components and to obtain reproducible results. When moving from one washing step to another, it is important to remove the previous solution completely. Turn the tubes upside down and remove all liquid from the inside wall of the tubes using clean tissue paper. This has to be none swiftly, before the cell pellet loses its consistency.
7. It is crucial that the homogenization conditions are standardized as far as possible. Most obvious factors that give variability include: (a) the composition of the solutions used, (b) the age and general condition of the cells affecting (e.g., the efficiency of their resuspension prior to homogenization), (c) washing of the cells, (d) the time between addition of HS and the initiation of homogenization (we use 5 min), and (e) the manual performance of the homogenization itself. Optimally, the cells should pass through the clearance of the homogenizer one by one. If resuspension of the cells is not adequate and cell clumps remain (or are generated during homogenization), more resistance can be felt during the homogenization. However, although the appearance of the nuclei in the homogenates can vary considerably between experiments, complete preservation of rounded structure is not a must. The nuclei can appear distorted without any concomitant release of DNA and can still be effectively pelleted during the centrifugation.
8. Iodixanol gradients have been previously used for the separation of ER, Golgi, and plasma membranes from Vero cells *(20)*. In addition to their lower viscosity relative to sucrose or glycerol gradients, iodixanol offers the advantage that gradients can be prepared that are isotonic thoughout their length, reducing the osmotic effects on cellular components. This is of importance regarding the stability of protein complexes that bind to the cytoplasmic surface of membranes. Our previous analytical differential centrigugation *(23)* and live cell-imaging studies *(18)* demonstrated the size hereerogeneity of the IC elements. Therefore, velocity sedimentation that fractionate components based on their size and shape was a method of choice for the separation of the vacuolar, tubular, and vesicular elements of the IC. Indeed, these gradients can resolve the two IC markers p58 and Rab1 (**Fig. 2B**), which are preferentially enriched in the vacuolar and tubular domains of the IC, respectively *(18)*, but display largely overlapping distributions in sucrose density gradients (R.S. and J.S., unpublished data). We have applied this method succesfully to analyze the distributions of other proteins residing in the ER–IC–

Golgi system (B.B.H. and J.S., unpublished data). Moreover, it may be applicable for the separation of vacuolar and tubular domains of endosomes. Finally, combination of this velocity sedimentation step with density gradient centrifugation or immunoisolation makes it possible to prepare purified membrane fractions that can be subjected to comprehensive compositional analysis.

Acknowledgments

The authors are grateful to Eivind Rødahl (Haukeland University Hospital, Bergen) for the kind gift of the PC12 cell subclone and Bruno Goud (Institut Curie, Paris) for providing the antiserum against Rab1. This work was financially supported by The University of Bergen, The Norwegian Cancer Society and The Functional Genomics (FUGE) Program of the Research Council of Norway.

References

1. Rothblatt, J., Novick, P., and Stevens. T.H. (1994) *Guidebook to the secretory pathway*. Oxford University Press, New York, NY.
2. Schekman, R.W. (1994) Regulation of membrane traffic in the secretory pathway. *Harvey Lect*. **90**, 41–57.
3. Simpson, J.C., Neubrand, V.E., Wiemann, S., and Pepperkok, R. (2001) Illuminating the human genome. *Histochem. Cell Biol*. **115**, 23–29.
4. Gilchrist, A., Au, C.E., Hiding, J., et al. (2006) Quantitative proteomics analysis of the secretory pathway. *Cell* **127**, 1265–1281.
5. Mogelsvang, S. and Howell, K.E. (2006) Global approaches to study Golgi function. *Curr. Opin. Cell Biol*. **18**, 438–443.
6. Pepperkok, R. and Ellenberg, J. (2006) High-throughput fluorescence microscopy for systems biology. *Nat. Rev. Mol. Cell Biol*. **7**, 690–696.
7. Griffiths, G., Parton, R.G., Lucocq, J., et al. (1993) The immunofluorescent era of membrane traffic. Trends Cell Biol. **3**, 214–219.
8. Kuismanen, E. and Saraste, J. (1989) Low temperature-induced transport blocks as tools to manipulate membrane traffic. *Methods Cell Biol*. **32**, 257–274.
9. Klausner, R.D., Donaldson, J.G., and Lippincott-Schwartz, J. (1992) Brefeldin A: insights into the control of membrane traffic and organelle structure. *J. Cell Biol*. **116**, 1071–1080.
10. Thyberg, J. and Moskalewski, S. (1999) Role of microtubules in the organization of the Golgi complex. *Exp.Cell Res*. **246**, 263–279.
11. Rodriguez-Boulan, E. and Powell, S.K. (1992) Polarity of epithelial and neuronal cells. *Annu. Rev. Cell Biol*. **8**, 395–427.
12. Simons, K., Dupree, P., Fiedler, K., et al. (1992) Biogenesis of cell-surface polarity in epithelial cells and neurons. *Cold Spring Harb. Symp. Quant. Biol*. **57**, 611–619.

13. Horton, A.C. and Ehlers, M.D. (2004) Secretory trafficking in neuronal dendrites. *Nat. Cell Biol.* **6**, 585–591.
14. Sabatini, D.D. (2005) In awe of subcellular complexity: 50 years of trespassing boundaries within the cell. *Annu. Rev. Cell Dev. Biol.* **21**, 1–33.
15. Tooze, S,A. (1998) Biogenesis of secretory granules in the trans-Golgi network of neuroendocrine and endocrine cells. *Biochim. Biophys. Acta.* **1404**, 231–244.
16. Martin, T.F. and Grishanin, R.N. (2003) PC12 cells as a model for studies of regulated secretion in neuronal and endocrine cells. *Methods Cell Biol.* **71**, 267–286.
17. Chanat, E., Dittie, A.S., and Tooze, S.A. (1998) Analysis of the sorting of secretory proteins to the regulated secretory pathway. A subcellular fractionation approach. *Methods Mol Biol.* **88**, 285–324.
18. Greene, L.A. and Tischler, A.S. (1976) Establishment of a noradrenergic clonal line of rat adrenal pheochromocytoma cells which respond to nerve growth factor. *Proc. Natl. Acad. Sci. U S A.* **73**, 2424–2428.
19. Sannerud, R., Marie, M., Nizak, C., et al. (2006) Rab1 defines a novel pathway connecting the pre-Golgi intermediate compartment with the cell periphery. *Mol. Biol. Cell* **17**, 1514–1526.
20. Majoul, I. V., Bastiaens, P. and Soling, H. D. (1966) Transport of an external Lys-Asp-Glu-Leu (KDEL) protein from the plasma membrane to the endoplasmic reticulum: studies with cholera toxin in Vero cells. *J. Cell Biol.* **133**, 777–89.
21. Balch, W.E. and Rothman, J.E. (1985) Characterization of protein transport between successive compartments of the Golgi apparatus: asymmetric properties of donor and acceptor activities in a cell-free system. *Arch. Biochem. Biophys.* **240**, 413–425.
22. Peränen, J., Rikkonen, M. and Kääriäinen, L. (1993) A method for exposing hidden antigenic sites in paraformaldehyde-fixed cultured cells applied to initially unreactive antibodies. *J. Histochem. Cytochem.* **41**, 447–454.
23. Ying, M., Flatmark, T. and Saraste, J. (2000) The p58-positive pre-Golgi intermediates consist of distinct subpopulations of particles that show differential binding of COPI and COPII coats and contain vacuolar H+-ATPase. *J. Cell Sci.* **113**, 3623–3638.

20

Tracking the Transport of E-Cadherin to and From the Plasma Membrane

Matthew P. Wagoner, Kun Ling, and Richard A. Anderson

Summary

The epithelial to mesenchymal transition (EMT) is the breakdown of epithelial cell morphology that gives way to a more mobile, mesenchymal phenotype. Although this process is fundamental to the development of multicellular organisms, it is also a key occurrence in many diseases, including cancers of epithelial origin E-cadherin is a central component of adherens junctions (AJs), which act as structural and signaling hubs in epithelial cells that oppose EMT. The loss of E-cadherin from the plasma membrane is an early indication of EMT and a marker of poor prognosis in many cancers making the trafficking of E-cadherin an area of great interest. Recent work from the authors' laboratory has established the role of type Iγ phosphatidylinositol 4-phosphate 5-kinase (PIPKIγ) in the trafficking of E-cadherin by studying the surface accessibility of E-cadherin in endocytosis and recycling assays. Additionally, immunofluorescence data demonstrated that cells lacking PIPKIγ lost E-cadherin at the plasma membrane. The biochemical and microscopic techniques used to investigate the trafficking of E-cadherin are presented herein.

Key words: E-cadherin; endocytosis; PIPKIγ; plasma membrane targeting; recycling; transport; type Iγ phosphatidylinositol 4-phosphate 5-kinase trafficking.

1. Introduction

Adherens junctions (AJs) are integral components of epithelia that serve as both structural and signaling centers for the cell *(1,2)*. Cell–cell contacts formed by AJs maintain the polarity and function of epithelial cells, and the loss of AJs is recognized as an early event in the epithelial to mesenchymal transition (EMT; *3,4*), which is a hallmark of many epithelial cancers *(5–8)*. Cadherins are

Ca^{2+}-dependent homodimeric cell–cell adhesion receptors that are central to the structural stability and signaling capability of AJs. Cadherins interact with the p120-, β- and α-catenins. These catenins serve not only as scaffolds between these intercellular receptors and the cytoskeleton but also as immediate effectors governing actin dynamics, AJ stability and the Wnt, MAPK, NFκb, and sonic hedgehog pathways *(5,9,10)*. The loss of E-cadherin from the plasma membrane is an indicator of poor prognosis in cancers of epithelial origin, which makes its mechanisms of assembly and disassembly into AJs an area of considerable interest *(1)*.

Recent work in our lab has determined that type Iγ phosphatidylinositol 4-phosphate 5-kinase (PIPKIγ) regulates the transport of E-cadherin to and from the cell surface and thus the formation and disassembly of AJs (11). To study the role of PIPKIγ on the transport of E-cadherin to and from the plasma membrane, the authors' laboratory used biochemical and microscopic means to label and visualize both cell surface and internalized receptors *(11,12)*. The methods used to track E-cadherin assembly and endocytosis are described herein.

2. Materials

2.1. Cell Culture and Lysis

1. Madin-Darby Canine Kidney (MDCK) growth media: Dulbecco's Modified Eagle's Medium (DMEM; Cellgro) supplemented with 10% fetal bovine serum (FBS; Gibco), stored at 4°C.
2. The trypsin (0.25%) ethylenediamine-tetraacetic acid (EDTA; 2.2 mM) solution was in Hank's Buffered Salt Solution (HBSS) without sodium bicarbonate, calcium, or magnesium (Cellgro) stored in aliquots at –20°C for long-term storage, and 4°C for short-term (a few weeks) storage.
3. MDCK cells were cultured in a 24-mm transwell with 3 μm pore and a polycarbonate membrane insert from Corning.
4. Phosphate-buffered saline (PBS) solution is 137 mM NaCl, 2.7 mM KCl 10 mM Na_2HPO_4 and 1.8 mM KH_2PO_4, adjusted to pH 7.4 with HCl. It may be stored at room temperature as an autoclaved 10x stock and diluted as needed.
5. Lysis buffer: 50 mM Tris-HCl, pH 7.5, 150 mM NaCl, 0.5% NP-40, 1 mM EDTA, 1 mM phenylmethanesulphonyl fluoride (PMSF), and 10% glycerol with an EDTA-free protease inhibitor (PI) tablet (Roche). The PI tablets are stored at 4°C.
6. Sample buffer (5X): 60 mM Tris-HCl, pH 6.8, 25% glycerol, 2% sodium dodecyl sulfate (SDS), 70 mM β-mercaptoethanol and 0.03% bromophenol blue stored in aliquots at –20°C.
7. Disposable cell lifter (Fisher).

2.2. SDS-Polyacrylamide Gel Electrophoresis (SDS-PAGE)

1. The gel system we use is a Mini PROTEAN 3 Cell electrophoresis system from Bio-Rad.

2. Buffer A: 3 M Tris-HCl, pH 8.9 stored at room temperature.
3. Buffer B: 1 M Tris-HCl, pH 6.7 stored at room temperature.
4. Molecular-weight marker: Pageruler Prestained Protein Marker (Fermentas) stored at –20°C.
5. Running buffer (10X): 200 mM Tris, 1.92 M glycine, and 1% SDS in water and is stored at room temperature. Dilute to 1X in water.
6. Ammonium persulfate (APS) solution (10%) is made in water and may be stored at 4°C for weeks at a time.
7. Acrylamide/*bis* solution, (30%) 29:1 (3.3%C) is purchased from Bio-Rad and stored at 4°C.
8. *N,N,N′,N′*-Tetramethylethylenediamine (TEMED) is purchased from FisherBiotech and stored at 4°C.
9. Needle (27.5 gage) and 1-mL syringe (Becton Dickinson) may both be stored at room temperature.

2.3. Western Blotting

1. PBS-T is phosphate-buffered saline (PBS) with 0.1% Tween-20 (Sigma-Aldrich) which may be stored at room temperature. Check the solution regularly for bacterial growth, which will appear as insoluble floc.
2. Blocking buffer (BB) is 5% (w/v) nonfat dry milk in PBS-T (LabScientific). The dry milk is stored at room temperature, but the BB is stored at 4°C.
3. Transfer buffer (TB): 20 mM Tris-HCl, 200 mM glycine, 10% methanol in water and is stored at 4°C.
4. Purified mouse anti E-cadherin antibody is available from BD Transduction laboratories and is stored at –20°C.
5. Peroxidase-conjugated AffiniPure Donkey Anti-mouse immunoglobulin (I_gG) is available from Jackson ImmunoResearch and is stored at –20°C.
6. SuperSignal west pico chemiluminescent substrate is available from Pierce and is stored at room temperature.
7. Polyvinylidene fluoride (PVDF) is available from Millipore and is stored at room temperature.
8. Chromatography paper, 3MM, or Whatmann paper, is available from Whatmann (Maidstone).
9. Film: x-Ray Film from Research Products International.
10. NIH ImageJ version 1.36b is available from http://rsb.info.nih.gov/ij/.

2.4. Immunostaining

1. Corning Micro Slides (Plain) were purchased from Corning Glass Works.
2. Cover slips (22 × 22 × 0.15 mm) were purchased from Fisher.
3. Blocking solution: 3% bovine serum albumin (BSA) in PBS.

4. Paraformaldehyde (PFM)/PBS solution is made as an 8% solution in PBS and stored frozen at −20°C. Thaw as needed and dilute to 4% in PBS. Insoluble precipitate may be resuspended in a 60°C water bath, but be careful to allow the solution to return to room temperature before use.
5. The buffer used to permeabilize the cells is PBS with 0.2% Triton X-100.
6. Alexa Fluor 488 goat anti-mouse IgG (2 mg/mL stock; Molecular Probes) may be diluted 1:400 in blocking solution to 50 μg/mL.
7. 4,5-Diamidino-2-phenylindole (DAPI) is stored as a 10-mg/mL solution in water at 4°C, and may be diluted as needed.
8. The mounting medium used is Vectashield from Vector labs.

2.5. Reversible Biotinylation and Endocytosis

1. Sulfo-N-hydroxysulfosuccinimide (NHS) SS-biotin (Pierce) is diluted to 1 mg/mL in 4°C PBS shortly before use. Solid Sulfo-NHS SS-biotin is sensitive to moisture and should be stored at −20°C in a desiccator. When removing the compound from the freezer, remove the entire desiccator and allow it to come to room temperature before removing and opening the biotin. This will prevent condensation from forming on the cold Sulfo-NHS SS-biotin powder when the container is opened.
2. Sulfo-NHS-SS-biotin blocking reagent is 50 mM NH_4Cl in PBS with 1 mM $MgCl_2$ and 0.1 mM $CaCl_2$.
3. HBSS solution (Cellgro) with 0.5 mM ethyleneglycol-tetraacetic acid (EGTA) is stored at room temperature. The EGTA should be sterile filtered (0.2 μm from Nalgene) before adding to the HBSS.
4. Glutathione solution should be made with fresh glutathione (60 mM) and 0.83 M NaCl, 0.83 M NaOH and 1% BSA which may all be diluted from stock solutions.

3. Methods

E-cadherin trafficking to and from the plasma membrane may be tracked using biochemical and microscopic means, three methods of which are described in detail here. To follow the internalization of E-cadherin, first the surface accessible protein must be biotinylated, allowing those proteins to be selectively precipitated using streptavidin-coated beads. Endocytosis of E-cadherin is induced by using EGTA to strip the cellular environment of its Ca^{2+}, which cadherins require for their homotypic interaction. As the biotinylated E-cadherin is internalized, it is protected from the glutathione washes, which strip the disulfide linked biotin from the sulfo-NHS group that links it to any surface accessible E-cadherin. When the cell lysate is probed with streptavidin beads, only the E-cadherin that was internalized at the time of the glutathione wash will be precipitated.

Quantifying the amount of internalized E-cadherin at various time points against the level of total E-cadherin biotinylated will allow you to determine the effect of your protocol on the rate of E-cadherin endocytosis. The rate of E-cadherin recycling is determined by inducing endocytosis of E-cadherin, and biotinylating it as it returns to the plasma membrane following the calcium rescue.

Visualizing the intracellular localization of E-cadherin is equally important for understanding its endocytosis and recycling. This may be accomplished using indirect immunofluorescence, which can unveil a wealth of information beyond what the biotinylation data can provide. Co-staining of E-cadherin with your protein of interest can help determine where in the cell two proteins may be co-localizing and at what stage in the endocytic process.

3.1. Biotin Conjugated E-Cadherin Endocytosis Assay

1. MDCK cells are grown to confluence in 24-mm diameter transwells (Corning) while the protein of interest is expressed at appropriate levels according to your protocol.
2. Remove the media from the cells and wash them once with cold PBS. Treat the apical and basal surfaces of the cells with 1 mg/mL sulfo-NHS SS-biotin in cold PBS at 4°C for 30 to 60 min to biotinylate the exposed cell-surface proteins. Quench the free sulfo-NHS-SS-biotin by washing the cells in sulfo-NHS-SS-biotin blocking reagent twice for 5 min. This is followed by several washes with 4°C PBS.
3. Induce endocytosis of E-cadherin by replacing the media in the transwell with HBSS solution (Cellgro) with 0.5 mM EGTA for 0, 15, 30, 60, 90, and 120 min at 18°C (*see* **Note 2**).
4. Incubate the cells from each time point in two 20-min washes of glutathione solution at 4°C to remove the biotin conjugated to all proteins still at the cell surface (*see* **Notes 3** and **4**). One well from each condition used in your protocol should be excluded from this wash to be used as a positive control for total cell-surface biotinylated E-cadherin.
5. Lyse the cells in 500 μL lysis buffer containing PIs (and no reducing agents) and scrape them from the transwell using a disposable cell lifter (Fisher). Transfer the lysate to a 1.5 mL low-retention microcentrifuge tube and rotate the tube at 4°C for 1 h. Next, clear the lysate of insoluble debris by centrifuging at 16,000g (in an 18-place rotor Eppendorf 5415C) for 15 min and aspirating the supernatant into a fresh 1.5-mL centrifuge tube. Save 20 μL of the cleared lysate for each condition in a separate tube as an input sample.
6. While the lysates are clearing, equilibrate the streptavidin affinity gel (EZview Red streptavidin affinity gel from Sigma) by briefly vortexing the gel in 750 μL of lysis buffer and spinning it down for 1 min at 4,500g (in an Eppendorf 5415C). Repeat the above wash. For each affinity pull-down, use approx 10 μL settled bead volume of gel, the transfer of which may be made easier by using a cut pipet tip (*see* **Notes 5** and **6**). Aspirate the wash lysates and resuspend the gel with an equal volume of lysis buffer, creating a gel slurry.

7. Add 20 μL of the slurry to the cleared lysates and allow them to rotate together for 4 h to precipitate the biotin-labeled protein.
8. Spin-down the tubes for 1 min at 4,500g and aspirate the supernatant. Resuspend the gel in 1 mL of lysis buffer before spinning again. Repeat this wash three to four times.
9. Aspirate the final wash solution and dry the beads with a 27.5-gage needle and a 1-mL syringe (Becton Dickinson). Resuspend the gel in 60 μL of sample buffer and heat the tubes to 100°C for 10 min.

3.2. Biotin-Conjugated E-Cadherin Recycling Assay

1. MDCK stable cell lines are grown to confluence in 24-mm diameter transwells (Corning) while the protein of interest is expressed at appropriate levels according to the institution's protocol.
2. The endocytosis of E-cadherin is induced by changing the MDCK media to HBSS with 2 mM EGTA for 40 min at 37°C.
3. The E-cadherin is recycled back to the plasma membrane upon calcium rescue, which is achieved by changing the media back to DMEM supplemented with 10% FBS for 0, 5, 15, 30, 45, or 60 min.
4. At the times indicated, wash the cells twice with PBS at 4°C then add 1 mg/mL sulfo-NHS SS-biotin PBS for 60 min at 4°C to biotinylate the exposed cell-surface proteins. Quench the free Sulfo-NHS-SS-biotin by washing the cells in Sulfo-NHS-SS-biotin blocking reagent twice for 5 min. This is followed by several washes with PBS at 4°C.
5. The cells are lysed and biotin-labeled proteins precipitated as described in **steps 5** to **9** of the endocytosis assay, and analyzed as described here.

3.3. Western Blotting for E-Cadherin

3.3.1. Pouring the Gels

1. These instructions are based on the Mini PROTEAN 3 Cell system from Bio-Rad, and are easily adaptable to other gel systems.
2. The glass plates should be thoroughly washed with a water soluble detergent such as Alconox (Alconox Inc.) and rinsed clean with distilled water to avoid the buildup of impurities. Before pouring the gels the plates should be cleaned again with 95% ethanol and a Kimwipe (Kimberly-Clark).
3. A 7.5% gel of 1.5-mm thickness may be prepared with 5 mL of water, 1 mL of Buffer A, 2 mL 30% acrylamide/*bis* solution, 29:1, 80 μL of a 10% SDS solution, 80 μL of a 10% ammonium sulfate solution, and 5 μL TEMED. Once poured, the gel should be covered with a small amount of 95% ethanol (*see* **Notes 7–9**) *(13)*.
4. After the resolving gel has solidified (∼15 min) the ethanol should be poured off, and any remaining ethanol should be removed with a clean piece of 3 MM chromatography paper (Whatmann paper). Next, a 5% stacking gel may be prepared

with 1.4 mL water, 250 µL Buffer B, 330 µL acrylamide (30%), 20 µL ammonium sulfate (10%), 20 µL SDS (20%), 2 µL TEMED. The gel comb should be inserted immediately after the stacking gel is poured on top of the resolving gel. Once the gel has solidified the comb may be removed and the wells should be rinsed with distilled water.

3.3.2. Running the Gels

1. Load the gels into the electrophoresis module and fill the chamber and the tank with 500 mL running buffer.
2. Load a prestained molecular-weight marker onto the far left lane of the gel, followed by 20 µL of each sample into the SDS-PAGE gel along with a prestained molecular-weight marker. Run the gel at 100 V for 20 min, or until the dye front has passed through the stacking gel. The gel may now be run at 200 V until the dye front enters the running buffer.

3.3.3. Transferring the Protein to a Membrane

1. After the dye front has left the gel, remove the lid from the gel box and rinse all of the components in deionized water. In a tub of transfer buffer, submerge two sponges and two pieces of Whatmann paper per gel to be transferred. Force out any pockets of air in the sponges or the paper. While assembling the transfer sandwich it is important to make sure no bubbles are resting between layers. Place an open transfer cassette (black side down) in the tub with the buffer and layer a sponge and a piece of 3 MM paper in the cassette. Remove a gel from its glass plates and place it face down on the submerged Whatmann paper.
2. Designate a corner of the nitrocellulose membrane as the upper left-hand corner by cutting it off with clean scissors. Only handle the membrane with tweezers or forceps to prevent foreign proteins from being deposited on the membrane. Pretreat the PVDF by rinsing it in methanol for 30 s. Remove the membrane from the methanol and wash off the excess methanol in the TB. Place the membrane on the submerged gel face down so that the cut upper left-hand corner of the membrane is on your right. Layer the second wet piece of Whatmann paper on the membrane, followed by the second sponge. Squeeze out any air bubbles that may have developed between layers while you were assembling the sandwich.
3. Place the transfer cassettes in the transfer assembly in the proper orientation (black to black) along with an ice block and a stir bar to help maintain the temperature of the TB. Place the assembled transfer apparatus in a cold room (4°C) on a magnetic stir plate and set the plate to spin at a low speed. Set the power supply to transfer at 100 V for 90 min (*see* **Note 10**).

3.3.4. Blotting for E-Cadherin

1. Once the transfer is complete, the transfer sandwich may be opened, and excess portions of the membrane may be trimmed off. Move this membrane into a

blotting container containing enough PBS-T (PBS with 0.1% Tween-20) to submerge the blots. Set the box on a shaker or rocker, and allow it to wash for 5 min.
2. Replace the PBS-T with 5% milk PBS-T and block the membrane for 45 min. Replace the blocking buffer with 5% milk PBS-T with mouse anti-E-cadherin antibody (BD Transduction Laboratories) diluted 1:2,500 to a final concentration of 0.1 μg/mL and blot for 1 h on the rocker at room temperature.
3. Wash the blot in PBS-T three times for 5 min each on the rocker.
4. Decant the wash solution and add more milk-PBS-T with peroxidase-conjugated AffiniPure Donkey Anti-mouse IgG at 80 ng/mL at room temperature for 30 min.
5. Wash the blot in PBS-T three times for 5 min each on the rocker.
6. Lay the blots on Parafilm "M" (Pechiney Plastic Packaging) face up (cut corner on the upper left-hand side) and expose the blot to SuperSignal west pico chemiluminescent substrate (Pierce) for 5 min or another enhanced chemiluminescent reagent for an appropriate length of time. Next, place the membrane between two pieces of transparency film in a cassette and expose the blot to x-ray film in a dark room. Commonly the exposure times for this experiment will range between 10 and 60 s. Examples of blots assaying the endocytosis and recycling of E-cadherin are shown in **Figs. 1** and **2**.
7. These results may be quantified to determine the percentage of E-cadherin internalization by scanning densitometry of the films. The densitometry may be determined using programs such as ImageJ (NIH), allowing for the quantitative comparison of the E-cadherin in the glutathione-washed time points to the lane containing the E-cadherin that never endured the glutathione wash.

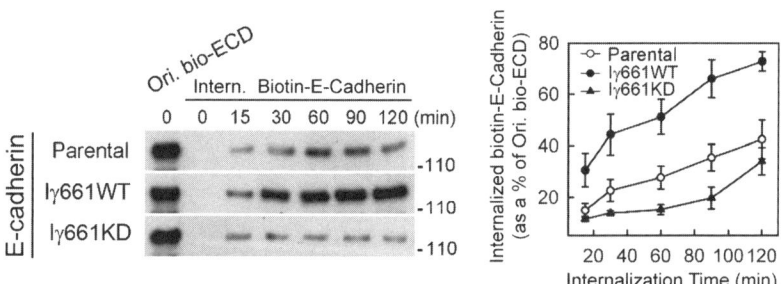

Fig. 1. Overexpression of wild-type PIPKIγ661 accelerates E-cadherin internalization, while expression of kinase dead (KD) PIPKIγ661 impedes it *(11)*. MDCK cells expressing PIPKIγ661 wild-type and kinase dead were biotinylated (Ori. Bio-ECD) with sulfo-NHS-SS-biotin followed by EGTA treatment to induce endocytosis. The cells were washed with glutathione to remove the disulfide linked biotin from cell surface accessible E-cadherin at intervals to monitor the rate of internalization. The cells were processed to determine total levels of biotinylated E-cadherin by western blotting. The results of the blotting were quantified (right) by densitometry using NIH ImageJ. The expression of active PIPKIγ661 clearly enhances the rapid endocytosis of E-cadherin while the kinase dead PIPKIγ661 does not.

Tracking the Transport of E-Cadherin

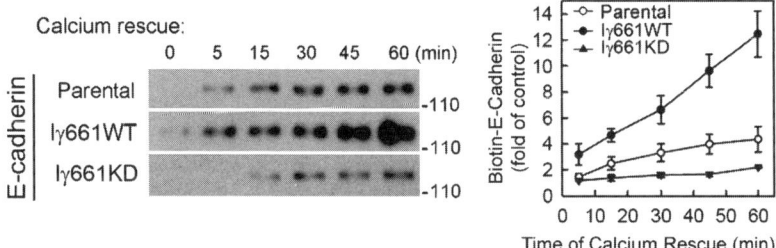

Fig. 2. Expression of PIPKIγ661 wild-type, but not kinase dead, accelerates E-cadherin recycling to the plasma membrane *(11)*. MDCK cells expressing PIPKIγ661 wild type and kinase dead were induced to internalize E-cadherin by EGTA treatment. E-cadherin recycling was then stimulated with calcium (return to normal medium) followed by cell surface biotinylation at discrete time points. The cells were then processed to determine total levels of biotin labeled E-cadherin by western blotting. The results of the blotting were quantified using scanning densitometry and NIH Image J. These results, taken with those in Fig. 1, support a model for PIPKIγ661 playing a key role in E-cadherin endosomal trafficking to and from the plasma membrane.

3.4. Immunofluorescent Staining of MDCK Cells for E-Cadherin

1. Culture MDCK cells on 22-mm cover slips in a 6-well plate. Throughout this experiment, coverslips should always be handled with very fine-tipped forceps to avoid altering the cells on the slide. To seed the cells in the 6-well plate, first sterilize the cover slips by submerging them in ethanol, which should be flamed before placing the slip on the plate. The cells should be grown to confluence before transfection.
2. Treat the cells according to your protocol and then proceed with the fixation and staining steps detailed here.
3. Rinse the cover slips by transferring them with tweezers to coplin jars containing 37°C PBS. Rinse the cover slips a second time in a second coplin jar containing fresh 37°C PBS.
4. Transfer the cover slips to another coplin jar containing a 4% PMF/PBS solution for 15 min at room temperature.
5. Rinse the cover slips twice as before with room temperature PBS.
6. Extract the cover slips by transferring them to a coplin jar with PBS plus 0.2% Triton X-100 for 15 min.
7. Rinse the cover slips twice in room temperature PBS as before, then incubate them in a third coplin jar of PBS for 5 min.
8. Block the cells in 3% BSA/PBS (blocking solution) solution for 1 h at room temperature in a coplin jar.
9. Dilute the mouse anti-E-cadherin antibody (250 μg/mL from BD Transduction Labs) 1:400 to a final concentration of 625 ng/mL in blocking solution. If you are also staining for other proteins, add additional primary antibodies to the same tube. Place the cover slips face-up on a hydrophobic surface in a humidity chamber and

add 200 μL of the primary antibody solution to each cover slip. Incubate at 37°C for 1 h.
10. Remove the cover slips from the humidifier and wash them in room temperature PBS with 0.1% Triton X-100 three times for 5 min.
11. Dilute the Alexa Fluor 488 goat anti-mouse IgG (2 mg/mL stock) 1:400 in blocking solution to 50 μg/mL. Place the cover slips face-up in the humidifier and add 200 μL of the secondary antibody solution to each cover slip. Incubate the cover slips in the humidifier for 30 min at 37°C (*see* **Note 11**).
12. After 30 min add 50 μL of 10 μg/mL DAPI to the secondary antibody solution on each cover slip for 2 min (optional).
13. Remove the cover slips from the humidifier and wash them in room temperature PBS with 0.1% Triton X-100 three times for 5 min.
14. Dry the cover slips by tapping the edges on a Kimwipe. Place a drop of vectashield mounting media (around 20 μL) on the slide and gently lower the cover slip onto the slide, making sure there are no bubbles between the two.
15. Place the slide face down on a large Kimwipe and apply firm, even pressure to the back of the slide, using a second Kimwipe to prevent your gloves from smudging the glass. This will press off any excess mounting media between the cover slip and the slide.
16. Using clear nail polish, seal the edges of the cover slip to the slide.
17. The slides may then be viewed by indirect immunofluorescence with emission and absorption spectra appropriate for the utilized fluorophores. Examples of the observed staining phenotype may be seen in **Fig. 3**.

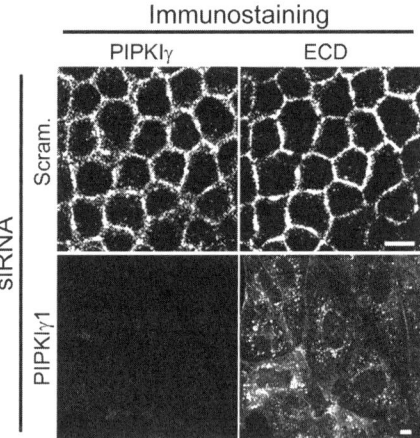

Fig. 3. PIPKIγ knockdown results in the mislocalization of E-cadherin in MDCK cells *(11)*. PIPKIγ was knocked down by the addition of PIPKIγ siRNA via the calcium phosphate precipitation method. Immunofluorescent staining for E-cadherin (ECD) indicates that it colocalizes with PIPKIγ. Staining of the knockdown cells indicate that loss of the kinase results in the mistargeting of E-cadherin. These data support the hypothesis that PIPKIγ is an important mediator of E-cadherin trafficking to the plasma membrane in epithelial cells.

4. Notes

1. It is recommended that only distilled, deionized water be used as a reagent in the assays described herein. Water that is to be used with cell culture work must be autoclaved for 40 min at 121°C 15 psi. All other reagents to be used in cell culture must be either autoclaved or sterile filtered through a 0.2-μm filter (Nalgene).
2. Keeping the cells at 18°C prevents the endocytosed E-cadherin from entering the lysozomal degradative pathway.
3. Sulfo-NHS SS-biotin is impermeable to cells and will therefore only label the proteins on the apical surface of the cell. Because of the disulfide bond linking the biotin to the labeled proteins, the biotin will be released from any proteins at the cell surface upon exposure to a reducing agent, such as the glutathione.
4. The glutathione solution should be made fresh before each use. The NaOH, NaCl, and BSA may all be added from concentrated stocks, but the glutathione itself should be resuspended before each use to ensure its effectiveness.
5. The binding affinity of EZview Red Streptavidin affinity gel is approx 10 μg of biotin per milliliter of settled gel. The 10 μL of affinity gel is therefore sufficient to bind 100 ng of biotin, which should be adequate to precipitate all of the biotin-labeled protein.
6. Beads should be pipeted using a pipet tip that is either of a naturally wide orifice or one that has had its most distal 5 mm cut off. This will allow the beads to flow more easily in and out of the tip. Because the beads tend to stick to the pipet tip, care should be taken to minimize the amount of contact they have with the tips.
7. Layering ethanol onto the freshly poured resolving gel creates a smooth interface between the resolving and stacking phases. This also serves to protect the gel from ambient oxygen, which inhibits efficient polymerization of acrylamide.
8. To avoid the occurrence of bubbles beneath and between the teeth of the gel comb, it should be washed and rinsed similar to the way the gel plates were. When inserting the comb, it should be lowered between the plates at an angle so that the rightmost tooth is dragged across the length of the unpolymerized gel as it is displacing it. This will also help to prevent the occurrence of bubbles.
9. To ensure the wells are properly formed, more than enough unpolymerized stacking gel should be added to the gel plate. Adding the comb may cause the excess gel to overflow the plates. Care should be taken when inserting the comb so that this excess unpolymerized acrylamide does not squirt out of the plates onto bare skin, as unpolymerized acrylamide is a skin-permeable neurotoxin.
10. For more information on casting gels of various volumes and densities **ref. 13** is an excellent resource for this and other areas requiring technical expertise.
11. Keeping the transfer cool lowers the resistance to current provided by the buffer. The resistance can be a problem as the buffer heats up and the amperage demands on the power supply increase to maintain the same voltage. If a cold room is not available, transfers may be performed on a bench-top provided they are kept cool by being packed in ice, or by some other means.
12. We have found the Alexa flurophore-conjugated secondary antibodies to be exceptionally effective and photostable *(14)*.

Acknowledgments

The authors would like to thank Nick Schill and Professor Richard Anderson for their editing, advice, and encouragement. This work was supported by R01 GM057549-08 and R01 CA104708-03.

References

1. Perez-Moreno, M. and Fuchs, E. (2006) Catenins: keeping cells from getting their signals crossed. *Dev. Cell.* **11**, 601–612.
2. Weis, W. I. and Nelson, W. J. (2006) Re-solving the cadherin-catenin-actin conundrum. *J. Biol. Chem.* **281**, 35593–35597.
3. Thiery, J. P. (2003) Epithelial-mesenchymal transitions in development and pathologies. *Curr. Opin. Cell Biol.* **15**, 740–6.
4. Huber, M. A., Kraut, N., and Beug, H. (2005) Molecular requirements for epithelial-mesenchymal transition during tumor progression. *Curr. Opin. Cell. Biol.* **17**, 548–558.
5. Bremnes, R. M., Veve, R., Hirsch, F. R., and Franklin, W. A. (2002) The E-cadherin cell-cell adhesion complex and lung cancer invasion, metastasis, and prognosis. *Lung Cancer.* **36**, 115–124.
6. Chen, H. C., Chu, R. Y., Hsu, P. N., et al. (2003) Loss of E-cadherin expression correlates with poor differentiation and invasion into adjacent organs in gastric adenocarcinomas. *Cancer Lett.* **201**, 97–106.
7. Liu, M., Lawson, G., Delos, M., et al. (2003) Prognostic value of cell proliferation markers, tumour suppressor proteins and cell adhesion molecules in primary squamous cell carcinoma of the larynx and hypopharynx. *Eur. Arch. Otorhinolaryngol.* **260**, 28–34.
8. D' Souza-Schorey, C. (2005) Disassembling adherens junctions: breaking up is hard to do. *Trends Cell Biol.* **15**, 19–26.
9. Vasioukhin, V., Bauer, C., Degenstein, L., Wise, B. and Fuchs, E. (2001) Hyperproliferation and defects in epithelial polarity upon conditional ablation of alpha-catenin in skin. *Cell* **104**, 605–617.
10. Perez-Moreno, M., Davis, M. A., Wong, E., Pasolli, H. A., Reynolds, A. B., and Fuchs, E. (2006) p120-catenin mediates inflammatory responses in the skin. *Cell* **124**, 631–644.
11. Ling, K., Bairstow, S., Carbonara, C., Turbin, D., Huntsman, D., and Anderson, R. (2007) Type Iγ phosphatidylinositol phosphate kinase modulates adherens junction and E-cadherin trafficking via a direct interaction with μ1B adaptin. *J. Cell. Biol.* **176**, 343–353.
12. Le, T. L., Yap, A. S., and Stow, J. L. (1999) Recycling of E-cadherin: a potential mechanism for regulating cadherin dynamics. *J. Cell. Biol.* **146**, 219–232.
13. Sambrook, J. and Russell, D., (2001) *Molecular Cloning: A Laboratory Manual*. 3 ed. Cold Spring Harbor Laboratory Press, Cold Spring Harbor, NY.
14. Panchuk-Voloshina, N., Haugland, R. P., Bishop-Stewart, J., et al. (1999) Alexa dyes, a series of new fluorescent dyes that yield exceptionally bright, photostable conjugates. *J. Histochem. Cytochem.* **47**, 1179–1188.

21

Analysis of Nucleocytoplasmic Shuttling of NFκB Proteins in Human Leukocytes

Chandra C. Ghosh, Hai-Yen Vu, Tomas Mujo, and Ivana Vancurova

Summary

Controlled nucleocytoplasmic localization regulates activity of NFκB as well as other transcription factors. Analysis of the nucleocytoplasmic protein shuttling has been greatly facilitated by the use of leptomycin B (LMB), an inhibitor of CRM1-dependent nuclear export. The authors have previously shown that LMB inhibits NFκB activity in human neutrophils by increasing the nuclear accumulation of NFκB inhibitor, IκBα. In this chapter, the authors describe a protocol that uses LMB to study the nucleocytoplasmic shuttling of IκBα in human macrophage-like U937 cells, thus inhibiting NFκB activity. This protocol should be readily adaptable to analyze the nucleocytoplasmic shuttling of other proteins in human leukocytes.

Key words: IκBα; leptomycin B; leukocytes; nuclear accumulation; nuclear export; NFκB; transcription factors; Western blotting.

1. Introduction

The biological function of many transcription factors is regulated by their nuclear translocation. Proteins are transported in and out of the nucleus through nuclear pores by either passive diffusion (for proteins smaller than 30 kDa) or by an active transport. The active import of nuclear proteins is mediated by nuclear localization sequences rich in basic amino acid residues, while the nuclear export is mediated by nuclear export signals (NESs) *(1–3)*. Dysfunctional nucleocytoplasmic protein shuttling has been involved in numerous diseases *(4,5)*.

Transcription factor NFκB comprises a family of proteins that serve as crucial regulators of genes involved in host immune and inflammatory responses, proliferation, and differentiation *(6,7)*. In addition to pro-inflammatory proteins such as tumor necrosis factor (TNF)-α and interleukin (IL)-8, NFκB induces transcription of anti-apoptotic genes *(6,7)*. Increased activation of NFκB has been associated with the pathogenesis of numerous inflammatory disorders as well as with human cancers *(8–11)*. Therefore, inhibition of NFκB activity represents an important therapeutic target. In most cell types, the NFκB proteins (p50 NFκB1, p65/Rel A, c-Rel, NFκB2/p52, and Rel B) are retained in the cytoplasm through their association with the inhibitory protein IκBα. Following cell stimulation with pro-inflammatory and stress signals, IκBα is degraded, and NFκB proteins translocate to the nucleus where they bind to NFκB responsive promoters and activate transcription *(12)*. Thus, in this "classical" model of NFκB regulation, NFκB activity is regulated by the nuclear translocation of NFκB subunits. One of the first genes induced after NFκB is activated is the inhibitor *IκBα*, since the IκBα promoter contains the NFκB-responsive region *(13)*. In most cell types, the newly synthesized IκBα continuously shuttles between the nucleus and cytoplasm, however, its nuclear export seems predominant over the nuclear import, resulting in mostly cytoplasmic localization of IκBα *(14–16)*.

Leptomycin B (LMB) is a *Streptomyces* secondary metabolite with antitumor and antifungal properties *(17)*. LMB has been widely used as a potent tool to study the nucleocytoplasmic shuttling of nuclear proteins *(18,19)*. LMB inhibits nuclear protein export by binding to the nuclear export receptor CRM1, thus preventing its interaction with NES, and disrupting the nuclear protein export machinery *(18,19)*. Proteins regulated by the CRM1-dependent mechanism include transcriptional regulators such as IκB proteins and proteins of the IκB kinase complex *(14,20)*, the Stat family of transcription factors *(20–23)* and the pro-inflammatory proteins cyclooxygenase-2 and TNF-α *(24,25)*.

It has been previously shown that LMB increases nuclear IκBα accumulation in human neutrophils, thus inhibiting NFκB activity and increasing neutrophil apoptosis *(26)*. It has also been shown that LMB is a potent inhibitor of TNF-α release from stimulated human neutrophils and monocytic cells *(25)*. A protocol that uses LMB to study nucleocytoplasmic shuttling of IκBα in human macrophage-like U-937 cells is described here. LMB increases the nuclear levels of IκBα in stimulated U-937 cells, thus inhibiting NFκB activity and NFκB-dependent transcription. The main points of this protocol are as follow: (a) cell culture and incubation of U-937 cells with LMB; (b) isolation of cytoplasmic and nuclear extracts; (c) analysis of nuclear and cytoplasmic proteins by Western blotting; and (d) analysis of NFκB activity by electrophoretic mobility shift assay (EMSA). This protocol should be readily adaptable to analyze the nucleocytoplasmic shuttling of other proteins in human leukocytes.

2. Materials

2.1. Cell Culture

1. U-937 cells obtained from the American Type Culture Collection.
2. RPMI medium (Gibco) supplemented with 10% fetal bovine serum (FBS), 2 mM glutamine, 1 mM sodium pyruvate, and penicillin-streptomycin solution.
3. Phorbol 12-myristate 13-acetate (PMA, Calbiochem, 524400) is dissolved at 1 mg/mL in dimethyl sulfoxide (DMSO) and stored at –80°C.
4. Lipopolysaccharide (LPS, Sigma, L-4391) is dissolved at 1 mg/mL in sterile water and stored at –80°C.
5. LMB (Sigma, L-2913; *see* **Note 1**).
6. Phosphate-buffered saline (PBS) pH 7.2.
7. 75-cm^2 culture flasks.
8. Standard 24-well plates with clear flat bottom.
9. Teflon cell scrapers.
10. 2 mL microcentrifuge tubes.

2.2. Preparation of Cytoplasmic and Nuclear Extracts

1. Relaxation buffer (RB): 10 mM Hepes, pH 7.5, 10 mM KCl, 3 mM NaCl, 3 mM MgCl$_2$, 1 mM ethylenediamine-tetraacetic acid (EDTA), 1 mM ethyleneglycol-tetraacetic acid (EGTA; *see* **Note 2**).
2. 10% Igepal CA-630 (Sigma): In a sterile 15-mL centrifuge tube, dilute 1 mL of 100% Igepal CA-630 in 9 mL of sterile water. Mix well by inverting the tube several times. Avoid vortexing, which could cause foaming. Store at room temperature. If turbidity or sediment develop during storage, a clear liquid can be obtained on heating to 40°C (*see* **Note 3**).
3. Nuclear extract buffer (NEB): 20 mM Hepes, pH 7.5, 25% glycerol, 500 mM KCl, 1 mM MgCl$_2$, 1% Igepal CA-630, 1 mM EDTA (*see* **Note 4**).
4. 1 M Dithiothreitol (DTT): Dissolve 154 mg of DTT (Cleland's Reagent) in 1 mL of sterile water. Aliquot and store at –20°C (*see* **Note 5**).
5. 100 mM Phenylmethylsulfonyl fluoride (PMSF): Dissolve 17 mg of PMSF in 1 mL of absolute ethanol. Store at –20°C (it will not freeze; *see* **Note 6**).
6. Protease inhibitor (PI) cocktail for mammalian cell extracts containing pepstatin A, bestatin, leupeptin, aprotinin, 4-[2-aminoethyl]-benzenesulfonyl fluoride [AEBSF] and *trans*-epoxysuccinyl-L-leucylamido[4-guanidino]butane (E-64), obtained from Sigma (P-8340; *see* **Note 7**).
7. 5X Sample buffer (5XSB): 62.5 M Tris-HCl, pH 6.8, 10% glycerol, 2% sodium dodecyl sulfate (SDS), 0.05% bromophenol blue, and 5% 2-mercaptoethanol (added just before use).
8. SB: 0.4 mL of 5XSB, 0.55 mL of deionized water, and 50 µL of 2-mercaptoethanol (added just before use).

2.3. SDS-Polyacrylamide Gel Electrophoresis (SDS-PAGE)

1. 90% Ethanol
2. Acrylamide/bis (30%): Dissolve 87.6 g of acrylamide and 2.4 g of $N'N'$-bis-methylene-acrylamide in 200 mL of deionized water. Adjust volume to 300 mL. Filter and store at 4°C in the dark (see **Note 8**).
3. 10% SDS: Dissolve 10 g of SDS in 90 mL of water with gentle stirring. Bring volume to 100 mL and filter. Store at room temperature.
4. 1.5 M Tris-HCl, pH 8.8: Dissolve 27.23 g of Tris-base in 80 mL of deionized water with gentle stirring. Adjust pH to 8.8 with 6 N HCl. Bring total volume to 150 mL, filter, and store at 4°C.
5. 0.5 M Tris-HCl, pH 6.8: Dissolve 6 g of Tris-base in 60 mL of deionized water. Adjust pH to 6.8 with 6 N HCl. Bring total volume to 100 mL, filter, and store at 4°C.
6. 10% Ammonium persulfate (APS): Dissolve 100 mg of APS in 1 mL water. Prepare fresh daily.
7. Tetramethylethylenediamine (TEMED; BioRad).
8. 10X Running buffer: Dissolve 30.3 g of Tris-base, 144 g of glycine, and 10 g of SDS in 800 mL of deionized water. Do not adjust pH with acid or base. Bring the total volume to 1,000 mL, and store at 4°C.
9. Prestained molecular-weight markers (BioRad).

2.4. Western Blotting

1. Nitrocellulose membrane Hybond C (Amersham).
2. Filter paper.
3. Transfer buffer (TB): 25 mM Tris-HCl, 192 mM glycine, 20% methanol, pH 8.3.
4. 10X Tris-buffered saline (TBS) buffer: 100 mM Tris-HCl, pH 7.5, 1.4 M NaCl, 15 mM $MgCl_2$.
5. TBST: TBS buffer containing 0.1% Tween-20.
6. TBSTM blocking buffer: TBST containing 5% nonfat dry milk (Carnation; see **Note 9**).
7. IκBα rabbit polyclonal antibody (Santa Cruz Biotechnology, sc-371) is diluted 1:250 in TBSTM.
8. Actin rabbit polyclonal antibody (Sigma, A5060) is diluted 1:2,000 in TBSTM.
9. Horseradish-labeled anti-rabbit immunoglobulin-G secondary antibody (Amersham) is diluted 1:2,000 in TBSTM.
10. ECL PLUS Western Blotting Analysis kit (Amersham, RPN2132).
11. Stripping buffer: 62.5 mM Tris-HCl, pH 6.7, 2 % SDS, 100 mM 2-mercaptoethanol.

2.5. Electrophoretic Mobility Shift Assay (EMSA)

1. NFκB double-stranded DNA oligonucleotide 5′-TTGTTACAAG<u>GGGACTTTC</u>CGCTG<u>GGGACTTTCC</u>AGGGAGGC-3′ containing two tandemly repeated NFκB binding sites (underlined; see **Note 10**).

2. T4 polynucleotide kinase (Promega, M4101).
3. γ-[32P]ATP, specific activity 6000 Ci (222TBq)/mMole (Perkin-Elmer; *see* **Note 11**).
4. 1% SDS/100 mM EDTA STOP solution.
5. MicroSpin G-25 columns (Amersham).
6. 10X Binding buffer (10XBB): 200 mM Hepes, pH 7.5, 5 mM EDTA, 5 mM DTT, 10% Igepal CA-630, 50% glycerol.
7. Acetylated bovine serum albumin (BSA; Sigma, B-2518): Dissolve content of the vial (10 mg) in 0.5 mL of sterile water to get stock concentration of 20 mg/mL. Aliquot and store at –20°C.
8. Poly-dI.dC (Pharmacia, 27-7880-01): Dissolve content of the vial (0.5 mg) in 0.5 mL of sterile water to get stock concentration of 1 mg/mL. Aliquot and store at –20°C.
9. 5X Tris-glycine-EDTA (TGE) buffer: Dissolve 30.3 g of Tris-base and 150 g of glycine in 900 mL of deionized water. Add 20 mL of 0.5 M EDTA, pH 8.0. Measure pH to make sure it is 8.4 to 8.5. Adjust volume to 1,000 mL, filter, and store at room temperature.
10. Acrylamide:*Bis* (30%): Dissolve 29.2 g of acrylamide and 0.8 g of *N'N'-bis*-methylene-acrylamide in 70 mL of deionized water. Adjust volume to 100 mL, filter, and store at 4°C in the dark.
11. 10% APS: Dissolve 100 mg of APS in 1 mL of water. Prepare fresh daily.
12. TEMED (BioRad, Hercules, CA).
13. 6 X DNA loading dye: Dissolve 25 mg of bromophenol blue in 7 mL of sterile water in 15-mL tube. Add 3 mL of 100% glycerol, mix well, aliquot in microcentrifuge tubes, and store at 4°C.
14. Whatmann DE-81 chromatography paper.

3. Methods

This section describes the protocol for analysis of nucleocytoplasmic shuttling of IκBα in human macrophage-like U-937 cells. This protocol is an adaptation of a standard procedure that has been successfully used to study NFκB regulation in the authors' laboratory and others *(25–30)*. Cells are treated with the CRM1-dependent inhibitor of nuclear export, LMB, and the nuclear and cytoplasmic fractions are isolated and analyzed by Western blotting. **Figure 1A** shows the increased nuclear levels of IκBα induced by LMB treatment of LPS-stimulated U-937 cells. This increased nuclear localization of IκBα in stimulated leukocytes results in the inhibition of NFκB activity (**Fig. 1B**). **Figure 2** illustrates our model of the NFκB inhibition in activated leukocytes by the LMB-induced increased nuclear accumulation of IκBα. The following steps can be accomplished in 4 to 5 d.

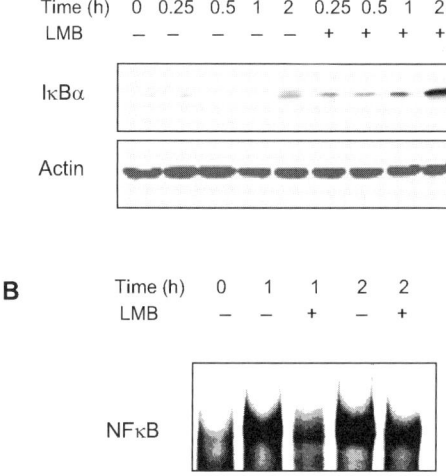

Fig. 1. Effect of leptomycin B (LMB) on nuclear levels of IκBα and NFκB activity in lipopolysaccharide (LPS)-stimulated U-937 cells. (**A**) U-937 cells were treated with phorbol 12-myristate 13-acetate (PMA; 10 ng/mL) for 24 h, and stimulated with LPS (1 μg/mL) with and without LMB (10 nM) for 0, 0.25, 0.5, 1, and 2 h. Nuclear levels of IκBα were analyzed by immunoblotting. Equal protein loading was confirmed by using actin antibody. Each lane contains approx 2×10^5 cells. (**B**) U-937 cells were treated with PMA (10 ng/mL) for 24 h, and stimulated with LPS (1 μg/mL) with and without LMB (10 nM) for 0, 1, and 2 h. NFκB activity was measured in nuclear extracts by electrophoretic mobility shift assay.

3.1. Cell Culture

1. Grow U-937 monocytic cells in RPMI medium supplemented with 10% FBS, 2 mM glutamine, 1 mM sodium pyruvate and penicillin-streptomycin solution to a concentration 1×10^6 cells/mL.
2. Induce differentiation of U-937 cells into macrophages by adding phorbol 12-myristate 13-acetate (PMA) at a final concentration 10 ng/mL. Mix well but gently. Transfer the cells to a 12-well plate; each well will get 2 mL of the above cell suspension.
3. Incubate the plate 1 d in tissue culture incubator (37°C, 5% CO_2 humidified atmosphere), to allow cell differentiation (induced by 24-h treatment with PMA).
4. After 24 h, add LMB at a final concentration of 10 nM and /or LPS at a final concentration of 1 μg/mL. Incubate the cells in tissue culture incubator for the desired times.
5. At the specific times, transfer the cells from the plate to 2-mL prechilled microcentrifuge tubes and follow the protocol for preparation of cytoplasmic and nuclear extracts (*see* **Subheading 3.2.**).

Analysis of Nucleocytoplasmic Shuttling 285

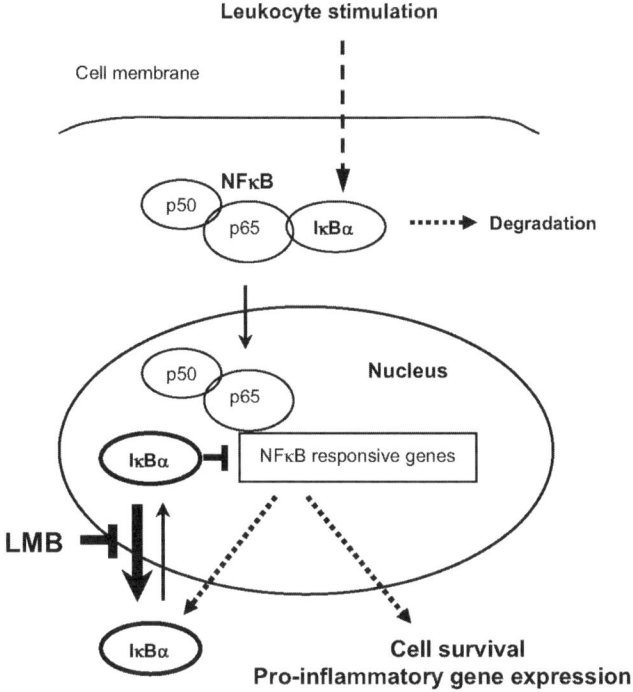

Fig. 2. Model of leptomycin B (LMB)-induced nuclear accumulation of IκBα in human leukocytes, resulting in the inhibition of NFκB activity.

3.2. Preparation of Cytoplasmic and Nuclear Extracts

All steps must be performed at 4°C or on ice, unless stated otherwise.

1. At the specified times, transfer the cells from the cell culture plates to 2-mL microcentrifuge tubes.
2. Centrifuge the tubes containing the cell suspension in a refrigerated centrifuge at 1,700g for 5 min.
3. Carefully aspirate and discard the supernatants, ewsuspend the cells in 1.5 mL of ice-cold PBS, and centrifuge at 1,700g for 5 min.
4. Carefully aspirate and discard the supernatants.
5. Add 0.15 mL of ice-cold RB containing 2 mM DTT, 2 mM PMSF and the PI cocktail (Sigma) used in a concentration 15 μL per 10^6 cells. Mix well but gently by pippeting up and down, and by tapping on the bottom of the centrifuge tube. The RB is hypotonic; the cells will swell in this buffer.
6. Incubate the cell suspensions on ice for 15 min.
7. Solubilize the plasma membrane by adding 0.05 volumes (7 μL per 0.15 mL of cell suspension) of 10 % Igepal CA-630. Vortex vigorously for 10 s to lyse the

cells. Do not leave the cells in RB containing Igepal CA-630 for longer than 15 min because this might also lyse the nuclear membranes.
8. Centrifuge for 5 min at 200g (see **Note 12**).
9. Carefully aspirate the supernatant (cytoplasmic extract [CE]), add 50 μL of 5XSB, and boil immediately for 7 min on a boiling water bath. Label as CE, and store at −70°C (see **Note 13**).
10. Once the CE is removed, the nuclear pellets will become more visible. Wash the nuclear pellets by adding 0.2 mL of RB containing the Sigma PIs 2 mM PMSF and 2 mM DTT.
11. Centrifuge as described in **step 8**, remove and discard the supernatant.
12. For analysis by Western blotting, resuspend each nuclear pellet in 50 μL of 2XSB, and boil immediately for 7 min on a boiling water bath. Label as nuclear extracts (NE), aliquot into several tubes and store at −70°C (see **Note 14**).
13. For analysis of NFκB DNA binding activity, resuspend the nuclear pellets in 50 μL of ice-cold NEB containing 2 mM DTT, 2 mM PMSF and the PI cocktail (Sigma) used in a concentration 15 μL / 10^6 cells. Vortex, incubate on ice 20 min, vortex again and centrifuge 10 min at 14,000g. Transfer the supernatants NE) to prechilled tubes, remove small amount (5 μL) for determining protein concentration, aliquot, and store at −70°C.

3.3. SDS-PAGE

To run SDS-PAGE, the authors use the BioRad Mini PROTEAN-3 System. For separation of IκBα, 12% gels are used.

1. Clean the glass plates with 90% ethanol and assemble according to the BioRad Mini-PROTEAN 3 Instructions. Make sure the plates are flush, to prevent gel leakage. Make a mark 2 cm off the top of the short plate.
2. Prepare resolving gel solution (10 mL) by mixing 3.4 mL of deionized water, 4.0 mL of Acrylamide/Bis solution (30%), 2.5 mL of 1.5 M Tris-HCl, pH 8.8 buffer, and 0.1 mL of 10% SDS.
3. Immediately before pouring the gels add 50 μL of 10% APS and 5 μL of TEMED. Swirl gently and pour the gels immediately.
4. Overlay the gels with water.
5. Allow the gels to polymerize for about 45 min.
6. Prepare 4% gel solutions for the stacking gels. Mix 6.1 mL of deionized water, 1.3 mL of acrylamide/*bis* solution, 2.5 mL of 0.5 mL Tris-HCl, pH 6.8, and 0.1 mL of 10% SDS.
7. Before casting the stacking gels, pour off water from the top of the resolving gels.
8. Add 50 μL of 10% APS and 10 μL of TEMED to the stacking gel solution. Swirl gently, and pour the solution between the glass plates on the top of the resolving gel. Insert combs immediately.
9. Allow the gels to polymerize for about 1 h.

10. Gently remove the combs and rinse the wells thoroughly with 1X running buffer that is prepared by diluting 50 mL of 10X running buffer with 450 mL of deionized water.
11. Assemble plates in the tank according to the BioRad Mini-PROTEAN 3 Cell Instruction Manual. Add about 125 mL of 1X running buffer to the inner chamber, and about 200 mL of 1X running buffer to the tank (lower chamber).
12. Load the samples into the wells with a pipet using gel-loading tips.
13. Run the gels at constant 120 V for about 50 min, or until the blue dye from the samples reaches the bottom of the gel.
14. Turn off the power, disassemble the gel apparatus, and carefully remove the gel from the glass plates. Follow the protocol for Western blotting (*see* **Subheading 3.4.**).

3.4. Western Blotting

To analyze the cytoplasmic and nuclear levels of IκBα, we first probe the membranes with IκBα antibody, which is then followed by actin antibody to confirm equal protein loading. The procedure described here uses the BioRad semi-dry transfer apparatus.

1. Soak two thick filter papers, nitrocellulose membrane (cut to the size of the filter paper) and the gel in TB for 15 min.
2. Assemble the transfer sandwich (from bottom to top): thick filter paper, membrane, gel, thick filter paper. Make sure there are no bubbles between membrane and the gel.
3. Once the lid is put on the transfer sandwich, activate the power. For one 12% gel, the transfer conditions are 18 V (constant), 35 min. For two 12% gels, the conditions are 20 V (constant), 50 min.
4. After the transfer is complete, turn off the power, and dissemble the transfer sandwich. The gel can be now discarded, and the colored prestained markers should be clearly visible on the membrane.
5. Transfer the membrane into blocking solution (TBSTM), and incubate 1 h with gentle rocking at room temperature (*see* **Note 15**).
6. Pour off the blocking solution, and incubate the membrane with IκBα primary antibody diluted 1:250 in TBSTM for 1 h at room temperature.
7. Pour off the primary antibody and wash the membrane five times in TBST (5 min each wash).
8. After washing, incubate the membrane for 1 h at room temperature with horseradish peroxidase (HRP)-labeled secondary anti-rabbit antibody diluted 1:2,000 in TBSTM.
9. Wash the membrane five times in TBST.
10. Pour off TBST from the membrane, place the membrane on a sheet of saran wrap, and develop the signal with ECL PLUS system.
11. Detect the signal with digital imager or autoradiographic film. The image shown in **Fig. 1A** illustrates IκBα nuclear levels detected by digital imager *Syngene*.

12. To confirm equivalent amounts of loaded proteins, strip the membrane for 30 min at 50°C with stripping buffer, and incubate it for 1 h at room temperature with actin antibody diluted 1:2,000 in TBSTM.
13. Develop the signal with HRP-labeled secondary antibody and ECL PLUS system as described above (*see* **Subheading 3.4.10**).

3.5. EMSA

The NFκB oligonucleotide used as a probe for EMSA is a 42-bp double-stranded construct (5′-TTGTTACAAGGGGACTTTCCGCTGGGGACTTTC CAGGGAGGC-3′) containing two tandemly repeated NFκB binding sites (underlined; *see* **Note 10**).

1. End label annealed double-stranded NFκB oligonucleotide by using T4 kinase. To accomplish this, mix the following components in a microcentrifuge tube on ice: 7 μL of water, 2 μL of 10X kinase buffer, 4 pmol of annealed double-stranded NFκB oligonucleotide, 2 μL (15 U) of T4 kinase, and 50 μCi [γ-^{32}P]ATP (*see* **Note 11**).
2. Incubate for 30 min at 37°C in a water bath.
3. Stop the reaction by adding 5 μL of 1% SDS/100 mM EDTA STOP solution, vortex, and spin briefly.
4. Purify the labeled oligonucleotide on a MicroSpin G-25 column.
5. Prepare 7.5% gel in TGE buffer by mixing 41 mL of water, 15 mL of 5XTGE buffer, 18.8 mL of acrylamide:*bis* (30%) solution, 0.6 mL of 10% APS and 60 μL of TEMED. Pour the gel and let it polymerize for 45 min. Rinse the wells with 1XTGE buffer using a 10 mL hypodermic syringe.
6. Pre-run the gel without samples in 1XTGE buffer at 150 V (*see* **Note 16**).
7. Prepare binding reactions by mixing the following components in microcentrifuge tubes on ice: 8 μL of water, 2 μL of 10X BB, 2 μL of poly-dI.dC stock solution, 1 μL of BSA stock solution, 5 μL of the NE containing approx 5 μg protein, and 1 μL of ^{32}P-labeled NFκB oligonucleotide (*see* **Note 11**).
8. Mix gently by tapping on the bottom of the tubes, pulse spin, and incubate 15 min at room temperature.
9. Load the samples on the pre-run gel.
10. Run the gel for 2.5 h at 180 V (*see* **Note 16**).
11. After electrophoresis, transfer the gel to Whatmann DE-81 paper, cover it with plastic wrap, and dry it.
12. Expose the dried gel to PhosphorImager (Molecular Dynamics). The image shown in **Fig. 1B** illustrates NFκB activity measured in nuclear extracts of U-937 cells stimulated with LPS in the absence and presence of LMB.

4. Notes

1. LMB (MW 540.7) can be purchased from Sigma, Calbiochem, or Biomol. LMB from Sigma is supplied as a methanol solution at a concentration of 5 μg/mL. LMB inhibits the nuclear export of NES-containing proteins at concentrations

10–20 nM *(18–28)*. For the final working concentration of 10 nM, the Sigma LMB stock should be diluted 900-fold. The LMB stock solution should be stored at $-20°C$, and when opened, kept on ice to minimize evaporation.
2. RB can be prepared in 100 mL amount, and stored at 4°C. Just before use, take out the amount required for the day, and add DTT (2 mM final concentration), PMSF (2 mM final concentration), and the Sigma PI cocktail mix (15 μL per 10^6 cells); keep on ice.
3. Igepal CA-630 is a nonionic, nondenaturing detergent that is chemically indistinguishable from Nonidet P-40. It solubilizes plasma membranes to release the nuclei.
4. NEB can be prepared in 10 mL amount, and stored at 4°C. Just before use, take out the amount required for the day, and add DTT, PMSF, and the Sigma PI cocktail mix; keep on ice.
5. DTT is a strong reducing agent. It is used to prevent formation of intra- and intermolecular disulfide bonds of proteins. DTT should be added to RB and NEB in the final working concentration 2 mM just before use.
6. PMSF is unstable in aqueous environment. It is essential that it is dissolved in absolute alcohol (ethanol, methanol, or isopropanol), and is added to the RB and NEB in the final working concentration 2 mM just before use.
7. This PI cocktail contains PIs with broad specificity for the inhibition of serine, cysteine, aspartic proteases, and aminopeptidases. It should be stored at –20°C, and added to the RB and NEB just before use, in the final concentration 15 μL per 10^6 cells.
8. Unpolymerized acrylamide is extremely toxic when ingested, and absorbs easily into the skin, so use care and wear gloves while handling solutions that contain it. Once the acrylamide is polymerized it is no longer absorbable, but care still should be taken when handling the gel.
9. Although blocking with 5% Carnation nonfat dry milk in TBST works well in most cases, alternatively 1% protease-free BSA in TBST can be used.
10. This oligonucleotide was designed to detect NFκB activity in human leukocytes *(31–33)*. Alternatively, commercially available double-stranded NFκB oligonucleotides can be used (such as from Santa Cruz Biotechnology).
11. Perform the ^{32}P-labeling steps in an area designated for use of radioactive materials. Follow the safety guidelines for work with radioactivity.
12. It is helpful to always centrifuge the tubes in one position (e.g., the cap snaps facing toward the center of the rotor); this way you can always expect the pellets to be at the same place.
13. Once the cytoplasmic extracts are collected, keep them on ice and boil them immediately with 5XSB to prevent protein degradation. Once boiled, the samples are more stable and can be stored at –70°C for several weeks. To prevent repeated thawing of the samples, store them in smaller aliquots.
14. Similarly as with the CE, boil the NE with 2XSB immediately to prevent protein degradation. Once boiled, the samples are more stable and can be stored at –70°C for several weeks. Avoid repeated thawing by storing the samples in smaller aliquots.

15. The Western blotting can be interrupted at this point, and the membranes can be incubated in TBSTM overnight at 4°C.
16. The EMSA gel can be run either at room temperature with cooling using running water, or at 4°C.
17. It is convenient to prepare a common binding mix by mixing all components (for desired amount of reactions) except for the NEs and labeled oligonucleotide.

Acknowledgments

This work was supported by NIH grant GM079581 and by the St. John's University Faculty Research Award to I.V.

References

1. Paine, P. L., Moore, L. C., and Horowitz, S. B. (1975) Nuclear envelope permeability. *Nature* **254**, 109–114.
2. Breeuwer, M. and Goldfarb, D. S. (1990) Facilitated nuclear transport of histone H1 and other small nucleophilic proteins. *Cell* **60**, 999–1008.
3. Gorlich, D. and Mattaj, I. W. (1996) Nucleocytoplasmic transport. *Science* **271**, 1513–1518.
4. Yashiroda, Y. and Yoshida, M. (2003) Nucleo-cytoplasmic transport of proteins as a target for therapeutic drugs. *Curr. Med. Chem.* **10**, 741–748.
5. Peters, R. (2006) Introduction to nucleocytoplasmic transport: molecules and mechanisms. *Methods Mol. Biol.* **322**, 235–258.
6. Baeuerle, P, A. and Baltimore, D. (1996) NFκB: ten years after. *Cell* **87**, 13–20.
7. Ghosh, S. and Karin, M. (2002) Missing pieces in the NFκB puzzle. *Cell* **109**, S81–96.
8. Barnes, P. J. and Karin, M. (1997) NFκB – a pivotal transcription factor in chronic inflammatory diseases. *New Engl. J. Med.* **336**, 1066–1071.
9. Baldwin, A. S. (2001) The transcription factor NFκB and human disease. *J. Clin. Invest.* **107**, 3–6.
10. Yamamoto, Y. and Gaynor, R. B. (2001) Therapeutic potential of inhibition of the NFκB pathway in the treatment of inflammation and cancer. *J. Clin Invest.* **107**, 135–142.
11. Tak, P. P. and Firestein, G. S. (2001) NFκB: a key role in inflammatory diseases. *J. Clin. Invest.* **107**, 7–11.
12. Karin, M. and Neriah, Y. B. (2000) Phosphorylation meets ubiquitination: The control of NF-κB activity. *Annu. Rev. Immunol.* **18**, 621–663.
13. Arenzana-Seisdedos, F., Thompson, J., Rodriguez, M. S., Bachelerie, F., Thomas, D., and Hay, R. T. (1995) Inducible nuclear expression of newly synthesized IκBα negatively regulates DNA-binding and transcriptional activities of NFκB. *Mol. Cell. Biol.* **15**, 2689–2696.
14. Arenzana-Seisdedos, F., Turpin, P., Rodriguez, M., et al. (1997) Nuclear localization of IκBα promotes active transport of NF-κB from the nucleus to the cytoplasm. *J. Cell Sci.* **110**, 369–378.

15. Rodriguez, M., Thompson, J., Hay, R. T., and Dargemont, C. (1999) Nuclear retention of IκBα protects it from signal-induced degradation and inhibits NFκB transcriptional activation. *J. Biol. Chem.* **274**, 9108–9115.
16. Sachdev, S., Bagchi, S., Zhang, D. D., Mings, A. C., and Hannink, M. (2000) Nuclear import of IκBα is accomplished by a Ran-independent transport pathway. *Mol. Cell Biol.* **20**, 1571–1582.
17. Yoshida, M., Nishikawa, M., Nishi, K., Abe, K., Horinouchi, S., and Beppu, T. (1990) Effects of leptomycin B on the cell cycle of fibroblasts and fission yeast cells. *Exp. Cell. Res.* **187**, 150–156.
18. Kudo, N., Wolff, B., Sekimoto, T., et al. (1998) Leptomycin B inhibition of signal-mediated nuclear export by direct binding to CRM1. *Exp. Cell. Res.* **242**, 540–547.
19. Kudo, N., Matsumori, N., Taoka, H., et al. (1999) Leptomycin B inactivates CRM1/exportin 1 by covalent modification at a cysteine residue in the central conserved region. *Proc. Natl. Acad. Sci. U S A.* **96**, 9112–9117.
20. Verma, U. N., Yamamoto, Y., Prajapati, S., and Gaynor, R.B. (2004) Nuclear role of IκB Kinase-γ/NFκB essential modulator (IKK γ/NEMO) in NFκB-dependent gene expression. *J. Biol. Chem.* **279**, 3509–3515.
21. Banninger, G. and Reich, N. C. (2004) STAT2 nuclear trafficking. *J. Biol. Chem.* **279**, 39199–39206.
22. Bhattacharya, S. and Schindler, C. (2003) Regulation of Stat3 nuclear export. *J. Clin. Invest.* **111**, 553–559.
23. Zeng, R., Aoki, Y., Yoshida, M., Arai, K., and Watanabe, S. (2002) Stat5B shuttles between cytoplasm and nucleus in a cytokine-dependent and -independent manner. *J. Immunol.* **168**, 4567–4575.
24. Jang, B.C., Munoz-Najar, U., Paik, J. H., Claffey, K., Yoshida, M., and Hla, T. (2003) Leptomycin B, an inhibitor of the nuclear export receptor CRM1, inhibits COX-2 expression. *J. Biol. Chem.* **278**, 2773–2776.
25. Miskolci, V., Ghosh, C., Rollins, J., et al. (2006) TNFα release from peripheral blood leukocytes depends on a CRM1-mediated nuclear export. *Biochem. Biophys. Res. Commun.* **351**, 354–360.
26. Castro-Alcaraz, S., Miskolci, V., Kalasapudi, B., Davidson, D., and Vancurova, I. (2002) NFκB regulation in human neutrophils by nuclear IκBα: correlation to apoptosis. *J. Immunol.* **169**, 3947–3953.
27. Huang, T. T., Kudo, N., Yoshida, M., and Miyamoto, S. (2000) A nuclear export signal in the N-terminal regulatory domain of IκBα controls cytoplasmic localization of inactive NFκB/IκBα complexes. *Proc. Natl. Acad. Sci. U S A.* **97**, 1014–1019.
28. Chen, Y., Wu, J., and Ghosh, G. (2003) KappaB-Ras binds to the unique insert within the ankyrin repeat domain of IκBβ and regulates cytoplasmic retention of IκBβ x NFκB complexes. *J. Biol. Chem.* **278**, 23101–23106.
29. Ganesh, L., Yoshimoto, T., Moorthy, N. C., et al. (2006) Protein methyltransferase 2 inhibits NFκB function and promotes apoptosis. *Mol. Cell Biol.* **26**, 3864–3874.
30. Berchtold, C. M., Wu, Z. H., Huang, T. T., and Miyamoto, S. (2007) Calcium-dependent regulation of NEMO nuclear export in response to genotoxic stimuli. *Mol. Cell Biol.* **27**, 497–509.
31. Vancurova, I., Miskolci, V., and Davidson, D. (2001) NFκB activation in TNFα-stimulated neutrophils is mediated by protein kinase Cδ. Correlation to nuclear IκBα. *J. Biol. Chem.* **276**, 19746–19752.

32. Vancurova, I., Bellani, I., and Davidson, D. (2001) Activation of NFκB and its suppression by dexamethasone in polymorphonuclear leukocytes: Newborn *versus* adult. *Pediatr. Res.* **49**, 257–262.
33. Miskolci, V., Castro-Alcaraz, S., Nguyen, P., Vancura, A., Davidson, D., and Vancurova, I. (2003) Okadaic acid induces sustained activation of NFκB and degradation of the nuclear IκBα in human neutrophils. *Arch. Biochem. Biophys.* **417**, 44–52.

22

Imaging pHluorin-Based Probes at Hippocampal Synapses

Stephen J. Royle, Björn Granseth, Benjamin Odermatt, Aude Derevier, and Leon Lagnado

Summary

Accurate measurement of synaptic vesicle exocytosis and endocytosis is crucial to understanding the molecular basis of synaptic transmission. The fusion of a pH-sensitive green fluorescent protein (pHluorin) to various synaptic vesicle proteins has allowed the study of synaptic vesicle recycling in real time. Two such probes, synaptopHluorin and sypHy, have been imaged at synapses of hippocampal neurons in culture. The combination of these reporters with techniques for molecular interference, such as RNAi allows for the study of molecules involved in synaptic vesicle recycling. Here the authors describe methods for the culture and transfection of hippocampal neurons, imaging of pHluorin-based probes at synapses and analysis of pHluorin signals down to the resolution of individual synaptic vesicles.

Key words: Endocytosis; exocytosis; hippocampal synapses; imaging; neurons; synaptic vesicle; synaptopHluorin; sypHy.

1. Introduction

Neurons transmit information at chemical synapses via release of neurotransmitter from synaptic vesicles (*1*). The exocytosis and subsequent endocytosis of vesicular membrane and proteins has been intensely studied by neurobiologists for more than three decades (*2*). One of the most widely used preparations for studying synaptic vesicle cycling is hippocampal neurons in culture. Optical methods for measuring exocytosis and endocytosis of synaptic vesicles have come to the fore owing to the inaccessibility of small synapses to electrophysiological recording methods (3). Of these optical methods, one of the most direct is the imaging of a pH-sensitive green fluorescent protein (GFP), termed *pHluorin*, developed by Miesenböck and others (4,5). The pHluorin can be targeted to the interior of synaptic vesicles by fusion to proteins such as synaptobrevin (synaptopHuorin) (4), synaptophysin (sypHy) (6), synaptotagmin (7) or VGLUT1 (8). As illustrated in **Fig. 1**, the pHluorin is quenched by the low pH inside the vesicle (pH ~5.5). Upon stimulation, vesicles fuse with the plasma membrane, protons are lost and the fluorescence of pHluorin increases as the pH moves to 7.4. This is seen as an upward deflection in the sypHy trace recording the fluorescence of sypHy in a small region over a synaptic bouton. The fluorescence declines again following endocytosis and vesicle re-acidification. Re-acidification is fast

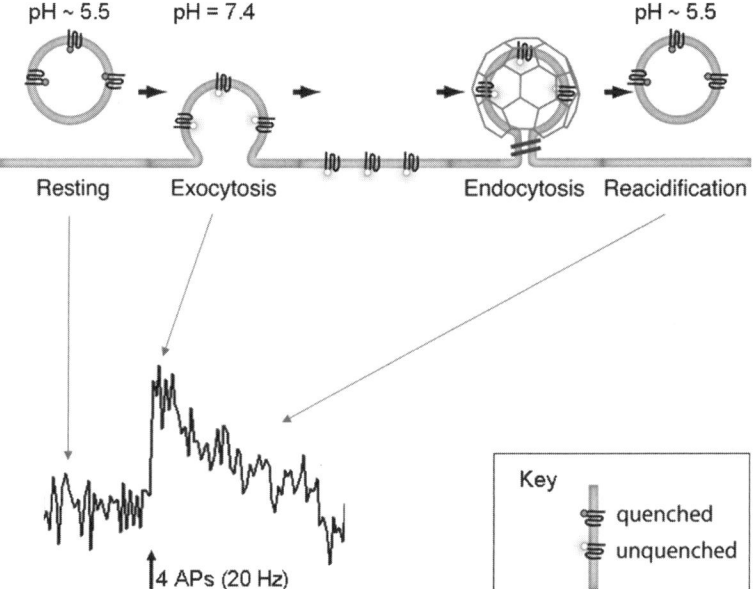

Fig. 1. Schematic of imaging exocytosis and endocytosis with pHluorin-based probes. SypHy comprises superecliptic pHluorin (pH-sensitive green fuorescent protein) fused to the second intravesicular loop of synaptophysin. The pHluorin is quenched by the low pH inside the vesicle (pH ~5.5). Upon stimulation (black arrow), vesicles fuse with the plasma membrane, protons are lost and the fluorescence of sypHy increases as the pH moves to 7.4. This is seen as an upward deflection in the sypHy trace (red). The fluorescence declines again following endocytosis and vesicle re-acidification. Reacidification is fast (τ ~4 s), so the decline in sypHy signals gives us a measure of endocytosis in virtually real time.

($\tau \sim 4$ s), so the decline in sypHy signals gives us a measure of endocytosis that is close to real time *(6,9)*.

The imaging of pHluorin-based probes at hippocampal synapses has been used to describe the kinetics of endocytosis *(6,10,11)* and to investigate the molecules involved in vesicle retrieval *(6)*. In our hands, sypHy is the best currently available pHluorin-based probe. It is better localized to vesicles and has a greater signal-to-noise ratio compared with synaptopHluorin *(6)*. We have optimized conditions for the culture of healthy hippocampal neurons and for their transfection with reasonable efficiency. Furthermore, we have developed methods to image pHluorin-based probes with high temporal resolution, the analysis of which allows for the resolution of exocytosis at the level of an individual synaptic vesicle.

2. Materials
2.1. Primary Culture of Hippocampal Neurons

1. Cover slips (16 mm diameter, thickness 1, BDH/VWR 406/0189/2); Poly-D-lysine hydrobromide (70,000–150,000 MW, Sigma P0899).
2. Laminin (1 mg/mL, Sigma L2020). Dissection medium: 490 mL EBSS (no calcium, magnesium or phenol red, Gibco 14155-048), 5 mL HEPES (1 M, pH 7.4, Sigma H0887), 5 mL penicillin-streptomycin (Gibco 15140-122). Store at 4°C. Standard cell culture plasticware: tissue culture flasks (75 cm^2 and 25 cm^2), 6-well plates (Costar), 15-mL tubes. Culture medium A: 412.5 mL minimal essential medium (MEM; no phenol red, Gibco 51200-046), 10 mL Glucose (1 M in MEM); 5 mL penicillin-streptomycin, 5 mL sodium pyruvate (100 mM, Sigma S8636), 12.5 mL HEPES (1 M, pH 7.4), 5 mL N2 supplement (Gibco 17502-048), 50 mL horse serum (heat inactivated, Gibco 26050-070), sterile filter, store in the dark at 4°C.
3. Litter of E18 rat pups, approx six pups (12 hippocampi) are required for six plates (72 cover slips). Dissection kit: 2 × watchmakers forceps, dissection scalpel, fine scissors, angled scissors, small scissors, and small spatula; all bake-sterilized. Dissection microscope.
4. Papain PAP2 powder (Worthington PAPAIN-022).
5. Culture medium B: 500 mL Neurobasal (no phenol red, Earle's salts added, Gibco, 12348-017); 10 mL B27 supplement (Gibco 17504-044), 1.25 mL L-glutamine (200 mM, Gibco 25030-024), 5 mL penicillin-streptomycin. Store in the dark at 4°C.

2.2. Transfection of Hippocampal Neurons

1. Culture medium B.
2. Good-quality midiprep DNA of plasmids for expression (under a cytomegalovirus promoter [CMVp]) of monomeric red fluorescent protein (mRFP) or sypHy

(~0.5 mg/mL, A260/A280 = 1.8). Optional: chemically synthesized short, interfering RNA (siRNA, Qiagen, 40 μM). Lipofectamine 2000 (Invitrogen, 11668-019). MEM.

2.3. Imaging pHluorin-Based Probes at Synapses

1. Normal extracellular solution (NES): 136 mM NaCl, 2.5 mM KCl, 10 mM HEPES, 1.3 mM $MgCl_2$, 10 mM glucose, 2 mM $CaCl_2$, 0.01 mM 6-cyano-7-nitroquinoxaline-2,3-dione disodium salt (CNQX; Sigma) and 0.05 mM DL-2-amino-5-phosphonopentanoic acid (DL-APV; Tocris), pH 7.4.
2. Imaging chamber (Warner instruments RC-25F), customized by the attachment of two parallel platinum wires, 5 mm apart, for electric field stimulation held in place by beeswax (**Fig. 2**).
3. Imaging set up: Nikon Diaphot 200 microscope with a 40× (1.3 numerical aperture [NA]) oil immersion objective and a Xenon 100 W epifluorescence unit. The filter set for pHluorin comprises: 475AF40 excitation filter, 505DRLP dichroic mirror and 535AF45 emission filter (Omega Filters). Using these filters with no attenuation, the illuminating light leaving the objective is approx 300 mW. To visualize mRFP, the filter set comprises: 560AF55 excitation filter, 595DRLP dichroic mirror and 645AF75 emission filter (Omega Filters). Images are captured at a depth of 16-bit using a Photometrics Cascade 512B charge-coupled device (CCD) camera controlled by an Apple Macintosh G4 computer running IPLab (Scanalytics). Where required a Uniblitz VMM-D3 shutter driver controlled by a second Mac running IgorPro (Wavemetrics) is used. Stimulation is via a Grass S48 stimulator and SIU5 stimulus isolation unit.

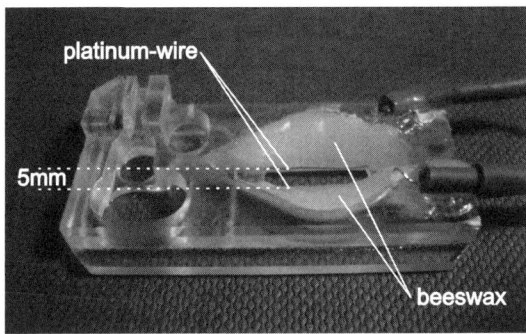

Fig. 2. The imaging chamber. Photograph of the chamber used for imaging hippocampal cultures. The chamber (RC-25F, Warner instruments) is modified by the addition of two platinum wires 5 mm apart. The wires are sealed in place by beeswax. These wires are then connected to the electrical stimulator. Cover slips of transfected neurons are attached to the underside lip of the holder and the holder is mounted into a P-1 platform (Warner instruments). Normal extracellular solution is delivered via tube into the right side of the chamber and a suction pipe withdraws solution to the left.

3. Methods

3.1. Primary Culture of Hippocampal Neurons

1. Autoclave sufficient number of cover slips, rinse in sterile ddH_2O. Put cover slips into a Petri dish with 20 mL of poly-D-lysine (50 μg/mL). Incubate at room temperature, best used within 2 wk.
2. On day 1: rinse each cover slip in ddH_2O and then place in the appropriate cell culture trays. 2×16 mm Cover slips are placed side by side in each well of a 6-well plate, leave for 1 h to dry in the culture hood. Apply to each cover slip 150 μL of laminin (20 μg/mL) in sterile dissection media, laminin should sit on top of the cover slip and not seep onto the well. Leave overnight in primary cell culture incubator (humidified, 37°C and 95% air/5% CO_2). Preincubate culture media A in a vented 75-cm^2 tissue culture flask (*see* **Note 1**).
3. On day 2: The dissection, trituration, and quantification steps are done in a laminar flow hood, all other culture work is done in a class II cell culture hood. Sterile procedures are used throughout. Uterus from pregnant rat is placed in 50 mL of dissection media in a 200-mL beaker on wet ice. All dissection steps are done in Petri dishes filled with cold dissection media. One by one, get the pups out by cutting the skin of the uterus with small scissors, open the membraneous sac surrounding the pup with forceps and then remove the pup (the umbilical cord will still be attached). Cut the head off and place in a Petri dish. When all the heads are removed, dissect out the brain. We find it easiest to make a small incision in the center of the nape of the neck and pull the skin and skull forward and over the brain. Once the skull is off, lift out the brain with the spatula place in media on ice. Bisect the hemispheres, peel the meninges away carefully and remove the hippocampi under a dissection microscope. It is very important to avoid contamination of the tissue with blood cells. Place dissected hippocampi in media on ice (*see* **Note 2**).
4. Before you dissect the last brain, get a vial of PAP2 from the fridge and add dissection media to 10 U/mL and put in 15-mL falcon tube in a 37°C water bath. Following dissection of all hippocampi, transfer to the tube. Incubate for approx 15 min, until pieces become fluffy around the edges. Let the pieces settle to the bottom of the tube (or spin for 2 min at 300g) and then remove as much digestion solution as you can. Wash with 10 mL of cold culture medium, do not disturb the pellet. Add 2 mL cold culture medium A. Triturate the pieces with two flame-polished pipets of successively smaller bore. Around five to six passes with each is normally enough, the mixture should be cloudy with no big pieces left. Use hemocytometer to quantify yield, don't count dead (granulated cells). Typically a yield is around 2.5×10^6 cells per milliliter. Diluting in cold culture medium A, a typical plating density is 0.2 to 0.25×10^6 cells per milliliter (using 150 μL per cover slip). Remove the laminin from the cover slips, but don't let it dry. Plate the neurones one tray at a time. Do not to let the cell suspension spill off the cover slip. Place the trays in the incubator. Aapproximagtely 1 or 2 h later add 2 mL of prewarmed culture medium A to each well (*see* **Note 3**).

5. On day 3, replace 1 mL of media with fresh prewarmed culture medium A. Then feed the cells approximately every 3 d by replacing 50% of media with prewarmed culture medium B.

3.2. Transfection of Hippocampal Neurons

1. Neurons are typically transfected after 8 d *in vitro* (DIV). Two hours prior to transfection, remove half of the culture medium and add to the well an equal volume of prewarmed culture medium B. Save the removed medium in a 25-cm^2 tissue culture flask and add an equal volume of fresh culture medium B, keep in incubator (*see* **Note 4**).
2. For one well of a 6-well plate, mix 2 μg sypHy and 2 μg mRFP DNA with 50 μL MEM in a 1.5 mL tube (tube A). In tube B, mix 2 μL of lipofectamine2000 with 50 μL MEM, incubate at room temperature for 5 min. Add 50 μL from tube B to tube A and mix. Incubate for 20 min in the dark at room temperature. Add mixture dropwise to the cultures, 100 μL per well. Return cells to incubator. After 2 h, remove all media and replace with the preconditioned media kept from **step 1**. This method typically yields 2 to 20 transfected neurons per cover slip. Multiply volumes according to the number of wells required. The neurons are fed alongside the remaining cultures to keep them healthy (*see* **Note 5**).

3.3. Imaging pHluorin-Based Probes at Synapses

1. Imaging is carried out on neurons 14 to 21 DIV. Mount cover slips into the imaging chamber. Perfuse cells at approx 0.2 mL per minute with NES at 23±2 °C using a Gilson minipuls 3 peristaltic pump. Apply suction to the opposite side of the chamber to remove surplus (*see* **Note 6**).
2. Find a transfected neuron by looking for mRFP expression in the RFP channel. This minimizes photobleaching of sypHy that would occur during the search for transfected neurons in the GFP channel. In the quenched (resting) state, very little sypHy fluorescence can be seen by eye (*see* **Note 7**).
3. We evoke action potentials using 20 mA pulses of 1 ms duration. The amount of current that is required depends on the imaging chamber and should be tested empirically by preloading untransfected cultures with fluo3-AM ester (Molecular Probes) and identifying thresholds for the Ca^{2+} signal associated with action potential firing.
4. For detection of the low fluorescence signals from single vesicles, sources of noise must be minimized. The back illuminated CCD chip in our camera was cooled to −25°C, pixels were binned 2 × 2 and acquired at the lowest speed setting (5 MHz). High-quality recordings are obtained with no illumination attenuating filters or shutters and electron multiplication gain set to 2,067 with digitalization conversion gain 3 e$^-$/analog-to-digital unit. The perfusion system should be turned off during image acquisition. For each cover slip, acquire two data sets. The first

consists of 10 single action potentials (APs) at 45 s intervals (2,325 images in total) and the second consists of the same 10 single APs followed by a final train of 40 APs at 20 Hz (3000 images in total). Only a subset (128 × 128 pixels) of the chip was used to reliably obtain all frames in data sets acquired at 5 Hz (195 ms exposure time). Each frame is triggered by a computer running IgorPro. The same program triggers the stimulation pulses (*see* **Notes 8–10**).

5. For experiments imaging responses to trains of action potentials, a "time-lapse" protocol is used. The camera exposure times and illumination are synchronized using an Uniblitz VMM-D3 shutter driver with an on and off set of 10 ms in order to minimize photobleaching. Images of sypHy (512 × 512 pixels) were acquired at 0.5 Hz with 250 ms of exposure at 25% illumination. In a typical experiment, 40 s (20 images) of baseline are followed by 4 or 40 stimuli at 20 Hz. Recovery is imaged for a further 120 s (240 images).

3.4. Analysis of pHluorin Signals

Data sets are analysed using the IPLab (version 3.9.4) and/or IgorPro (version 5.04 or later) software running in either MacOS X or WindowsXP. Performance is similar for both Mac and PC systems and the most important aspect for computing performance seems to be the amount of RAM. At least 1 GB is needed in either case. IPLab is a versatile and easy to use software package for image acquisition and analysis, IgorPro can do the same things as IPLab but much faster, the user interface is however more demanding and requires more time to learn.

1. Identify synapses by finding points where the fluorescence intensity increases when stimulated by 40 APs at 20 Hz. Such hot-spots were found by subtracting an average baseline signal of 5 to 20 frames immediately before stimulation to 5 to 20 images directly after.
2. Regions of interest (ROIs) are centered over each synapse. Each ROI is aligned to accommodate the largest amount of basal fluorescence surrounding the point of greatest change identified above (by eye or by custom-written scripts in IgorPro). The size of the ROIs should be as small as possible to maximize the signal-to-noise ratio, but large enough to contain both the site of fusion and endocytosis, otherwise the fluorescence measurements will not represent the true kinetics of synaptic vesicle cycling. SynaptopHluorin is very mobile and mixes with a general membrane pool after vesicle fusion *(12,13)*. To allow for pHluorin movement due to this phenomenon, ROIs generally need to be bigger for synaptopHluorin than for sypHy *(6)*. We suggest trying different ROI sizes and work with the smallest size that gives consistent kinetics. The authors find that ROIs of 1.6 × 1.6 μm or more work fine for sypHy and twice that size for synaptopHluorin.
3. The illumination intensity required for imaging single synaptic vesicles rapidly leads to photobleaching of the unquenched surface pool of pHluorin molecules (**Fig. 3**). The data traces can be corrected by subtracting a double exponential

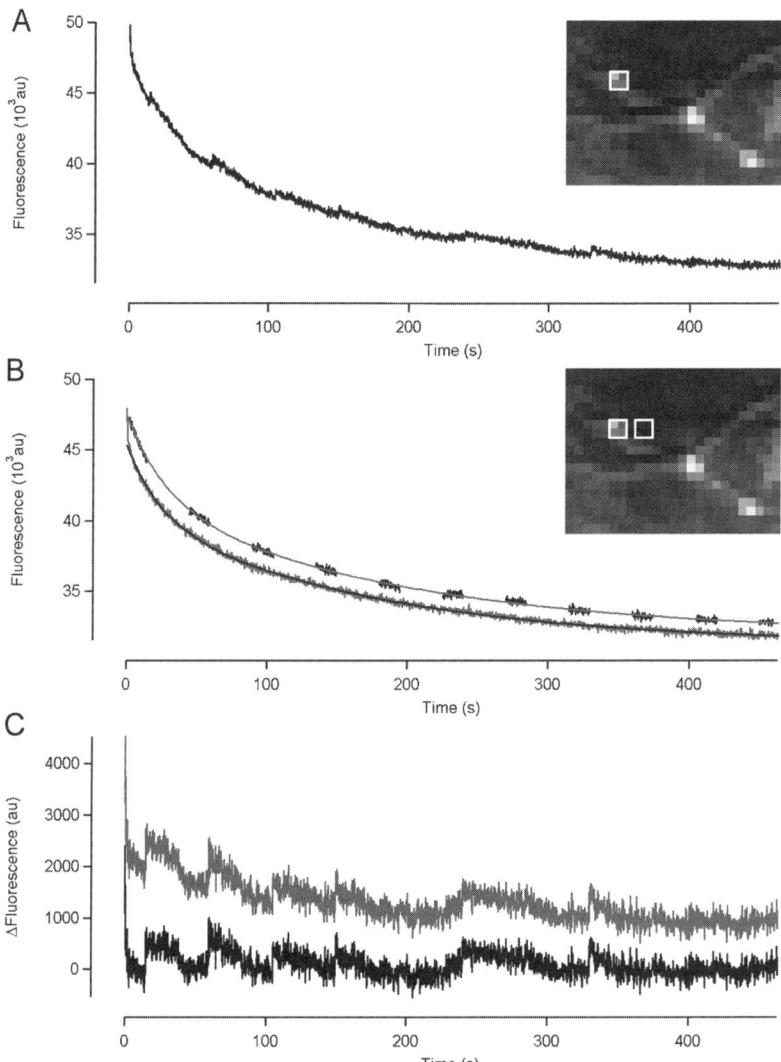

Fig. 3. Correction for photobleaching in single vesicle imaging experiments. (**A**) Fluorescence intensity trace (5 Hz acquisition rate) from a 2 × 2 pixel region of interest ROI; 1.6 × 1.6 μm, *see* inset) covering a synapse expressing sypHy. Every 45 s an action potential (AP) is evoked by field stimulation (10 times in total). (**B**) Bi-exponential curve fits to the fluorescence intensity trace at the synapse with 30 s following the AP blanked (grey fit to black data) or local background (black fit to gray data, see inset for ROI localization). (**C**) Traces after correction by subtracting the biexponential functions in B from the data in A. Using the function obtained at the synapse gives the trace shown in black, the gray trace is corrected using the fit to the local background.

decaying function with values from either fitting to the original trace with the 30 s immediately subsequent the stimulus blanked out or to the local background (the ROI shifted to a neighboring region without synapses, **Fig. 3**).
4. Synaptic vesicle release evoked by single APs can easily be identified as rapid jumps in fluorescence time-locked to the stimulation pulses. Endocytosis is much slower than exocytosis and to reliably identify the kinetic components, many events (>100) need to be averaged before curve fitting. The fluorescence decay after a single AP can be modeled as two consecutive and irreversible reactions with first-order kinetics: endocytosis (k_e) followed by reacidification and quenching of sypHy (k_r):

$$F(t) = \frac{[A]_0}{k_r - k_e}(k_r(1 - e^{-k_e t}) - k_e(1 - e^{-k_r t}))$$

Under our experimental conditions (10 mM HEPES buffer and 23°C) the rate constant of vesicle reacidification is approx 0.25 s^{-1}. When fitted to our data, the rate constant for endocytosis is approx 0.067 s^{-1} (*6*).
5. Nonphasic release of synaptic vesicles after an AP could affect the apparent time constant of the fluorescence decay. Therefore, each response to single APs should be individually examined by eye and traces containing nonphasic events discarded from further analysis.

4. Notes

1. We use separate cell culture incubators for primary neuronal cultures and standard cell lines. This is to prevent possible contamination. The culturing protocol yields a mixed neuron/glia culture. Culture media are without phenol red as its color interferes with imaging experiments. Phenol red can be supplemented in one well of a plate if it is necessary to assay pH changes.
2. Rat pups are used at E18. Neuronal survival is good at this age whereas at earlier ages, the meninges are troublesome to remove. Neurons from rat pups older than E18 do not survive well.
3. A common cause for poor neuronal survival is the plating density of neurons being too dense or sparse. We found that changing from culture medium A to B after 3 DIV allows a wider range of densities to be used, versus leaving the cells in medium A. The dissection time should be rapid: less than 2 h between removal of the embryos and plating. The substrate is another suspicious factor if neuronal cultures are unhealthy. It is very important that laminin is aliquoted upon receipt. Aliquots are stored in a –80 C freezer. Laminin should be thawed slowly in a refrigerator for 1 h prior to use, thus preventing cross-linking of laminin molecules.

4. Transfection of neurons is a black art: we found that replacing the medium with culture media B instead of culture media A greatly improves transfection efficiency, but the reason for this is unknown. Other changes were found to improve transfection, such as incubating lipofectamine-DNA complexes in the dark, they remain in our protocol more out of superstition than a definite scientific requirement.
5. To interfere with molecules involved in synaptic vesicle cycling. Plasmids for expression of dominant-negative constructs that are mRFP-tagged can be co-transfected with sypHy, or siRNAs can be co-transfected along with mRFP and sypHy. For co-transfection with siRNA, in addition to tubes A and B, also prepare tube C, with 12 μL of 40 μM siRNA and 50 μL Neurobasal. Prepare tube D with 4 μL lipofectamine2000 and 50 μL neurobasal. Add 50 μL from tube D to tube C. Following incubations as above, add 100 μl of AB mix and 100 μL of CD mix to the well. Either keep the mixture with the cells or replace medium after 24 h. Transfections were on 11 DIV neurons and experiments performed at 72 h post-transfection.
6. When modifying the chamber the wires are placed 5 to 10 mm apart and as close to the cover slip as possible. The perfusion solution is crucial for stimulating the neurons electrically. In early experiments we used the solution described in *(11)*, this resulted in unreliable stimulation. We believe the reason for this is culture-dependent as a different culturing protocol is used by this group. Vacuum-grease is applied to the edges of the cover slip and chamber to improve the seal between the glass and the chamber/holder (**Fig. 2**).
7. For the design of sypHy, we found that it was essential to position the pHluorin into the second intravesicular loop of synaptophysin, in between two cysteine residues that potentially form a disulphide bridge *(14)*. Attempts to fuse pHluorin to the N-terminus or the C-terminus of synaptophysin did not target the pHluorin probe into synaptic vesicles. A precursor version of sypHy that lacked the fourth transmembrane domain and the C-terminus, was also incorrectly targeted. On the other hand a sypHy-double version with two pHluorin probes in tandem inserted into the second intravesicular loop showed brighter fluorescence at the synapses before stimulation but did not result in a better signal-to-noise ratio. We therefore opted to work with the original sypHy construct.
8. We do not reduce background noise by photobleaching synaptopHluorin or sypHy before applying test stimuli, as suggested in *(10)*. In our hands, extensive pre-bleaching damages the cells.
9. Although cooling the CCD more than –25°C would decrease the dark current, spurious charge electrons become more frequent and skew the background noise distribution toward more electrons.
10. Our imaging setup is on a vibration-isolation table (VH3036-OPT, Newport) covered by a dark Faraday cage. When imaging single vesicles, experiments are carried out with no room lighting and with computer monitors switched off to reduce ambient light. The cage is fitted with a heavy black cloth curtain to prevent any remaining light leaking in. An electric fan is essential to circulate air to reduce the build up of heat generated from the camera and lamp in the cage.

Acknowledgments

Superecliptic pHluorin and monomeric RFP cDNAs were kind gifts from Drs James Rothman and Roger Tsien. This work was supported by the Medical Research Council (MRC), the Swedish Research Council and the Human Frontiers Science Program (HFSP).

References

1. Katz, B. (1969) *The Release of Neural Transmitter Substances*. Liverpool University Press, Liverpool.
2. Royle, S. J., and Lagnado, L. (2003) Endocytosis at the synaptic terminal. *J Physiol* **553**, 345–355.
3. Ryan, T. A. (2001) Presynaptic imaging techniques. *Curr Opin Neurobiol* **11**, 544–549.
4. Miesenbock, G., De Angelis, D. A., and Rothman, J. E. (1998) Visualizing secretion and synaptic transmission with pH-sensitive green fluorescent proteins. *Nature* **394**, 192–195.
5. Sankaranarayanan, S., De Angelis, D., Rothman, J. E., and Ryan, T. A. (2000) The use of pHluorins for optical measurements of presynaptic activity. *Biophys J* **79**, 2199–2208.
6. Granseth, B., Odermatt, B., Royle, S. J., and Lagnado, L. (2006) Clathrin-mediated endocytosis is the dominant mechanism of vesicle retrieval at hippocampal synapses. *Neuron* **51**, 773–786.
7. Diril, M. K., Wienisch, M., Jung, N., Klingauf, J., and Haucke, V. (2006) Stonin 2 is an AP-2-dependent endocytic sorting adaptor for synaptotagmin internalization and recycling. *Developmental cell* **10**, 233–244.
8. Voglmaier, S. M., Kam, K., Yang, H., et al. (2006) Distinct endocytic pathways control the rate and extent of synaptic vesicle protein recycling. *Neuron* **51**, 71–84.
9. Atluri, P. P. and Ryan, T. A. (2006) The kinetics of synaptic vesicle reacidification at hippocampal nerve terminals. *J Neurosci* **26**, 2313–2320.
10. Gandhi, S. P. and Stevens, C. F. (2003) Three modes of synaptic vesicular recycling revealed by single-vesicle imaging. *Nature* **423**, 607–613.
11. Sankaranarayanan, S. and Ryan, T. A. (2000) Real-time measurements of vesicle-SNARE recycling in synapses of the central nervous system. *Nat Cell Biol* **2**, 197–204.
12. Fernandez-Alfonso, T., Kwan, R., and Ryan, T. A. (2006) Synaptic vesicles interchange their membrane proteins with a large surface reservoir during recycling. *Neuron* **51**, 179–186.
13. Wienisch, M. and Klingauf, J. (2006) Vesicular proteins exocytosed and subsequently retrieved by compensatory endocytosis are nonidentical. *Nat Neurosci* **9**, 1019–1027.
14. Johnston, P. A. and Sudhof, T. C. (1990) The multisubunit structure of synaptophysin. Relationship between disulfide bonding and homo-oligomerization. *J Biol Chem.* **265**, 8869–8873.

23

Analysis of Receptor Tyrosine Kinase Internalization Using Flow Cytometry

Ning Li, Kristen S. Hill, and Lisa A. Elferink

Summary

The internalization of activated receptor tyrosine kinases (RTKs) by endocytosis and their subsequent down regulation in lysosomes plays a critical role in regulating the duration and intensity of downstream signaling events. Uncoupling of the RTK cMet from ligand-induced degradation was recently shown to correlate with sustained receptor signaling and increased cell tumorigenicity, suggesting that the corruption of these endocytic mechanisms could contribute to increased cMet signaling in metastatic cancers. To understand how cMet signaling for normal cell growth is controlled by endocytosis and how these mechanisms are dysregulated in metastatic cancers, we developed flow cytometry-based assays to examine cMet internalization.

Key words: cMet; endocytosis; flow cytometry; hepatocyte growth factor; Internalin B; receptor tyrosine kinase.

1. Introduction

Receptor tyrosine kinases (RTKs) are a family of single-pass transmembrane proteins that control a wide variety of cellular events in multicellular organisms including cell proliferation, differentiation, survival, and migration. Escape from the negative regulatory mechanisms that normally inactivate RTK signaling, promote neoplastic growth and cell transformation *(1–6)*. Tumor cell growth and metastasis are the result of increased signaling from multiple receptors, including the cMet RTK. cMet activation by binding to its physiological ligand hepatocyte growth factor (HGF) or to the Internalin B (InlB) protein of

Listeria monocytogenes, promotes receptor endocytosis and triggers the activation of multiple downstream signaling cascades *(7–10)*. In normal cells, cMet signaling is critical for cell growth, motility, and organ development, as well as their regenerative response to acute injury (reviewed in **ref. 11**). Conversely, dysregulated cMet signaling has been shown to promote cell proliferation, increased cell motility, and invasive morphogenetic responses correlating closely with the increased metastatic potential of several human cancers (reviewed in **refs. 11** and *12)*. Receptor internalization from the cell surface represents a key first step in the inactivation process, by targeting the internalized receptor for lysosomal degradation. Corruption of these mechanisms contributes to prolonged cMet signaling *(5,6,13)* suggesting that defects in cMet endocytic trafficking could be a contributing factor to increased cMet levels and signaling in some human metastatic cancers.

Flow cytometry is a qualitative and rapid method for counting, examining, and sorting the fluorescent characteristics of single cells. Flow cytometry offers several advantages for examining the endocytic trafficking properties of RTKs over several techniques including confocal microscopy, (co)immunoprecipitations, and cell surface biotinylation assays. First, flow cytometry is a rapid, high-throughput approach that can be used to quantify thousands of events in a subpopulation of cells. Second, it measures the signal intensity of all cells within a population and is therefore less susceptible to investigator bias. Third, flow cytometry allows the study of rare cell subsets that comprise too small of a population for analysis by conventional biochemical methods. Finally, the parametric nature of flow cytometry makes it amenable for measuring multiple events simultaneously within individual cells of heterogeneous populations. This chapter describes two flow cytometry-based assays we have developed for measuring residual cell surface levels of cMet in response to ligand and the direct internalization of cMet/ligand complexes under steady-state conditions.

2. Materials
2.1. Cell Culture

1. Cell line: Human mammary epithelial cells (T47D) stably expressing full-length human cMet (T47D/cMet; a generous gift of M. Park, McGill University) are maintained in Dulbecco's Modified Eagle's Medium (DMEM) containing 10% fetal bovine serum (FBS), supplemented with 400 μg/mL G418 as explained elsewhere *(7,8)*.
2. Phosphate-buffered saline (PBS): To make a 10xPBS stock solution, dissolve 80 g NaCl, 2 g KCl, 11.5 g Na_2HPO_4, and 2 g KH_2PO_4 in distilled water and adjust the final volume to 1 L, store at room temperature. PBS is prepared using the 10xPBS stock diluted in distilled water and stored at room temperature.
3. FBS: aliquot and store at −20°C.

Analysis of Receptor Tyrosine Kinase Internalization

4. Fluorescence-activated cell sorting (FACS) buffer: PBS prechilled to 4°C and supplemented with 2% FBS immediately before use. Prechill to 4°C before use.
5. Acid stripping buffer: DMEM supplemented with 0.2% bovine serum albumin (BSA) and with the pH adjusted to pH 3.5 using HCl.
6. 4% Formaldehyde: A 4% formaldehyde solution in PBS is prepared immediately before use from a 16% formaldehyde stock (Ted Pella, Inc., Cat. no. 18505).
7. 5 mM ethylenediamine-tetra-acetic acid (EDTA)-PBS: Calcium free PBS (Mediatech, Inc. Cat. no. 21-031-CV) supplemented with 5 mM EDTA pH 8.0.
8. Goat anti-HGF receptor (HGFR): R&D Systems Inc. (Cat. no. AF276), aliquot and store at −20°C.
9. Chicken anti-goat Alexa^{488}-conjugated secondary antibody: Invitrogen (Cat. no. A21467), aliquot and store at 4°C.
10. Recombinant human HGF(rhHGF): Pepro Tech Inc. (Cat. no. 100-39), aliquot and store at −20°C.

2.2. Preparation and Labeling of Recombinant InlB

1. Recombinant InlB: pKI22 expressing recombinant InlB containing an N-terminal His$_6$ tag was expressed in BL21 DE3 cells and isolated by Ni$^+$ chromatography as previously described *(10)*. A protein fraction enriched in recombinant InlB was desalted (HiPrep 26/10 Desalting column, Amersham Pharmacia Cat. no. 17-5087-01), purified by ion-exchange chromatography (HiTrap SP Amersham Pharmacia, Cat. no. 17-1152-01) and eluted with PBS containing 0.4 to 0.5 M NaCl *(10)*. Pure recombinant InlB was Amicon concentrated to 2 to 3 mg/mL of protein (BCATM Protein Assay Kit, Pierce Cat. no. 23225), exchanged into PBS and stored in aliquots at −80°C.
2. Alexa^{488}-antibody labeling kit: from Invitrogen (Cat.no. A-20181), store at −80°C.
3. P6 micro bio-spin chromatography columns: Bio-Rad Laboratories (Cat.no. 732-6221), store at 4°C.
4. Cover slips: VWR, selected micro-cover glasses 12 mm round, no. 1 thickness (Cat. no. 48366-251).
5. Nunclon*△ 4-well sterile tissue culture dishes (VWR, Cat. no. 62407-068).
6. 0.45 M sucrose in DMEM: solution is made by dissolving 7.7 g of sucrose (Fisher Scientific, Cat. no. S5-3) in 50 mL of DMEM without serum.
7. Alexa^{594}-Transferrin (Tfn): molecular probes (Invitrogen, Cat. no. T-13343).
8. Glass slides: FISHER finest Premium Microscope Slides Frosted $3'' \times 1'' \times 1$ mm (Cat. no. 12-544-3).
9. Mounting reagent: FluorSave™ Reagent (Calbiochem, Cat. no. 345789).
10. Zeiss LSM 510 Meta microscope (Carl Zeiss, Inc.).

3. Methods

In this section, we describe two flow cytometry-based assays to examine cMet internalization. We previously reported that the *Listeria* protein InlB mimics HGF-induced cMet internalization through clathrin-coated pits and receptor degradation in lysosomes *(7)*. Accordingly, InlB is an effective ligand to use with HGF to examine cMet endocytosis under normal conditions, a prerequisite for studies that examine the role of receptor endocytosis as a determinant of cMet signaling in metastatic cancers. To obtain reproducible and reliable information, it is often necessary to utilize complementary approaches that distinguish receptor internalization from ligand uptake. The first method we describe measures the relative amount of residual cMet remaining on the cell surface following ligand-induced receptor internalization. Following treatment with ligand for predetermined times, the cells are removed from the plates and stained under nonpermeable conditions using an antibody specific for the extracellular domain of cMet.

The second method examines the internalization of fluorescently labeled HGF or InlB under steady-state conditions. In this assay, serum starved cells are incubated at 37°C in media containing fluorescently labeled ligand to trigger the internalization of receptor/ligand complexes. Noninternalized ligand is removed from the cell surface using an acid-washing protocol, the cells are then harvested and processed for flow cytometry. The level of internalized fluorescently labeled ligand is used as an indicator for receptor internalization. Conditions for ligand labeling with the fluorescent dye Alexa^{488} are described.

3.1. Internalization Assay Using cMet Cell Surface Staining

1. All of the solutions required for cell treatment are prepared in advance as required. T47D/cMet cells are allowed to adhere to 60-mm plates at a density of 6×10^5 cells/plate for each data point (*see* **Note 1**). Additional control plates for negative controls are also required (*see* **step 12**).
2. The following day when the cells are 80% confluent, the cells are serum starved for 5 h by incubation at 37°C in prewarmed DMEM (without serum; *see* **Note 2**).
3. To initiate receptor internalization, the cells are incubated in prewarmed DMEM containing 100 ng/mL HGF or InlB at 37°C for 5, 10, or 15 min. One plate is incubated in DMEM without ligand for 15 min at 37°C as a control.
4. Receptor internalization is halted by shifting the cells to ice followed by two washes in ice-cold PBS for 10 min each at 4°C with gentle shaking.
5. Noninternalized ligand is stripped from the cell surface using freshly prepared, ice-cold acid stripping buffer (DMEM/0.2% BSA, pH 3.5 adjusted with HCl), three times for 5 min each on a shaking platform (*see* **Note 3**). The cells are then washed with ice-cold PBS three times for 5 min each with gentle shaking.
6. Incubate the cells in prechilled FACS buffer containing 3μg/mL of goat anti-HGFR antibody at 4°C for 1 h (*see* **Notes 4** and **5**).
7. After washing the cells three times with ice-cold PBS for 5 min each on a shaking platform, the cells are incubated in prechilled FACS buffer containing 8 μg/mL of the appropriate Alexa^{488}-conjugated secondary antibody for 1 h on ice.

8. Following three 5 min washes with ice-cold PBS, the cells are removed from each plate using 1 mL of 5 mM EDTA-PBS prewarmed to 37°C for 10 min (*see* **Notes 6** and **7**).
9. Transfer each cell suspension to individually labeled microfuge tubes and pellet the cells by centrifugation at 1,100g for 3 min at 4°C.
10. Following three washes in ice-cold FACS buffer, resuspend the cells in FACS buffer at a density of 10^6 cells per milliliter. If the sample will not be analyzed immediately, add an equal volume of 4% formaldehyde to fix the cells and store in the dark at 4°C.
11. Prior to analysis by flow cytometry using a BD FACSCanto™ instrument, the fixed cells are washed three times with ice-cold FACS buffer by centrifugation at 1,100g for 2 min each. Typically 20,000 cells are acquired for each time point (*see* **Note 8**).
12. Data analysis to quantify changes in the mean surface receptor fluorescence values was performed using FACS DIVA software version 5.0.1 (BD Biosciences). The mean fluorescence intensity (MFI) of the cells for each data set is compared to negative control plates stained with either the secondary antibody only or with a nonspecific primary antibody and the secondary antibody. The relative percentage of residual cell surface cMet at each time point (t_x) is calculated relative to the MFI of cells without internalization (t_0) as (MFI t_x – MFI control antibody only/MFI t_0 – MFI control antibody only) × 100. An example of the results is shown in **Fig. 1**.

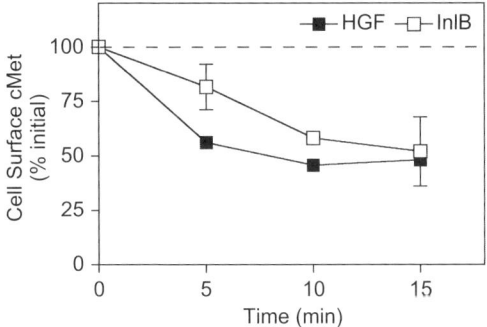

Fig. 1. cMet internalization in response to HGF or InlB. T47D/cMet cells were left untreated with ligand or incubated in media containing 100 ng/mL of HGF or InlB for 5, 10, or 15 min at 37°C. The cells were chilled to 4°C and cMet remaining on the cell surface was specifically labeled with anti-HGFR antibody using non permeabilized conditions and quantified by flow cytometry. Results represent the mean fluorescence intensities (MFI) normalized to untreated control cells under each experimental condition from duplicate experiments. Values represent the means for each data set with standard error. The results indicate that treatment with HGF or InlB results in a decrease in surface cMet levels, consistent with receptor internalization.

3.2. Conjugation of Alexa^{488} to HGF and InlB

1. Add 200 μL of a 2 mg/mL solution of InlB or HGF (*see* **Note 9**) with 20 uL of freshly prepared 1.0 mM sodium bicarbonate to increase the pH of the reaction mixture to approx 8.0 for optimal ligand labeling.
2. Transfer the entire reaction mixture to a single reaction vial containing the appropriate, unconjugated fluorophore. Although we typically use Alexa^{488} for our labeling studies; we have also had good results using Alexa^{594}.
3. Incubate the reaction for 1 h at room temperature in the dark with occasional shaking, followed by an overnight incubation in the dark at 4°C mixing end-over-end (*see* **Note 10**).
4. The following morning, a P6 micro bio-spin column is pre-equilibrated with PBS by washing three times with 500 μL of PBS by centrifugation at 1,100g for 2 min. Transfer the pre-equilibrated tube to a new prelabeled microfuge tube.
5. The labeling reaction must be clarified by centrifugation at 10,000g for 10 min at 4°C to remove excess aggregates. The resulting supernatant from the labeling reaction is then carefully transferred to the pre-equilibrated P6 micro bio-spin column and centrifuged at 1,100g for 4 min at room temperature. The combination of a clarification step with the spin column ensures the removal of unconjugated dye that would otherwise contribute to nonspecific, high background fluorescence.
6. The fluorescently labeled ligand is recovered in the column eluent. Dilute a sample of the labeled ligand and measure the absorbance of the conjugated solution at 280 nm and 494 nm in a cuvette with a 1 cm path length. The concentration of incorporated label into the ligand is determined using the following equation: Protein concentration (M) = [A_{280}-(A_{494} × 0.11)] × dilution factor/ molecular extinction coefficient. The molecular extinction coefficient for HGF and InlB is measured using the following equation: $\varepsilon = A_{280}/cl$ where c is the concentration of the protein (moles/L) and l is the length of the light-absorbing samples (cm).
7. The labeled ligand is aliquoted and stored at −80°C (*see* **Note 11**).

3.3. Confocal Microscopy to Alexa^{488}-Labeling

1. Cover slips that have been precleaned with 70% ethanol under sterile conditions are placed into 4-well plates in preparation for cell plating. When the cover slips have air dried, plate duplicate sets of T47D/cMet cells at a density of 5× 10^4 cells per well and allow them to adhere overnight.
2. The following morning cells are serum starved for 5 h at 37°C by incubation with prewarmed DMEM (without serum).
3. One set of cover slips is washed with prewarmed serum free DMEM for 15 min at 37°C. The second set of cover slips is incubated in prewarmed DMEM containing 0.45 M sucrose for 15 min at 37°C to produce a hypertonic condition that inhibits endocytosis (*see* **Note 12**).

4. The appropriate cover slips are then incubated in prewarmed DMEM with or without 0.45 M sucrose supplemented with 100 ng/mL Alexa[488]-labeled InlB and 5 μg/mL Alexa[594]-Tfn for 15 min at 37°C (*see* **Note 13**).
5. The plate containing the cover slips is rapidly transferred to ice and washed three times with ice-cold PBS for 5 min each on a shaking platform to remove excess ligand.
6. The PBS is gently removed by aspiration and 500 μL of 4% formaldehyde is added to each well for 10 min at room temperature.
7. Following fixation the formaldehyde is discarded (into a hazardous waste container) and the cells are washed three times at room temperature with PBS to remove residual formaldehyde.
8. Each cover slip is removed from the well using forceps and carefully air dried to remove excess solution. One drop of FluorSave™ Reagent is placed on a clean glass slide and the cover slip is inverted into it so that the surface with the attached cells is directly in contact with the mounting reagent.
9. The slides are allowed to set overnight in the dark to avoid photobleaching of the fluorophore and stored at 4°C in the dark prior to imaging.
10. Images are captured using a Zeiss LSM 510 Meta confocal microscope and a 63x Oil objective. An example of the results is shown in **Fig. 2**.

3.4. Steady-State Internalization of Fluorescently Labeled Ligand

1. T47D/cMet cells are allowed to adhere onto 60-mm plates at a density of 6×10^5 cells/plate. The cells are serum starved for 5 h by incubation at 37°C in DMEM only.
2. The following morning, the cells are incubated in prewarmed DMEM supplemented with 100 ng/mL Alexa[488]-labeled HGF or Alexa[488]-labeled InlB at 37°C for 5, 10, or 15 min to promote cMet internalization. Controls include serum starved cells that were treated with media lacking ligand.
3. The cells are rapidly placed on ice to halt receptor trafficking and then washed three times with ice-cold PBS for 5 min each on a shaking platform to remove excess ligand.
4. Noninternalized ligand is removed from the cell surface using three, 5 min washes in ice-cold acid stripping buffer followed by three 5 min washes in ice-cold PBS. Plates are routinely covered with aluminum foil during the washes to minimize bleaching of the fluorophore.
5. The washed cells are then removed from each plate by incubation in 1 mL of prewarmed 5 mM EDTA-PBS at 37°C for 10 min and transferred into prelabeled microfuge tubes.
6. The cells are recovered by centrifugation at 1100*g* for 3 min at 4°C, followed by three washes in ice-cold FACS buffer.

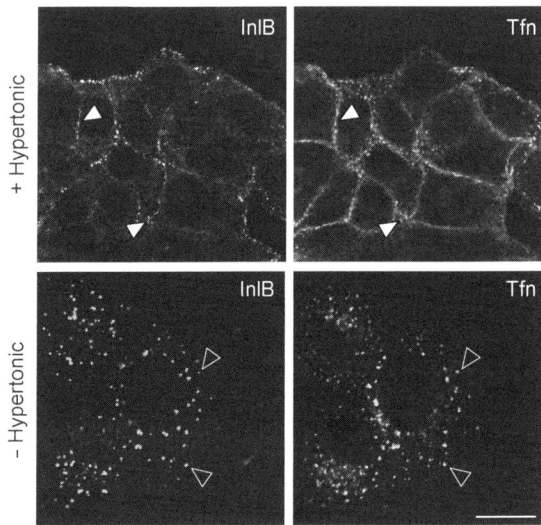

Fig. 2. Alexa[488]-labeled InlB internalization is blocked by hypertonic shock. T47D/cMet were cells preincubated in DMEM (– hypertonic) or containing 0.45 M sucrose (+ hypertonic) for 15 min at 37°C. The cells were then incubated in the corresponding media supplemented with 100 ng/mL of Alexa[488]-labeled InlB and 5 μg/mL Alexa [594]-labeled Transferrin (Tfn) for 15 min at 37°C and then processed for confocal microscopy. Representative examples are shown (scale, 10 μm). Hypertonic treatment results in the colocalization of InlB and Tfn on the cell surface (closed arrows) consistent with the inhibition of cMet internalization. Conversely, internalized InlB colocalizes with internalized Tfn (open arrows) in non hypertonic cells and appear as punctate spots reminiscent of early endosomes.

7. The cells are then gently resuspended in FACS buffer to a density of 10^6 cells per milliliter by trituration. If the sample will not be analyzed immediately, the cells are fixed by the addition of an equal volume of 4% formaldehyde and stored in the dark at 4°C.

8. The MFI of the cells is typically measured by flow cytometry using a BD FACSCanto™ instrument and FACS DIVA software version 5.0.1 (BD Biosciences). Fixed cells are first washed with ice-cold FACS buffer three times as described above prior to analysis. Typically 20,000 cells are acquired for each time point. To quantify changes in the value of internalized MFI (MFI_{int}), the MFI of the cells at each time point (t_x) is compared to control cells that were untreated with ligand ($MFI_{Control}$) and calculated using the following equation: $MFI_{int} = MFI\ t_x - MFI_{Control}$ (see **Note 14**). An example of the results is shown in **Fig. 3**.

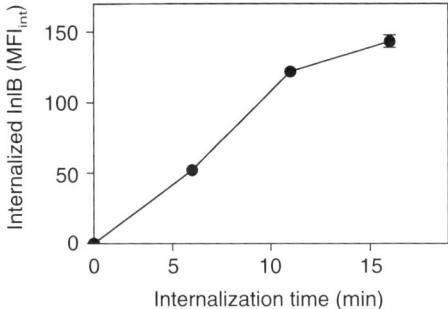

Fig. 3. Internalization of Alexa^{488}-labeled InlB. T47D/cMet cells were treated in DMEM with or without 100 ng/mL of Alexa^{488}-labeled InlB at 37°C for the indicated times. Following incubation the cells were chilled to 4°C, noninternalized ligand was removed by acid washing and the cells processed for flow cytometry. Results represent the internal mean fluorescence intensity (MFI$_{int}$) normalized to untreated control cells under each experimental condition from duplicate experiments. Values represent the means for each data set with standard error. The results show a rapid increase in internal InlB fluorescence consistent with the internalization of the cMet receptor in response to InlB.

4. Notes

1. Cell density is a critical determinant for the success of this approach. High cell densities could affect the efficiency of ligand binding to the receptor and hence downstream cell-signaling responses. For example, cell density has been shown to affect the internalization of the epidermal growth factor receptor (EGFR) and the formation of the endocytic Cbl-endophilin complex in HEK 293 cells *(14)*.
2. Serum starvation has the advantage that it can synchronize cell-signaling responses and receptor internalization events. Since the optimal time for serum starvation varies considerably between different cell lines and for different receptor systems, this step needs to be empirically determined. Growth hormones such as HGF and EGF normally exist in serum resulting in high levels of basal receptor activation and hence receptor internalization in cells cultured in media supplemented with serum. While growth factors such as EGF can be readily removed from sera by dialysis, the large size of HGF necessitates the use of cell incubations under serum-free conditions. We routinely access the efficacy of our serum starvation conditions for different cell lines using Western analysis with antibodies specific for two key tyrosine residues in the cMet kinase domain (Y1234, Y1235) that are phosphorylated in response to ligand binding *(7,8,10)*. Another potential concern is crosstalk between different RTK systems. For example, in human hepatocellular (HepG2 and HuH7) and pancreatic (DAN-G) carcinoma

cells activation of the EGFR by EGF also results in the transactivation and internalization of cMet *(15)*. Accordingly, it is best to test for receptor crosstalk by Western analysis prior to experimentation.
3. Acid stripping cells by washing with low pH solutions readily dissociates most ligands from their receptors *(4,13,16)*. However the optimal acid stripping conditions for removing cell surface associated ligand without damaging the cells must be empirically determined for each cell line and for each receptor/ligand combination. Some cell types do not tolerate low pH washes (e.g., pH 3.5). In these situations, we have washed cells using solutions with a slightly higher pH (e.g., pH 4.0) for longer time periods, taking care to monitor the response of the cells morphologically. We routinely monitor the effectiveness of the acid washing conditions for the removal of cell surface associated fluorescent ligand by confocal microscopy or flow cytometry.
4. Although fetal bovine sera or goat sera are generally good blocking reagents for these types of studies, it is preferable to use serum from the same species as the fluorophore-conjugated secondary antibody as a blocking reagent, to reduce any nonspecific binding events and lower background fluorescence accordingly. The use of BSA as a blocking agent may introduce other nonspecific binding sites that could increase background fluorescence.
5. The primary and secondary antibodies should be used at saturating concentrations to ensure complete labeling of cell surface cMet. The working concentration for each antibody is determined empirically. It is important to note that the saturating concentration of each antibody may vary between cell lines due to differences in the cell surface levels of cMet.
6. We routinely use 5 mM EDTA-PBS to detach adherent cells from plates and preserve the integrity of the cell surface receptors. However, some tightly adherent lines require the use of 0.25% Trypsin-EDTA in HBSS (Cellgro, Cat. no. 25-053-CI) for detachment. Since trypsin is a protease, trypsin treatment will result in the proteolytic cleavage of the extracellular domain of cMet and hence will result in reduced levels of cell surface staining. In these situations, we use flow cytometry to monitor the uptake of fluorescently labeled ligand as an alternative approach (*see* **Subheading 3.4.**).
7. Alternatively, ice-cold 5 mM EDTA-PBS can be used to harvest the cells prior to staining. This approach has the advantage that it utilizes less antibody during the labeling reaction. Following isolation, the cells are pelleted in microfuge tubes by centrifugation at 1100*g* for 2 min at 4°C and resuspended in FACS buffer at a concentration of 10^6 cells per milliliter prior to antibody staining. We routinely confirm cell density on a sample of the cells using a hemocytometer. If the cell density is too great it will affect the efficiency of subsequent cell surface staining. In this situation, the cells can be diluted to the correct cell density using additional FACS buffer. If the cell density is insufficient, the cells can be concentrated by centrifugation and resuspended in a smaller volume of FACS buffer.

8. Live cells can be distinguished from dead cells using the cell impermeant stain propidium iodide (PI). When bound to DNA the fluorescence of PI is enhanced 20 to 30 fold. Accordingly, dead cells and not live cells will stain strongly with PI. Incubate cells with 5 μg/mL PI (use 1 mg/mL stock solution diluted into FACS buffer) at room temperature for 15 min immediately prior to flow cytometry analysis.
9. Purified proteins must be resuspended in a buffer free of ammonium ions or primary amines to avoid competition with the amine groups of the reactive dye and inefficient protein labeling.
10. Under our ligand-labeling conditions, we estimate that each molecule of ligand is labeled with four to six molecules of dye without any apparent loss in biological activity. Since the kinetics of the labeling reaction could vary for different proteins, it is important to empirically determine the optimal labeling conditions for each protein. Over labeling could result in loss of protein function, reduce the affinity of the protein for its target or cause protein aggregation and hence nonspecific binding.
11. We have found that the conjugates are stable for at least 6 mo when aliquoted and stored at $-80°C$. We do not add sodium azide to the conjugates as this may confer some cell toxicity in subsequent experiments.
12. The treatment of cells with hypertonic media inhibits clathrin-mediated receptor internalization *(17)*. Although its precise mechanism of action remains unclear, hypertonic shock has been shown to cause a decrease in the size and number of clathrin-coated pits at the plasma membrane *(17)*. It is important to note that the inhibitory effect of hypertonic media on endocytosis is rapidly reversible. Thus ligand must be added to the cells in hypertonic media in order to obtain optimal inhibition of receptor endocytosis.
13. Since Tfn and its receptor are internalized exclusively via clathrin-coated pits, they are routinely used as specific markers for this endocytic route.
14. These methods can be used to study the roles of different molecules in RTK internalization by transfecting plasmids expressing dominant negative mutant proteins. Optimally it is best to use compatible fluorescent markers to identify the subpopulation of transfected cells that do not interfere with the fluorophores chosen to monitor ligand internalization or receptor retention on the cell surface. When studying a rare event in the cell population, we often increase the resuspension volume of the sample to compensate for the 200 μL void volume associated with the BD FACSCanto™ instrument.

Acknowledgments

This work was supported by grants from the National Science Foundation (IBN-343739) and the National Institutes of Health (CA-112605 and CA-119075) to L.A.E. We would like to thank Mark Griffin in the UTMB at Galveston Flow Cytometry and Cell Sorting Core Facility and Marta Lorinczi for their technical assistance.

References

1. Booden, M. A., Eckert, L. B., Der, C. J., and Trejo, J. (2004) Persistent signaling by dysregulated thrombin receptor trafficking promotes breast carcinoma cell invasion. *Mol Cell.* **24**, 1990–1999.
2. Oosterhoff, J. K., Kuhne, L. C., Grootegoed, J. A., and Blok, L. J. (2005) EGF signalling in prostate cancer cell lines is inhibited by a high expression level of the endocytosis protein REPS2. *Int. J. Cancer.* **113**, 561–567.
3. Li, L. and Cohen, S. N. (1996) Tsg101: a novel tumor susceptibility gene isolated by controlled homozygous functional knockout of allelic loci in mammalian cells. *Cell* **85**, 319–329.
4. Longva, K. E., Blystad, F. D., Stang, E., Larsen, A. M., Johannessen, L. E., and Madshus, I. H. (2002) Ubiquitination and proteasomal activity is required for transport of the EGF receptor to inner membranes of multivesicular bodies. *J. Cell. Biol.* **156**, 843–854.
5. Peschard, P., Fournier, T. M., Lamorte, L., et al. (2001). Mutation of the c-Cbl TKB domain binding site on the Met receptor tyrosine kinase converts it into a transforming protein. *Mol. Cell.* **8**, 995–1004.
6. Peschard, P. and Park, M. (2003) Escape from Cbl-mediated downregulation: a recurrent theme for oncogenic deregulation of receptor tyrosine kinases. *Cancer Cell* **3**, 519–523.
7. Li, N., Xiang, G. S., Dokainish, H., Ireton, K., and Elferink, L. A. (2005) The Listeria protein internalin B mimics hepatocyte growth factor-induced receptor trafficking. *Traffic* **6**, 459–473.
8. Shen, Y., Naujokas, M., Park, M., and Ireton, K. (2000) InlB-dependent internalization of Listeria is mediated by the Met receptor tyrosine kinase. *Cell* **103**, 501–510.
9. Veiga, E., and Cossart, P. (2005) Listeria hijacks the clathrin-dependent endocytic machinery to invade mammalian cells. *Nature Cell Biol.* **7**, 894–900.
10. Ireton, K., Payrastre, B., and Cossart, P. (1999) The Listeria monocytogenes protein InlB is an agonist of mammalian phosphoinositide 3-kinase. *J. Biol. Chem.* **274**, 17025–17032.
11. Birchmeier, C., Birchmeier, W., Gherardi, E., and Vande Woude, G. F. (2003) Met, metastasis, motility and more. *Nature Rev. Mol. Cell Biol.* **4**, 915–925.
12. Gao, C. F., and Vande Woude, G. F. (2005) HGF/SF-Met signaling in tumor progression. *Cell Res.* **15**, 49–51.
13. Abella, J. V., Peschard, P., Naujokas, M. A., et al. (2005) Met/Hepatocyte growth factor receptor ubiquitination suppresses transformation and is required for Hrs phosphorylation. *Mol. Cell Biol.* **25**, 9632–9645.
14. Schmidt, M. H., Furnari, F. B., Cavenee, W. K., and Bogler, O. (2003) Epidermal growth factor receptor signaling intensity determines intracellular protein interactions, ubiquitination, and internalization. *Proc. Natl. Acad. Sci. USA.* **100**, 6505–6510.
15. Fischer, O. M., Giordano, S., Commoglio, P. M., and Ulrich, A. (2004) Reactive oxygen species mediate Met receptor transactivation by G protein-coupled receptors and the epidermal growth factor receptor in human carcinoma cells. *J. Biol. Chem.* **279**, 28970–28978.

16. Klausner R. D., Ashwell G., van Renswoude J., Harford J. B., Bridges K. R. (1983) Binding of apotransferrin to K562 cells: explanation of the transferrin cycle. *Proc Natl Acad Sci USA.* **80**, 2263–6.
17. Heuser, J E., and Anderson, R. G. W. (1989) Hypertonic Media Inhibit Receptor-mediated Endocytosis by Blocking Clathrin-coated Pit Formation. *J.Cell Biol.* **108**, 389–400.

24

Measurement of Receptor Endocytosis and Recycling

Jane M. Knisely, Jiyeon Lee, and Guojun Bu

Summary

Receptor trafficking is essential to the delivery of nutrients and to the proper regulation of signaling pathways in mammalian cells. Numerous transmembrane receptors undergo clathrin-mediated endocytosis, followed by sorting in the early endosome. The low-density lipoprotein (LDL) receptor-related protein (LRP) is a multiligand endocytic receptor and a member of the LDL receptor family. At the cell surface, it binds to and continuously internalizes numerous ligands including lipoproteins, proteases, protease inhibitors, growth factors, and β-amyloid precursor protein via clathrin-mediated endocytosis. Its rapid endocytosis rate allows efficient clearance of extracellular and transmembrane ligands. Once internalized into the early or sorting endosome, LRP ligands are delivered to lysosomes to be degraded, whereas LRP is efficiently recycled to the plasma membrane. Herein, the authors describe quantitative methods to measure the endocytosis and recycling capacity of receptors, using LRP as a model receptor.

Key words: Endocytosis assay; flow cytometry; iodination; LRP; quenching antibody; recycling assay; TCA precipitation.

1. Introduction

Transmembrane receptors undergo sorting at multiple steps in the endocytic pathway. After trafficking through the secretory pathway, most cargo receptors bind to their cognate ligands at the cell surface, are internalized via a clathrin-mediated process, and are then subject to sorting in the early or sorting endosome. This step separates receptors that are destined to be degraded in the multivesicular bodies/lysosomes from those that will be recycled to the cell surface to undergo additional cycles of ligand binding and endocytosis *(1)*.

Cytoplasmic adaptor proteins have long been known to participate in clathrin-mediated sorting at the plasma membrane *(2)*, and more recently, the role of such adaptor proteins in sorting at the recycling step has been recognized *(3,4)*.

One well-studied cargo receptor known to undergo both rapid endocytosis and recycling to the cell surface is the low density lipoprotein (LDL) receptor-related protein (LRP), an extremely large member of the LDL receptor family *(5)*. LRP binds and internalizes numerous ligands, including lipoproteins, proteases, protease inhibitors, growth factors, and bacterial and viral proteins *(5)*. In addition, LRP has also been shown to function as a signaling receptor, and several cytoplasmic adaptor proteins (Dab1, PSD95, etc.) bind its 100 amino acid cytoplasmic tail *(6–8)*. The function of LRP in both endocytosis and signal transduction requires a tight regulation of the cell surface levels of LRP.

The 100 amino acid cytosolic tail of LRP contains two NPxY motifs, two di-leucine motifs and an YxxL motif. The YxxL and distal di-leucine motifs are the primary endocytosis signals *(9)*. The rapid endocytosis of LRP enables its ligands to be internalized and delivered to the lysosome for degradation. Subsequently, LRP is recycled to the plasma membrane. It has been shown that the proximal NPxY motif binds to SNX17, a member of Sorting Nexin family, and promotes LRP recycling *(3)*. Because of its large size, it is difficult to manipulate and transfect full-length LRP cDNA into cells. To circumvent this issue, the authors' laboratory has generated a set of LRP minireceptors (mLRP1-4), each consisting of one of the four extracellular ligand-binding domains fused to the transmembrane domain and cytoplasmic tail of LRP *(10)*. Each minireceptor contains an HA epitope at the N-terminus, allowing for easy detection of the transfected receptor.

Detailed here are two quantitative assays designed to follow the trafficking of cargo receptors through the endocytosis and recycling processes. The HA-tagged domain IV-containing LRP minireceptor (mLRP4) is used as an example in these protocols. Presented first is the kinetic analysis of endocytosis, which involves the use of ^{125}I-labeled ligand or antibody (*see* **Note 1**) to quantitatively determine the endocytic rate of wild-type and tailless (lacks the cytoplasmic tail) mLRP4. Second, a flow cytometry-based recycling assay, adapted from Austin et al. *(3,11,12)*, is used to demonstrate the different recycling rates displayed by wild-type mLRP4 and an mLRP4 mutant deficient in SNX17 binding.

2. Materials

2.1. Endocytosis Assay

2.1.1. Preparation of Iodogen Tubes

1. 12 × 75-mm glass (borosilicate) tubes (Fisher).
2. Iodogen powder (Pierce no. 28600).

3. Desiccator filled with dri-rite (with a rack for tubes inside).
4. Nitrogen gas tank (Airgas).
5. 5¾" pasteur pipet (Fisher).
6. Chloroform (Sigma).

2.1.2. Iodination of Ligand or Antibody

1. Handheld survey meter for γ-radiation.
2. Ring/clamp stand.
3. PD10 desalting column (GE Healthcare no. 17-0851-01).
4. Aerosol-barrier tips.
5. 50-mL conical centrifuge tubes.
6. Empty lead pig for storage of iodinated protein.
7. PBST buffer: phosphate-buffered saline + 0.01% Tween-20 (Sigma); stored at room temperature.
8. Cytochrome c (Sigma), 20 mg/mL solution in ddH$_2$O, stored at 4°C.
9. D-Tyrosine (Sigma): 10 mg/mL in ddH$_2$O (note: this is a saturated solution, so it will be cloudy; tap to resuspend before use), stored at 4°C.
10. 0.5 M sodium phosphate buffer, pH 7.4, stored at room temperature.
11. Carrier-free ^{125}I isotope (100 mCi/mL; Perkin Elmer Life Science).
12. 50 μg of protein to be iodinated, preferably in a volume of 90 μL or less.

2.1.3. Trichloroacetic Acid Precipitation to Determine the Percent of Incorporated Isotope

1. Cold PBS (pH 7.4).
2. 50 mg/mL solution of bovine serum albumin (BSA; EMD Bioscience), prepared in PBS, sterile filtered, and stored at 4°C.
3. 100% w/v TCA in ddH$_2$O (TCA, Sigma).
4. Gamma tubes (Fisher no. 14-959-5).

2.1.4. Kinetic Analysis of Endocytosis

1. Chinese Hamster Ovary (CHO) cell culture medium: Ham's F-12 supplemented with 10% defined fetal bovine serum (FBS; Sigma) and 2 mM L-Glutamine (Invitrogen), stored at 4°C.
2. Mirus TransIT CHO transfection kit (Mirus Bioscience).
3. Opti-Mem low serum media (Invitrogen).
4. Trypsin-ethylenediamine-tetraacetic acid (EDTA) (Invitrogen)
5. Ligand-binding buffer: 0.6 g/L BSA in Dubelcco's Modified Eagle's Medium (DMEM), sterile filtered and stored at 4°C.
6. Stop/strip solution: 0.2 M acetic acid, pH 2.6, 0.1 M NaCl, or PBS pH 2.0, stored at 4°C.

7. Low sodium dodecyl sulfate (SDS) lysis buffer: 62.5 mM Tris-HCl, pH 6.8, 0.2% SDS, 10% (v/v) glycerol.

2.2. Recycling Assay

2.2.1. Conjugation of Alexa Fluor 488 to Anti-HA Antibody

1. Dialysis buffer: 0.1 M sodium carbonate, pH 9.3, sterile filtered and stored at 4°C.
2. Snakeskin dialysis tubing (along with dialysis clips) or Slide-A-Lyzer Cassettes with buoys (Pierce).
3. Anti-HA immunoglobulin (Ig)G (Covance).
4. Alexa Fluor 488 Protein Labeling Kit (Invitrogen Cat. no. A10235), consisting of the following:

    a. Alexa Fluor 488 reactive dye (Component A)
    b. Sodium bicarbonate (Component B, MW = 84)
    c. Purification resin (Component C)
    d. Elution buffer (Component D)
    e. Purification columns
    f. Column funnels
    g. Foam column holders
    h. Disposable pipets
    i. Collection tubes
    *The kit can be stored under the conditions listed. For optimal storage conditions of individual components, refer to the labels on the vials or bags.

5. Bio-Rad protein assay dye reagent (Bio-Rad).
6. Disposable cuvettes (Fisher).

2.2.2. Kinetic Analysis of Recycling

1. U87 culture medium: DMEM (Sigma), supplemented with 10% defined FBS, 1 mM sodium pyruvate (Invitrogen), and 350 μg/mL G418.
2. G418 disulfate (Sigma) stock solution: 175 mg/mL in ddH$_2$O, sterile filtered and stored at –20°C.
3. Trypsin-EDTA (Invitrogen).
4. Alexa Fluor 488 conjugated antibodies recognizing extracellular domain of receptor: light sensitive, wrap with foil and store at –20°C or –80°C.
5. Ligand-binding buffer: *see* **Subheading 2.1.4**.
6. Quenching antibody: anti-Alexa Fluor 488 IgG, stored at –20°C or –80°C.
7. Flow cytometry buffer (PFN): PBS supplemented with 1.5% heat-inactivated FBS and 0.1% NaN$_3$, sterile filtered and stored at 4°C.
8. Cell dissociation solution (Sigma).
9. PBS.
10. Fluorescence-activated cell sorter tubes (VWR no. 60818-306).

3. Methods

3.1. Endocytosis Assay

For this assay, antibody to the extracellular region of the receptor or ligand is iodinated using the iodogen method. CHO cells are transfected with wild-type and mutant constructs to be tested, split to five 12-well plates, and assayed the next day. Specifically, iodinated antibody/ligand is added to the cells on ice and allowed to bind to cell surface receptors at 4°C. Next, each plate is warmed to 37°C for different amounts of time to initiate internalization of the receptor. After each warm up, the plates are placed back on ice and exposed to an acidic solution to strip the antibody/ligand from the cell surface. The cells are solubilized, and both the acid wash (cell surface) and lysate (internalized) fractions are counted in a γ-counter. The percent internalized at each time point is equal to the amount of internalized ligand divided by the total amount of cell-associated ligand (cell surface plus internalized), and can be plotted against time to obtain a kinetic internalization curve, as well as a half time for endocytosis.

3.1.1. Preparation of Iodogen Tubes

1. Make iodogen solution (1 mg/mL in chloroform). Keep covered with parafilm.
2. Connect Pasteur pipet to N2 outflow hose, start N2 flow. Pressure should not be too harsh (check on a tube with water inside, outflow should not disturb the surface tension).
3. Pipet 10 µL iodogen solution into the bottom of glass tube (do this in batches of two to three).
4. Dry the droplet with the N2 flow. Be sure to keep the droplet centered on the bottom—aim for a small "button" of dried iodogen on the inside bottom of the tube. Repeat with remaining tubes.
5. Coated tubes will retain activity for 2 yr if stored in a desiccator at room temperature.

3.1.2. Iodination of Ligand or Antibody to Follow Internalization

1. Cut open the tip of a PD10 desalting column to start drainage. Set up on the ring/clamp stand.
2. Set up one 50-mL conical tube in a Styrofoam stand under the PD10 column so as to collect drainage (be sure to secure everything so that the tube does not tip over).
3. Equilibrate the column with 20 to 25 mL PBST (do this while you are setting up all other materials). Once equilibrated, maintain about 500 µL of PBST on top of column until ready to load sample.

4. Perform iodination according to the radiation guidelines of your institution, preferably in a designated hood. Always wear disposable gloves and a protective lab coat when working with radioactivity, and don a lead apron for extra protection. Also, be sure to use aerosol barrier tips to prevent internal contamination of pipetors.
5. Put the ring/clamp stand and the ^{125}I isotope in the hood on absorbent underpads; also set up waste bag in hood.
6. In the hood, mix 50 μg of the protein to be iodinated (antibody or ligand, *see* **Notes 1** and **2**) and 10 μL (1 mCi) of ^{125}I isotope in the iodogen-coated tube. Bring total volume to 100 μL with phosphate buffer. Tap the tube gently to mix. Incubate at room temperature for 10 min.
7. After 10 min incubation, empty the flow-through from the PD10 column. Then quench the reaction by adding 20 μL D-tyrosine to the iodogen tube, being careful not to touch the sides. Tap gently for 30 s, and then add 20 μL of cytochrome *c* and mix by gentle tapping. It is important to try not to splash the mixture up the sides of the tube, as you will likely contaminate your pipetor.
8. Add the quenched mixture evenly and dropwise to the top of the PD10 column.
9. Add 300 μL of PBST buffer evenly and drop wise to the top of the column, and then add 4 to 5 mL of PBST to the column. Colored band should migrate in a horizontal band as it progresses downward.
10. Start collecting fluid in the labeled screw-cap vial when the colored band reaches the bottom of the column. Your collection should have most of the color in it. Stop collecting before the color fades to avoid collecting any free isotope.
11. Fill the vial to 1 mL with PBST, invert to mix. Store in the empty lead pig at 4°C until use (*see* **Note 3**).
12. Dispose of radioactive waste properly. Be sure to survey your equipment for radioactivity before putting it away.

3.1.3. TCA Precipitation to Determine the Percent of Incorporated Isotope

1. Pipet into triplicate Eppendorf tubes with tight closures: 598 μL PBS, 200 μL 50 mg/mL BSA, 200 μL 100% TCA, and 2 μL iodinated protein. Close lids, vortex, and incubate on ice for 10 min.
2. Centrifuge tubes at 16,000*g* for 10 min at 4°C. Pipet 500 μL of supernatant into a γ-tube, discard the remaining supernatant.
3. Cut the bottom of the tube, which contains the protein pellet, into another γ-tube.
4. Count both sets of tubes in a γ-counter.
5. To determine percent of unincorporated isotope: $\frac{\text{supernatant} \times 2}{\text{pellet}} \times 100$

Average the values from triplicate samples. If the percent incorporation is less than 95%, the iodination should be repeated, taking care not to collect any free isotope migrating at the end of the colored band.

3.1.4. Kinetic Analysis of Endocytosis

1. Plate CHO cells in 6-well plates such that they are 80 to 90% confluent for transfection. Plate four wells for each construct to be assayed (*see* **Note 4**).
2. The next day, transfect cells using the Mirus TransIT CHO kit, according to instructions. Specifically, for each well, combine 200 μL Opti-Mem with 8 μL CHO TransIT reagent in a polystyrene tube. Mix and incubate 15 min at room temperature. Next, add 2 μg maxiprep DNA of the construct to be tested, mix and incubate 15 min at room temperature. Add 3 μL CHO Mojo reagent, mix and incubate 15 min at room temperature. Finally, pipet mixture dropwise onto cells in complete medium, rocking to evenly distribute transfection complexes.
3. At 4 to 12 h post-transfection, trypsinize three wells for each construct, retaining the fourth for assessment of transfection efficiency. Resuspend cells into a total of 18 mL media/trypsinized cells, plate 1 mL/well in triplicate wells of 12-well plates.
4. The next day, perform endocytosis assay. Prewarm 50 mL ligand-binding buffer in a 37°C water bath. Rinse cells twice in ice-cold ligand-binding buffer. (Note: be careful to add the washes to the sides of the wells, so as not to disrupt the monolayers.)
5. Add ^{125}I-ligand at 5 nM final concentration or ^{125}I-antibody at 1 nM final concentration in cold ligand-binding buffer (0.5 mL per well). Allow binding to proceed at 4°C for 40 min with gentle rocking.
6. Remove unbound ligand and then wash cell monolayers three times with cold binding buffer.
7. Add cold stop/strip solution (ligand) or cold PBS pH 2.0 (antibody) to the zero time-point plate, and keep on ice.
8. For the remaining plates: place each one in a 37°C water bath and add 0.6 ml of ligand-binding buffer prewarmed to 37°C quickly but gently to the monolayers to initiate internalization. After each time point (1, 2, 5, and 10 min), quickly place each plate on ice, remove the ligand-binding buffer, and add 0.5 mL cold stop/strip solution (ligand) or PBS pH 2.0 (antibody).
9. Strip ligand/antibody from the cell surface by incubation of cell monolayers with cold stop/strip solution (ligand) or PBS pH 2.0 (antibody) for 20 min (0.5 mL for 10 min, twice). Collect this cell surface fraction to labeled γ-tubes, and count radioactivity in a γ-counter.
10. Solubilize cells with low-SDS lysis buffer. Collect this internalized fraction and count.
11. The sum of internalized and cell surface ligand represents the maximum potential internalization. Calculate the fraction of internalized ligand at each time point, and plot the average of triplicate wells on a scatter plot (**Fig. 1**). The half-time of endocytosis refers to time required to internalize half of the acid-washable ^{125}I-ligand/antibody initially bound to the cell surface.
12. If desired, use the fourth well of transfected cells to assess the transfection efficiency by Western blot. Alternatively, a green fluorescent protein expression construct can be cotransfected into this well, and the transfection efficiency can be easily ascertained using an inverted microscope equipped with the proper filter.

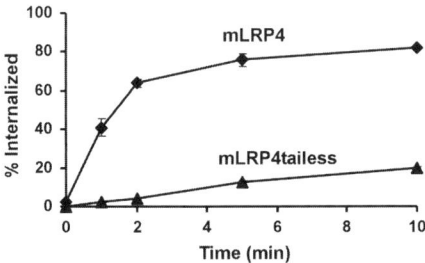

Fig. 1. Example of data derived from the kinetic analysis of endocytosis. The LRP minireceptor, mLRP4, undergoes rapid endocytosis, whereas a minireceptor lacking the cytoplasmic tail internalizes very slowly.

3.2. Recycling Assay

After a short incubation with fluorescently labeled antibody, U87 cells are chased at 37°C for different period times. The chase allows some receptor to be delivered to late endosome/lysosome while some recycle back to the plasma membrane. Recycled receptors will be recognized by anti-Alexa Fluor 488 IgG, which specifically recognizes Alexa Fluor 488, thereby quenching fluorescence. Thus, fluorescence measured after each chase by flow cytometry will represent the amount of receptors remaining inside the cell. Accordingly, the amount of receptors that recycle to the plasma membrane can be calculated by subtracting the remaining fluorescence from the total (0 min) at each time point. Percent of recycled receptor after each chase can be plotted as a function of chase time. As a control experiment, another set of cells can be chased without quenching antibody to determine any nonspecific disappearance of receptor.

3.2.1. Conjugation of Alexa Fluor 488 to Anti-HA Antibody (Slightly Modified From the Manufacturer's Instructions)

1. Dialyze anti-HA IgG against 2 L 0.1 M sodium carbonate (pH 9.3) overnight at 4°C. Place a stirring bar in the beaker so buffer can be kept in uniform solution.
2. The next day, replace buffer with a fresh 2 L and further dialyze for 4 h.
3. Remove protein from dialysis. Measure protein concentration using Bio-rad reagent. Dilute Bio-rad reagent in a 1:5 ratio with ddH$_2$O. Mix 10 µL of dialyzed anti-HA antibody with 1 mL of diluted Bio-rad reagent in a disposable cuvette and measure the absorbance at 595 nm in a spectrophotometer. Blank sample should contain 10 µL 0.1 M sodium carbonate. Calculate the concentration of protein according to the standard curve made by BSA.
4. Prepare a 1 M solution of sodium bicarbonate by adding 1 mL of ddH$_2$O to the provided vial of sodium bicarbonate (component B). Vortex or pipet up and down until fully dissolved.

5. Adjust antibody concentration to 2 mg/mL in PBS or 0.1 M sodium bicarbonate. If the concentration is lower than 2 mg/mL, *see* **Note 5**.
6. To 0.5 mL of the 2 mg/mL protein solution, add 50 µL of the 1 M bicarbonate prepared in **step 4**.
7. Allow a vial of reactive dye to warm to room temperature. Transfer the protein solution from **step 6** to the vial of reactive dye. This vial contains a magnetic stir bar. Cap the vial and invert a few times to fully dissolve the dye. Stir the reaction mixture for 1 h at room temperature. Because preparation of the purification column takes approx 15 min, you may wish to begin pouring the column.
8. Purify the labeled antibody by following the steps listed here:

   a. Assemble the column and position it upright. Attach a funnel to the top of a column. Gently insert the column through the X-cut in one of the provided foam holders to avoid damaging the column. Using the foam holder, secure the column with a clamp to a ringstand. Carefully remove the cap from the bottom of the column.

   b. Prepare elution buffer by diluting the room temperature 10X stock (component D) 10-fold in ddH$_2$O. Typically, less than 10 mL will be required for each purification.

   c. Using one of the provided pipets, stir the purification resin (component C) thoroughly to ensure a homogeneous suspension. Pipet the resin into the column, allowing excess buffer to drain away into a small beaker or other container. Resin should be packed into the column until the resin is approx 3 cm from the top of the column.

   d. Allow the excess buffer to drain into the column bed. Do not worry about the column drying out because the matrix will remain hydrated. Make certain the buffer flow through the column is even prior to adding the reaction mixture. If the flow of buffer is slow or stalled, repack the column. Remove the column funnel to load the sample. Carefully load the reaction mixture from **step 6** onto the column. Allow the mixture to enter the column resin. Rinse the reaction vial with approx 100 µL of elution buffer and apply to the column. Allow this solution to enter the column.

   e. Replace the funnel and slowly add elution buffer (prepared in **Subheading 2.2.**), taking care not to disturb the column bed. Continue adding elution buffer until the labeled protein has been eluted (typically about 30 min).
   Important: Collect, and retain as fractions, all of the eluted buffer.

   f. As the column runs, periodically illuminate the column with a handheld ultraviolet lamp. You should observe two colored bands, which represent the separation of labeled protein from unincorporated dye. Collect the first colored band, which contains the labeled protein, into one of the provided collection tubes. If desired, a foam holder can be used to support the collection tube. Add elution buffer to the column as necessary. Do not collect the slower moving band, which consists of unincorporated dye.

   g. Once the fraction containing the labeled protein has been successfully collected, all other fractions of eluted buffer may be discarded. In rare instances where there is no discernable band corresponding to labeled protein, the retained fractions can be used to recover any unlabeled protein.

9. Measure the absorbance of the conjugate solution at 280 nm and 494 nm (A_{280} and A_{494}) in a cuvette with a 1-cm pathlength. (Note: Dilution of the sample may be necessary.)
10. Calculate the concentration of conjugated protein in the sample:

$$\text{concentration (M)} = \frac{[A_{280} - (A_{494} \times 0.11)] \times \text{dilution factor}}{203,000}$$

where $203,000 \, \text{cm}^{-1}\text{M}^{-1}$ is the molar extinction coefficient of a typical IgG and 0.11 is a correction factor to account for absorption of the dye at 280 nm.

11. Calculate the degree of labeling:

$$\text{moles dye per mole protein} = \frac{A_{494} \times \text{dilution factor}}{71,000 \times \text{protein concentration (M)}}$$

where $71,000 \, \text{cm}^{-1}\text{M}^{-1}$ is the approximate molar extinction coefficient of the Alexa Fluor® 488 dye at 494 nm. For IgGs, labeling with 4 to 9 moles of Alexa Fluor® 488 dye per mole of antibody is optimal.

3.2.2. Recycling Assay (**Fig. 2.**), see **Note 6**

1. Plate U87 cells stably expressing wild-type or mutant LRP minireceptor in 6-well plates on the day before the assay; three wells are for wild-type and three wells are for mutant receptor-expressing cells. Cells should be 90% confluent on the day of experiment.
2. On the day of the assay, warm up binding buffer in 37°C water bath and chill PBS on ice. Prepare antibody solution in warm binding buffer (2 µg/mL in binding buffer).

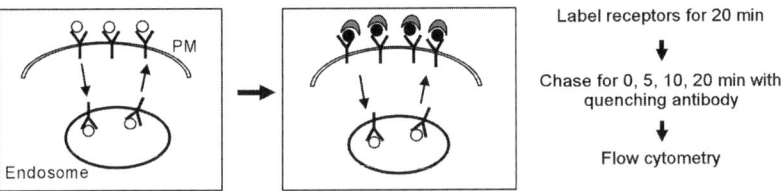

Fig. 2. Schematic representation of recycling assay. Receptors are labeled with fluorescent antibody (white circles), followed by a chase at 37°C for different amounts of time in the presence of the anti-Alexa 488 quenching antibody (black circles represent quenched antibody). The amount of fluorescence remaining, representing the amount of intracellular receptors, is measured by flow cytometry.

3 mL of antibody solution is required for one 6-well plate. For four plates, make 14 mL to have enough solution.
3. Quickly but gently wash cell monolayers twice with warm binding buffer. After the second wash, remove binding buffer completely by aspirating twice.
4. Add antibody solution into cells, 0.5 mL per well.
5. Incubate cells in antibody solution at 37°C for 20 min. During this time, prepare quenching antibody solution in warm binding buffer (24 μg/mL)
6. After incubation, remove antibody solution from wells and wash cells twice with warm binding buffer. After the second wash, remove binding buffer completely by aspirating twice.
7. Add 0.5 mL of quenching antibody solution to cells except for 0 min samples.
8. Incubate cells at 37°C for different period of times (e.g., 0, 5, 10, 20 min). For the 0 min time point, place the plate on ice immediately.
9. At each time point, place cells on ice, remove binding buffer, and wash cells twice with ice-cold PBS.
10. Chill PFN on ice. Remove PBS completely and add 1 mL of cell dissociation solution into each well and incubate at 37°C for 2 min.
11. Collect cells from the plate and transfer to FACS tubes.
12. Use 1 mL of ice-cold complete media per well to rinse remaining cells from the plate.
13. Centrifuge cells at 145 × g for 2 min at 4°C.
14. A small cell pellet should be seen at the bottom of tube. Carefully aspirate supernatant, add 2 mL of PFN to each tube, tap tube to resuspend pellet, and centrifuge at 145 × g for 2 min at 4°C. Repeat with a second wash step.
15. Gently resuspend cells in 500 μL of PFN and read in flow cytometer (BD FACS calibur).
16. Data interpretation (**Fig.** 3): Since anti-Alexa Fluor 488 IgG quenches fluorescence of Alexa Fluor 488 on the cell surface only, the data obtained from flow cytometry

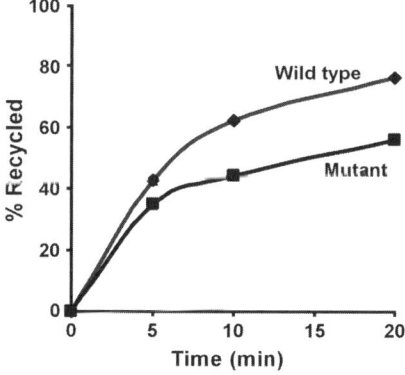

Fig. 3. Example of data derived from the kinetic analysis of recycling. Mutant receptor shows less efficient recycling compared to wild-type receptor. Of labeled wild-type receptors, 60% returned to the cell surface within 10 min, whereas only 50% of the mutant receptor was recycled after 20-min chase.

analysis represent the fluorescence from the receptor remaining within the cell (not the cell surface). Quenched fluorescence is calculated by subtracting the remaining fluorescence after each chase from the total initial signal, and can be plotted as percent of total after each chase (equal to recycled receptor).

$$\% \text{ of Recycled receptor} = \frac{\text{Total receptor (0 min)} - \text{intracellular receptor after each chase}}{\text{Total receptor (0 min)}} \times 100$$

4. Notes

1. Selection of ligand or antibody for endocytosis assay: The kinetic analysis of endocytosis can be performed with either antibody or ligand for the receptor of interest. Antibodies to epitope tags are preferred, while ligands and antibodies directed to the extracellular domain of the receptor may require more extensive testing to determine the degree of background binding. The iodogen method works best on antibodies that are in aqueous buffer (i.e., PBS or Tris-buffered saline), and the presence of glycerol inhibits labeling; therefore commercial antibodies stored in glycerol should be dialyzed into PBS. Furthermore, some commercial antibodies contain a carrier protein, such as BSA, which could increase the background level of radioactive binding to the cells and should be removed prior to iodination.
2. Limiting protein concentration: If the concentration of protein to be iodinated is such that 90 μL will not provide 50 μg, or if the protein is very precious, smaller amounts can be iodinated. The quantity of isotope can also be reduced accordingly, but the total 100 μL reaction volume should remain the same. The final concentration of iodinated antibody should be calculated to reflect any such changes.
3. Determination of iodinated protein concentration: Although some protein is certainly lost in the PD-10 column, the protein concentration is approximated as the amount of protein in the reaction divided by the final volume of the iodinated protein in the screw-cap vial. For the parameters given in the methods section, this would be 50 μg/1,000 μL = 50 ng/μL.
4. The use of alternative cell types for endocytosis assay: This protocol was designed for CHO cells, which are easily transfected using the Mirus TransIT CHO kit. The authors' laboratory has used this protocol successfully with stably transfected CHO cells, but has had limited success with transiently transfected HeLa cells.
5. Limiting protein concentration for fluorescently labeling antibody: Proteins at concentrations less than 2 mg/mL will not label as efficiently. If the protein cannot be concentrated to approx 2 mg/mL, use less than 1 mg protein per reaction to increase the molar ratio of dye to protein. In addition, using a dilute protein solution, especially at less than 1 mg/mL, will make it more difficult to efficiently remove the unconjugated dye from the dye-labeled protein with accept-

able yields, since the provided purification columns are designed to purify conjugates from a total volume of less than 1 mL. For reaction volumes greater than 1 mL, divide the solution of the conjugate and apply it to multiple purification columns or, to avoid further dilution of the conjugate, remove free dye by extensive dialysis.
6. Flow cytometry: Perform all steps on ice except for the cell dissociation step.

Acknowledgments

Work in the authors' laboratory is supported by grants from the National Institutes of Health, the Alzheimer's Association, and the American Health Assistance Foundation.

References

1. Maxfield, F.R. and McGraw, T.E. (2004) Endocytic recycling. *Nat. Rev. Mol. Cell Biol.* **5**, 121–132.
2. Bonifacino, J.S. and Traub, L.M. (2003) Signals for sorting of transmembrane proteins to endosomes and lysosomes. *Annu. Rev. Biochem.* **72**, 395–447.
3. van Kerkhof, P., Lee, J., McCormick, L., et al. (2005) Sorting nexin 17 facilitates LRP recycling in the early endosome. *Embo J.* **24**, 2851–2861.
4. Carlton, J., Bujny, M., Rutherford, A. and Cullen, P. (2005) Sorting nexins—unifying trends and new perspectives. *Traffic* **6**, 75–82.
5. Herz, J. and Strickland, D.K. (2001) LRP: a multifunctional scavenger and signaling receptor. *J. Clin. Invest.* **108**, 779–784.
6. Trommsdorff, M., Borg, J.P., Margolis, B. and Herz, J. (1998) Interaction of cytosolic adaptor proteins with neuronal apolipoprotein E receptors and the amyloid precursor protein. *J. Bio.l Chem.* **273**, 33556–33560.
7. Trommsdorff, M., Gotthardt, M., Hiesberger, T., et al. (1999) Reeler/Disabled–like disruption of neuronal migration in knockout mice lacking the VLDL receptor and ApoE receptor 2. *Cell* **97**, 689–701.
8. Lillis, A.P., Mikhailenko, I. and Strickland, D.K. (2005) Beyond endocytosis: LRP function in cell migration, proliferation and vascular permeability. *J. Thromb. Haemost.* **3**, 1884–1893.
9. Li, Y., Marzolo, M.P., van Kerkhof, P., Strous, G.J. and Bu, G. (2000) The YXXL motif, but not the two NPXY motifs, serves as the dominant endocytosis signal for low density lipoprotein receptor-related protein. *J. Biol. Chem.* **275**, 17187–17194.
10. Obermoeller-McCormick, L.M., Li, Y., Osaka, H., FitzGerald, D.J., Schwartz, A.L., Bu, G., et al. (2001) Dissection of receptor folding and ligand-binding property with functional minireceptors of LDL receptor-related protein. *J. Cel.l Sci.* **114**, 899–908.
11. Austin, C.D., De Maziere, A.M., Pisacane, P.I., Van Dijk, S.M., et al. (2004) Endocytosis and sorting of ErbB2 and the site of action of cancer therapeutics trastuzumab and geldanamycin. *Mol. Biol. Cell* **15**, 5268–5282.

12. Schapiro, F.B., Soe, T.T., Mallet, W.G. and Maxfield, F.R. (2004) Role of cytoplasmic domain serines in intracellular trafficking of furin. *Mol. Biol. Cell* **15**, 2884–2894.

25

Styryl Dye-Based Synaptic Vesicle Recycling Assay in Cultured Cerebellar Granule Neurons

Victor Anggono, Michael A. Cousin, and Phillip J. Robinson

Summary

Neurons transmit information by exocytosis of synaptic vesicles (SV), which contain neurotransmitter. Exocytosis is followed by efficient retrieval of the plasma membrane by endocytosis to generate a new SV. SV retrieval supports multiple cycles of synaptic transmission. Over the years, styryl dyes have been widely used to probe the mechanism of SV recycling in the processes of cultured neurons. The styryl dye method is complementary to electrophysiological measurements or genetic reporter approaches. Owing to their ease to culture, cerebellar granule neurons provide a robust neuronal model system for the assay. These cells are readily transfected with various DNA constructs to study the function of exocytic or endocytic proteins in SV recycling.

Key words: Cerebellar granule neuron; endocytosis; exocytosis; synaptic vesicle recycling; styryl dye; transfection.

1. Introduction

Communication between neurons is initiated via the release of a neurotransmitter by exocytosis of synaptic vesicles (SV) at presynaptic nerve terminals. To accommodate rapid and repeated rounds of release, SVs are retrieved from the plasma membrane by endocytosis, refilled with neurotransmitter and recycled. A functional block in synaptic vesicle endocytosis (SVE) leads to an activity-dependent rundown of SV exocytosis in isolated nerve terminals *(1)*. Hence, the recycling of SVs underlies all aspects of synaptic transmission, and is key to understanding the basis of synaptic plasticity *(2)*.

The development of the fluorescent molecules known as styryl dyes has allowed SV recycling to be investigated in real time in living neurons (for reviews, see **refs. 3–5**). Styryl dyes possess a hydrophilic head group and lipophilic tail group that allow them to stain membranes (because they fluoresce upon membrane binding) but not to penetrate them. Many dyes have been engineered that vary either in their tail length (which governs their dissociation kinetics from membrane) and/or in their number of double bonds (which regulate their fluorescent properties) (for review, see **ref. 6**). These commercially available dyes provide a straightforward assay for studying SV recycling by time-lapse fluorescence microscopy. This is achieved by loading the dye into retrieving SVs using a loading stimulus and then unloading the dye from the same SVs using an unloading stimulus. Using this protocol it is possible to gain information on total SV turnover in nerve terminals (the total amount of dye accumulated), the speed of SV fusion (the kinetics of dye unloading), the extent of SV exocytosis (the total amount of dye unloading), and the extent of SVE (the difference between accumulated dye and unloaded dye) *(3,4)*.

In this protocol, cerebellar granule neurons (CGNs) are used as the primary neuronal culture of choice. CGNs are simple to prepare, are very robust, and can be transfected with DNA constructs more readily than most other primary neurons *(1,7)*. Moreover, after 7 d *in vitro* (DIV), CGNs form functional glutamatergic synapses that are visible using electron microscopy and readily detectable using neurotransmitter release assays *(8)*.

2. Materials
2.1. Primary CGN Culture

1. Seven-day old Wistar rat pups.
2. Culture medium: Minimal Essential Medium (MEM) containing Earle's salt (Invitrogen) supplemented with 10% fetal bovine serum (Invitrogen), 25 mM KCl, 30 mM D-glucose, 2 mM L-glutamine 100 U/mL penicillin and 100 μg/mL streptomycin (Invitrogen). Note that the elevated KCl is required to depolarize the CGNs during growth.
3. Solution of trypsin (0.25%) and ethylenediamine-tetraacetic acid (EDTA) (1 mM) from Invitrogen.
4. Solution B: 0.3% bovine serum albumin (BSA, Sigma), 0.25% D-glucose, 1.5 mM MgSO$_4$. Prepare solution fresh in 100 mL of phosphate-buffered saline (without Ca^{2+} or Mg^{2+}, Invitrogen) and filter sterilize.
5. Solution T: 10 mL of trypsin-EDTA solution, 10 mL of solution B. Prepare solution fresh and filter sterilize.
6. Solution C: 500 U deoxyribonuclease (DNase, Sigma), 0.5 mg soybean trypsin inhibitor (SBTI, Sigma), 1.5 mM MgSO$_4$. Prepare solution fresh to a total volume of 10 mL with solution B and filter sterilize. The DNAse and SBTI are stored as single use 0.5 mL stock solutions at −20°C.

7. Solution W: 3.2 mL of solution C, 16.8 mL of solution B. Prepare solution fresh and filter sterilize.
8. Cushion solution: 4% BSA, 3 mM MgSO$_4$ in Earle's Balanced Salt Solution (Invitrogen). Store solution filter sterilized in 10 mL aliquots at $-20°C$.
9. Ara-C stock solution: cytosine β-D-arabinofuranoside (Ara-C, Sigma) (cytotoxic and carcinogenic agent which should be handled with care) is dissolved in ultra-pure water at 10 mM and stored in aliquots at $-20°C$. Ara-C working medium is prepared by dilution in culture media and sterilized.
10. Poly-D-lysine (Sigma) is dissolved in ultra-pure water (0.0015%), filter sterilized and stored in 45 mL aliquots at $-20°C$.
11. Microscope cover slips (22 mm, no. 1 thickness) and sterile 12-well and 6-well tissue culture plates.
12. Sterile screw cap tubes for centrifugation (15 mL) and sterile pipets and pipet tips.

2.2. Transfection

1. Plasmid DNA is purified with *Wizard Plus* SV maxipreps DNA purification system (Promega) and stored at $-20°C$ (*see* **Note 1**).
2. HEPES-buffered saline (2X, HeBS): 247 mM NaCl, 1.4 mM Na$_2$HPO$_4$, 15 mM D-glucose, 10 mM KCl, 42 mM HEPES, pH 7.14. Store in 1 mL aliqouts at $-20°C$ (*see* **Note 2**).
3. CaCl$_2$ solution: prepare 2.5 mM solution and filter sterilize. Store at room temperature.

2.3. SV Recycling Assay

1. Styryl dye (FM1-43 or FM4-64, Molecular Probes, Eugene, OR) is dissolved in water at 1 mM and stored in 20 μL aliquots at $-20°C$ (*see* **Note 3**).
2. Basic stock solution (10X): 1.2 M NaCl, 35 mM KCl, 4 mM KH$_2$PO$_4$, 50 mM NaHCO$_3$, 200 mM N-[Tris(hydroxymethyl)methyl]-2-ethanesulfonic acid (Sigma), 50 mM D-glucose, 12 mM Na$_2$SO$_4$. Store in 50 mL aliquots at $-20°C$.
3. MgCl$_2$ stock solution: prepare 1.2 M solution and store at room temperature.
4. CaCl$_2$ stock solution: prepare 0.5 M solution and store at room temperature.
5. NaCl stock solution: prepare 3 M solution and store at room temperature.
6. KCl stock solution: prepare 3 M solution and store at room temperature.
7. Basic solution: thaw an aliquot of 10X basic stock solution and dilute to 500 mL with ultra pure water then adjust to pH 7.4 with NaOH. To this, add 0.5 mL of 1.2 M MgCl$_2$ stock solution and 1.3 mL of 0.5 M CaCl$_2$ stock solution (for 1.2 mM MgCl$_2$/1.3 mM CaCl$_2$).
8. Na-basic solution: add 750 μL of 3 M NaCl stock solution to 45 mL of 1X basic solution (for 170 mM NaCl/3.5 mM KCl).
9. K-depol solution: add 750 μL of 3 M KCl stock solution to 45 mL of 1X basic solution (for 120 mM NaCl/53.5 mM KCl).
10. These instructions assume the use of an RC-21BR polycarbonate perfusion chamber (Warner) to house the cells in a 260 μL liquid flow through

environment enclosed between two cover slips 2.5 mm apart. The chamber requires a platform and a stage adaptor to mount onto a microscope (Warner). Also required are VC-6 six channel perfusion valve control systems (Warner) and a Dynamax peristaltic pump (Rainin) coupled to a standard Olympus IX81 epifluorescence microscope (Tokyo, Japan). The microscope is equipped with appropriate dichroic mirrors and emission filters for fluorescence imaging and a Hamamatsu Orca-ERG charge-coupled device digital camera (Hamamatsu) for image capture.

3. Methods

3.1. Primary CGN Culture

1. Place 100 autoclaved cover slips in 45 mL of sterile poly-D-lysine solution then placed on a rotating platform for 2 h to coat the cover slips. The poly-D-lysine solution is discarded and the coated cover slips are separated on a tissue paper (that has previously been sprayed with 70% ethanol to sterilise) using two fine forceps inside a laminar flow hood. Cover slips are dried in the laminar flow hood and then placed into sterile 12-well plates and put into the CO_2 incubator. They can also be stored for up to 1 month at 4°C for later use.
2. On the day, prepare the laminar flow hood with the following solutions and instruments: solution B, solution T, solution W, solution C, large scissors, small scissors, rough forceps, two fine forceps, a fine spatula, a scalpel blade and a 6-cm tissue-culture dish. All these instruments are sprayed with 70% ethanol and placed on a tissue paper previously sprayed with ethanol. Tape a bag onto the hood to store the rat carcasses.
3. Decapitate a 7-d-old rat pup with a large scissors then cut around the side of the head with the small curved scissors (taking care not to cut the brain) and then lift off the roof of the skull with the strong forceps. The cerebellum should be clearly visible and is lifted out using the fine forceps into a tissue culture dish containing solution B. This process is repeated for every rat used in the preparation (typically three to four rats; *see* **Note 4**).
4. Once the dissection is completed, the cerebella are transferred to a clean tissue culture dish and cut into approx 1-mm square blocks (*see* **Note 5**).
5. The chopped tissue is transferred into the sterile solution T using the fine spatula. Incubate the tube at 37°C for 20 min and gently agitate every 5 min (*see* **Note 6**).
6. In the interval, flame-polish three sterile glass pipets using a Bunsen flame. Narrow the neck of two pipettes using the flame to give one fine bore and one medium bore. Leave the final pipet untouched. Thaw an aliquot of the cushion solution in a 37°C water bath at this point.
7. After trypsinization, add 20 mL of solution W into the cell suspension and centrifuge at $1,000g$ for 1 min (*see* **Note 7**).
8. Gently decant the supernatant and resuspend the pellet in approx 1.5 mL of solution C using the wide-bore pipet. After this, triturate the cells thoroughly using the flame-polished pipets, starting with the untouched, followed by the medium- and narrow-bore pipets. This is an important step and the quality of the culture depends on producing a homogenous suspension at this stage.

9. After trituration, layer the cell suspension (usually this is 1.5 mL) on top of the cushion solution (10 mL in a sterile 15-mL tube) and centrifuge at 1,500g for 5 min. The cells will be pelleted through the cushion, leaving behind any cell debris generated during the trituration. The cushion solution should look cloudy.
10. The supernatant is carefully discarded by aspiration and the cell pellet is resuspended in 2 mL of prewarmed (37°C) culture medium. Take 10 µL of the cell suspension and dilute with 90 µL of culture medium and count the number with a haemocytometer under a light microscope.
11. A cell suspension is prepared at a density of 2×10^6 cells per millilitre and 75 µL is spotted on the center of each poly-D-lysine coated cover slip, in every well of a 12-well tissue culture plate to achieve a density of 1.5×10^5 cells per cover slip.
12. The culture plates are returned to the CO_2 incubator for 40 to 60 min to allow the cells to adhere.
13. Gently and slowly add 1.5 mL of culture medium into each well so as to minimize disturbance of the cells on the cover slip and return the culture plates to the CO_2 incubator.
14. The next day, change the medium on the cells using fresh culture medium containing Ara-C (10 µM). Cells are fed every 5 to 6 d by replacing the medium with Ara-C working medium (*see* **Note 8**).

3.2. Transfection by Calcium Phosphate Precipitation

1. Cultured CGNs are transfected with plasmid DNA between 6 and 10 DIV. Remove the medium from each well to a sterile tube and temporarily store this conditioned medium in the CO_2 incubator for later use. Keep the lid of the tube loose to allow gas exchange (*see* **Note 8**).
2. Wash the cells twice with 1 mL of prewarmed MEM and incubate the cells with 1 mL of MEM for 1 h in the incubator. (Proceed to **step 8**.)
3. During the incubation period, prepare the transfection solution in 1.5-mL microcentrifuge tubes. This is done in two separate reactions (**steps 4–7**). The final transfection solution will contain (based on one 25-mm well of a 12-well plate): 22.5 µL of 2X HeBS, 2.25 µL of 2.5 M $CaCl_2$, 1.5 µg of DNA and water (to bring $CaCl_2$ and DNA to a total volume of 22.5 µL; *see* **Note 9**).
4. Label each 1.5-mL microcentrifuge tube with the specific DNA that is to be transfected. Add enough water (so the final volume at this step will be 22.5 µL) to the tube corresponding to the specific target DNA, followed by 2.25 µL of $CaCl_2$ solution. Plasmid DNA (1.5 µg) is then added to the tube. Briefly vortex and spin the DNA/$CaCl_2$ solution.
5. Label another set of 1.5-mL microcentrifuge tubes and add 22.5-µL of 2X HeBS.
6. While the HeBS tube (from **step 5**) is being stirred on the vortexer, add the DNA/$CaCl_2$ solution (from **step 4**) dropwise to the HeBS tube and continue vortexing for at least 20 s or longer (*see* **Note 10**).
7. Briefly spin the tube from **step 6** (2 s on a Tomy PMC-060 Capsulefuge) to consolidate the dispersed solution following vortexing and incubate in the dark for 30 min at room temperature.

8. After 1 h of cell incubation in MEM, add 45 µL of the transfection solution (from **step 7**) to the appropriate well dropwise. Gently agitate by manual rocking the well to disperse the DNA precipitate and then incubate the plate for 25 min in the CO_2 incubator (*see* **Note 11**).
9. Wash the cells twice with 1 mL of prewarmed MEM and replace with conditioned culture medium saved from **step 1**. The use of conditioned medium is essential for the survival of the CGNs.
10. Inspect the transfected cultures between 1 and 5 d post-transfection by fluorescence microscopy. Examples of CGNs transfected with various DNA constructs are shown in **Fig. 1**.

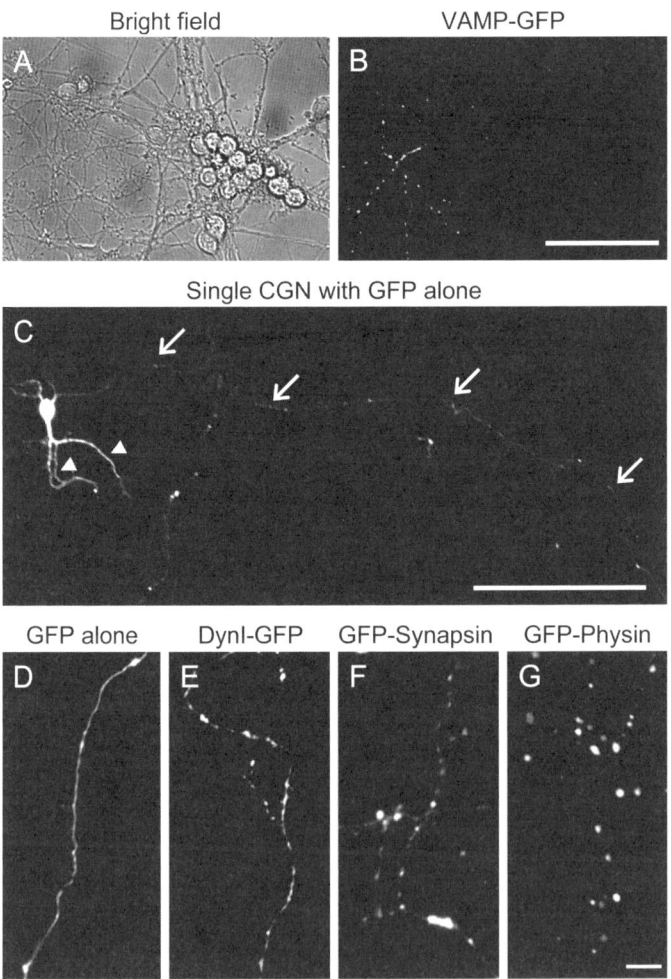

Fig. 1. (Continued)

3.3. SV Recycling Assay

1. The experiment is set up at the live cell-imaging microscope in the RC-21BR perfusion chamber. Before starting, ensure that no air bubbles are present in the tubing by priming all the syringes of the perfusion apparatus by running the appropriate buffer through them.
2. Lift out one cover slip containing CGNs from the incubator and place in a well of a fresh 6-well plate containing Na-basic solution. Incubate for at least 10 min to repolarize the CGNs (*see* **Note 12**).
3. During this time period prepare the dye-loading solution (10 μL of 1 mM FM1-43 or FM4-64 dye diluted into 1 mL of K-depol solution, final concentration of 10 μM). After the 10-min incubation (**step 2**), replace the Na-basic solution with dye-loading solution and incubate for 2 min (*see* **Note 13**).
4. The dye-loading solution is then aspirated and cells are washed with Na-basic solution twice, followed by 10 min incubation in Na-basic solution to repolarize the CGNs. This allows dye-loaded SVs to complete recycling and become competent for fusion.
5. During this time, pregrease a perfusion chamber with petroleum jelly on the top and bottom edges. The chamber should also have Na-basic solution primed through the inlet tube to ensure no air bubbles occur when perfusion begins.
6. Remove the CGN cover slip from the dish (**step 4**) and dry its base and edges to ensure a good seal with the chamber. Mount the cover slip on the base of the chamber. Quickly add 260 μL of Na-basic solution to fill the chamber. Then add another 22-mm clean cover slip on top to seal the chamber.

Fig. 1. (Continued) Transfection of primary cerebellar granule neurons (CGNs) in culture. (**A**) A bright field image from a primary culture of 9-d-old CGNs that have been transfected with the synaptic vesicle (SV) protein VAMP–GFP. (**B**) Monochrome VAMP–GFP image (green channel) shows puncta representing mature nerve terminals (synapses) along the neurites of the cultured neurons. (**C**) CGN transfected with plasmid DNA expressing GFP alone, which fills the cytoplasm. Note the morphology of a single granule neuron, which consists of a cell body, thick dendrites (arrow heads) and a long, thin horizontal axon (arrows). The long vertical axon originates from another neuron in the well. (**D–F**) Images of CGNs transfected with different green fluorescent protein (GFP)-tagged proteins. (**D**) GFP (empty vector) fluorescence is distributed evenly along the neurite of transfected CGN. (**E**) Dynamin I-GFP fluorescence is distributed along the axon and also shows prominent punctate staining at synapses along the axon of a transfected CGN. It is also present within the axon between synapses. (**F, G**) Both GFP–synapsin I (**F**) and GFP–synaptophysin I (**G**) fluorescence are observed as discrete punctate staining at synapses of transfected CGNs, representing sites of SV recycling. Scale bars, 50 μm (A–C) and 10 μm (D–G).

7. Screw down the chamber into the platform and check to see if it is watertight. Test using 100 μL of Na-basic solution gently pushed through the chamber. Check for leakage on the chamber top and bottom and check for unrestricted flow.
8. Mount the chamber on the microscope stage and connect it to the perfusion apparatus on the inlet and to the peristaltic pumps on the outlet. Perfuse the sample for 30 s at 2 to 5 mL per minute to ensure there are no leaks and that the flow is normal.
9. Perfuse the sample with Na-basic solution continuously to wash out any remaining dye and switch the microscope to red illumination (550 nm excitation and a long pass emission filter of 600 nm) to view FM4-64 puncta (in conjunction with green fluorescent protein [GFP]) or to green illumination (480 nm excitation and a band pass emission filter of 515-555 nm) for FM1-43 (in conjunction with red fluorescent protein [RFP]).
10. When working with transfected neurons, choose the correct filter set (green or red depending on either GFP or RFP) and search for transfected cells using the microscope. Take an image of the transfected cell with either 40X or 63X objectives prior to the FM dye time-lapse imaging. The combination of GFP and FM4-64 is often used to study the effect of endocytic protein mutants in the authors' laboratories (*see* **Note 14**).
11. Once a field of neurites has been identified, time-lapse imaging is initiated with an acquisition rate of an image every 3 s. After 20 s of imaging, continue imaging and start to perfuse K-depol solution (unloading solution) into the chamber for 30 s before returning to Na-basic solution. The K-depol solution will stimulate SV exocytosis and thus unload the dye accumulated during the loading step. An optional second stimulation can be applied at time 120 s with K-depol solution for 30 s to ensure complete unloading of accumulated dye. Switch back to Na-basic solution and collect images for a further 30 s.
12. Take the chamber off the stage and immediately clean the lens with a lens cleaning tissue. Dismantle the chamber and clean with normal tissue paper, making sure that the inlet and outlet are not blocked.
13. The acquired time-lapse images are quantified using imaging software such as Metamorph (Molecular Devices) or Simple PCI (Compix). Regions of interest (ROI, between 90 and 100), typically $1.5 \times 1.5\,\mu m^2$, are selected to overlap the largest portion of the fluorescent spot that is an individual synapse. Only select discrete fluorescent puncta that clearly represent single synapses. Record the fluorescence intensities of individual synaptic boutons (ROI) over time using the imaging software. Two datasets must always be obtained: the responses from the transfected neurons and one from the untransfected neurons in the same field. The responses from the untransfected neurons are used as the internal control.
14. The fluorescence intensity values of both groups of traces just prior to stimulation are normalized to an arbitrary number. Then both groups (transfected and untransfected) are averaged to obtain the mean and standard error for each time point.

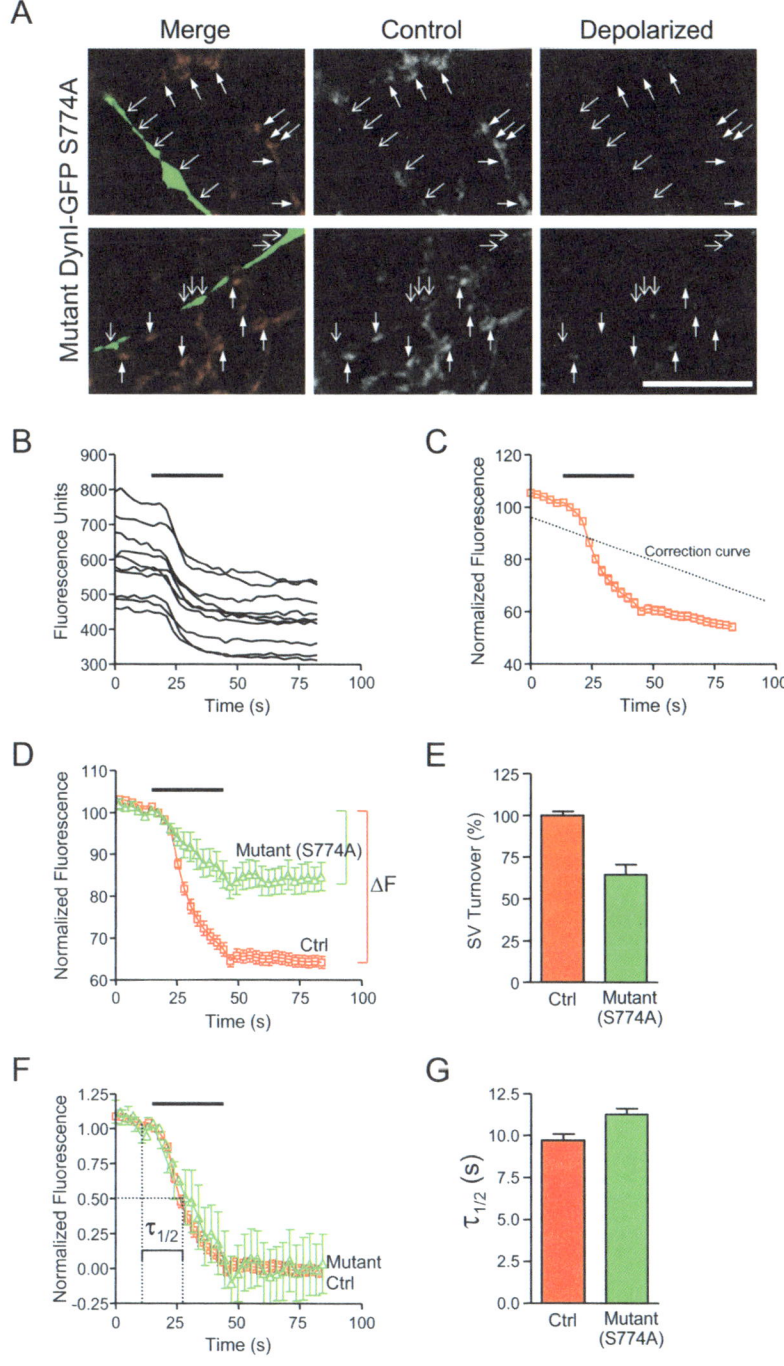

Fig. 2. (Continued)

15. Often the normalized curve includes a component that is due to photobleaching and can be corrected by subtraction with a correction curve. The correction curve can be generated as a function of straight line with a gradient value similar to the slope of the normalized traces.
16. Estimates of the total amplitude of SV turnover from a given bouton are obtained by calculating the magnitude of the difference in average fluorescence intensity (F) averaged over the first three time points before and after KCl stimulation.
17. The kinetics of FM dye unloading are estimated by normalizing the start and completion of unloading to 1 and 0 in the averaged traces. The time constant (τ) can be obtained by fitting the curve to a single exponential function. However using maximal KCl stimulation, the initial 50% of dye loss is linear, negating the need for transformation of the data (see **Note 15**). An example of a SV recycling assay is shown in **Fig. 2**.

Fig.2. (Continued) Synaptic vesicle (SV) recycling assay using FM4-64. (**A**) Primary cultures of cerebellar granule neurons (CGNs) were transfected with green fluorescent protein (GFP)-tagged Dynamin I S774A (green, this phosphomutant cannot be phosphorylated and blocks SV endocytosis, see **ref. 1**) and stimulated uptake of FM4-64 (red) are visualized by fluorescence microscopy (left panels). Two separate fields are shown as examples (upper and lower panels). The monochrome FM4-64 image shows puncta from either transfected (open arrow heads) or untransfected (filled arrow heads) neurites, representing sites of SV recycling (middle panels). CGNs were depolarized by a single 53.5 mM KCl stimulus to unload the accumulated FM4-64 (right panels). The fluorescence of all puncta was reduced. Note that puncta from transfected neurites all exhibit markedly less or no dye loading (middle panels, open arrow heads) suggesting a block in synaptic vesicle endocytosis (SVE). Scale bar, 10 µm. (**B**) Average fluorescence intensities of 10 individual puncta from untransfected neurons were plotted against time of stimulation. (**C**) The raw estimation of the total amplitude of SV recycling was quantified by averaging the fluorescence intensity of 60 puncta upon KCl stimulation within one experiment. Note that the change in fluorescence intensity includes a component that is due to photobleaching and this can be corrected by subtraction with a correction curve. (**D**) Representative corrected traces of FM4-64 unloading from the transfected (green triangles) and untransfected (red squares) synaptic puncta from a single experiment using mutant dynI-S774A. Accumulated FM4-64 can be calculated as F. (**E**) The extent of FM4-64 accumulation and thus SV turnover from at least three independent experiments is expressed as a percentage of that in untransfected neurons ± standard errors of the mean. (**F**) SV exocytosis was determined by examining the kinetics of FM4-64 unloading and the time constant (τ). Representative traces from a single experiment are shown. (**G**) SV exocytosis is unaffected in neurons overexpressing dynamin I S774A phosphomutant, an indication that the block in SV turnover shown in panel E is specific to SVE. Solid bars (in panels B–D, F) represent the period of KCl stimulation.

4. Notes

1. A high-purity DNA plasmid is required for the transfection procedure. Any contaminant may be toxic to the primary neurons. Plasmid purification kits that are endotoxin-free often produce superior quality DNA for transfection. It is advisable to use a high-quality plasmid DNA with an A260/280 ratio of 1.8 to 1.9 by ultraviolet spectroscopic measurement.
2. The pH of the 2X HeBS buffer should be optimized to achieve high transfection efficiency. Test batches of HeBS that have different pH values between 7.0 and 7.2 with 0.02 increments to determine the optimal pH value.
3. All styryl dyes go off quickly in water at room temperature ($\sim$1 h). When preparing the stock solution, immediately freeze 20 μL aliquots, which are enough for two experiments at a time. When using the dye keep it on ice in the dark at all times.
4. The decapitation procedure may require a licence in some countries, such as the United Kingdom; however the use of anaesthesia of CO_2 asphyxiation is severely damaging to the neurons and these euthanasia methods must not be employed.
5. It is best to cut the cerebella using a McIlwain tissue chopper (375 μm intervals at right angles). It is an expensive instrument and manual chopping with a scalpel blade is a reasonable alternative, which still allows good access for tryptic digestion.
6. The trypsinization of the cells is a key point of the preparation. If the cells are exposed to trypsin for too long they will not be viable. A good indication of over-trypsinized cells is when the cell fails to pellet properly, instead a whispy tail emanates from it after centrifugation. If this occurs, either reduce the time the cells are incubated with trypsin, or reduce the concentration of trypsin in solution T.
7. The SBTI inhibits the action of trypsin and the DNAse degrades the DNA released from the cell nucleus, thus stopping the cell suspension from becoming viscous.
8. Ara-C prevents division of non-neuronal cells, which if left unchecked, would overrun the cultures. When feeding the cells with new culture medium it is important to ensure that it is gassed in the CO_2 incubator for at least 1 h prior to use. Granule neurons become very sensitive to pH changes as they age in culture.
9. DNA concentration may vary depending on the plasmid construct and needs to be optimized. If doing co-transfection with two different plasmids, use a higher concentration of each DNA (e.g. ,start with 1 μg of each DNA). Always calculate extra volume when preparing the transfection solution for multiple wells.
10. A key step in obtaining a good calcium phosphate precipitate is the addition of the DNA/$CaCl_2$ mix to the HeBS solution. The HeBS solution must be continually vortexed during addition and for at least 20 s afterward. For larger volumes, air can be bubbled into the tube using a pipette attached to a pipetman to serve as an agitation method.
11. The amount of precipitate in the transfection solution depends on the calcium concentration. If the precipitate is too clumpy, try a lower concentration of $CaCl_2$. If it does not form fast enough try a higher concentration of $CaCl_2$.

12. CGNs that have been cultured for at least 9 to 10 d are normally used for this assay. It is important to repolarize the neurons because they are cultured in medium containing 25 mM KCl, which is depolarizing. If cells are not cultured in such elevated potassium, they may enter apoptosis. Hence, the cells should not be left for more than 60 min in Na-basic solution. If required, take three cover slips at a time from the incubator into a separate well in the 6-well plate.
13. The dye-loading solution contains 53.5 mM KCl for depolarization of the CGNs. The stimulus strength can be varied if required. Dye-loading time may vary from 10 s to 2 min depending on the nature of the experiments. Dye can also be loaded by using Na-basic instead of K-depol solution and immediately applying an electrical field stimulus in an RC-21BRFS field stimulation perfusion chamber (Warner).
14. The transfection efficiency in CGNs is very low, so identifying an ideal field of view may take time. However, the time taken to identify this field should be kept to a minimum because excess illumination kills the cells as a result of dye phototoxicity *(6)*. Ideally, one is looking for a transfected neuron that is not part of a clump of other neurons (as judged by observing in the DIC channel). This is important because SV recycling in individual neurites needs to be resolved after loading with dye.
15. Kinetic analysis is particularly useful when it is not possible to differentiate between SV exocytosis and SVE by other methods. An effect on SV turnover without any effect on destaining kinetics is a strong indication of an effect that is specific to SVE.

Acknowledgments

This work was supported by grants from the National Health and Medical Research Council of Australia (to P.J.R.) and the Wellcome Trust (GR070569, to M.A.C.), an Australian Bicentennial Scholarship (to V.A.), and a University of Sydney Postgraduate Award (to V.A.).

References

1. Anggono, V., Smillie, K. J., Graham, M. E., Valova, V. A., Cousin, M. A., and Robinson, P. J. (2006) Syndapin I is the phosphorylation-regulated dynamin I partner in synaptic vesicle endocytosis. *Nat. Neurosci.* **9**, 752–760.
2. Sudhof, T. C. (2004) The synaptic vesicle cycle. *Annu. Rev. Neurosci.* **27**, 509–547.
3. Cousin, M. A. and Robinson, P. J. (1999) Mechanisms of synaptic vesicle recycling illuminated by fluorescent dyes. *J. Neurochem.* **73**, 2227–2239.
4. Angleson, J. K. and Betz, W. J. (1997) Monitoring secretion in real time: capacitance, amperometry and fluorescence compared. *Trends Neurosci.* **20**, 281–287.

5. Cochilla, A. J., Angleson, J. K., and Betz, W. J. (1999) Monitoring secretory membrane with FM1-43 fluorescence. *Annu. Rev. Neurosci.* **22**, 1–10.
6. Betz, W. J., Mao, F., and Smith, C. B. (1996) Imaging exocytosis and endocytosis. *Curr. Opin. Neurobiol.* **6**, 365–371.
7. Tan, T. C., Valova, V. A., Malladi, C. S., et al. (2003) Cdk5 is essential for synaptic vesicle endocytosis. *Nat. Cell Biol.* **5**, 701–710.
8. Evans, G. J. O. and Cousin, M. A. (2007) Activity-dependent control of slow synaptic vesicle endocytosis by cyclin-dependent kinase 5. *J. Neurosci.* **27**, 401–411.

26

Use of Quantitative Immunofluorescence Microscopy to Study Intracellular Trafficking

Studies of the GLUT4 Glucose Transporter

Vincent Blot and Timothy E. McGraw

Summary

Insulin regulates the glucose uptake in muscle and adipose cells by acutely modulating the amount of the GLUT4 glucose transporter in the plasma membrane. The steady-state cell surface distribution of a membrane protein is an equilibrium between exocytosis and endocytosis. The authors study the effect of insulin on GLUT4 using quantitative immunofluorescence microscopy adapted to single-cell analysis. They use an HA-GLUT4-GFP reporter molecule as a surrogate of GLUT4 trafficking. Insulin induces an increase of GLUT4 in the plasma membrane by both increasing GLUT4 exocytosis and decreasing its endocytosis. Quantitative immunofluorescence techniques such as those described in this review can be used to study the trafficking of virtually any membrane protein.

Key words: endocytosis, exocytosis, GLUT4; insulin action; quantitative microscopy.

1. Introduction

Insulin regulates glucose homeostasis in part by controlling postprandial glucose disposal by fat and muscle cells. In those tissues, the GLUT4 facilitative glucose transporter is responsible for insulin action on glucose transport. Glucose transport is regulated through modulation of the amount of GLUT4 in the plasma membrane of adipocytes and myocytes: GLUT4 is retained within intracellular compartments of nonstimulated cells and insulin acutely stimulates its translocation to the plasma membrane *(1,2)*.

GLUT4 level in the plasma membrane is an equilibrium between exocytosis from intracellular compartments to the cell surface and endocytosis from the plasma membrane. In nonstimulated adipocytes, GLUT4 exocytosis is slower than its endocytosis and less than 5% of GLUT4 is in the plasma membrane. Insulin signaling results in an increase of GLUT4 exocytosis and an inhibition of GLUT4 endocytosis. At the new insulin-stimulated steady state, 50% of GLUT4 is in the plasma membrane of adipocytes.

How insulin controls GLUT4 trafficking parameters remains unknown. If we are to understand insulin-resistance syndromes that lead to type 2 diabetes, it is necessary to understand the link between insulin signaling and GLUT4 intracellular trafficking.

The authors have developed single-cell quantitative fluorescence microscopy techniques to study GLUT4 trafficking in 3T3-L1 adipocytes. As a surrogate for GLUT4 trafficking, they study an HA-GLUT4-GFP reporter chimera developed by S.W. Cushman *(3,4)*. This construct has an HA epitope inserted in its first exofacial loop and the green fluorescent protein (GFP) fused to its carboxyl terminus (**Fig. 1**). Binding of anti-HA antibodies to the HA epitope is proportional to the amount of HA-GLUT4-GFP protein in the plasma membrane and the GFP fluorescence is proportional to the total HA-GLUT4-GFP expressed per cell. Thus, the ratio of surface anti-HA-associated fluorescence to

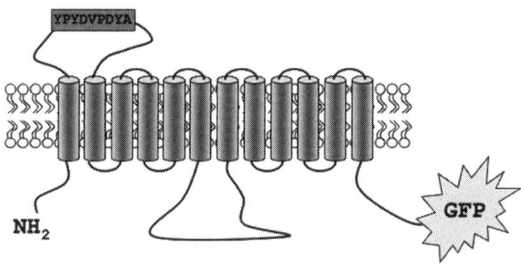

Fig. 1. Schematic representation of the HA-GLUT4-GFP chimera we used as a surrogate to study GLUT4 trafficking. GLUT4 is a 12 transmembrane domain protein with both its NH_2 and COOH termini in the cytoplasm. HA-GLUT4-GFP has an HA tag inserted in the first exofacial loop (red rectangle, the amino-acid sequence of the HA epitope is indicated in the one letter symbol) and a GFP protein fused at its carboxyl terminus.

total GFP fluorescence is a measure of the fraction of HA-GLUT4-GFP on the surface *(4)*.

Here, the authors describe the fluorescence assays used to study GLUT4 trafficking, with particular emphasis on how this quantitative approach allows for the detailed dissection of insulin regulation of GLUT4 trafficking.

2. Materials

2.1. Solutions Required for 3T3L1 Cells Culture and Experimental Manipulations

1. Use Dulbecco's Modified Eagle's Medium (DMEM) supplemented with 10% calf serum (CS) to maintain cultures of 3T3-L1 pre-adipocytes.
2. To differentiate 3T3-L1 pre-adipocytes into adipocytes use DMEM supplemented with 10% fetal bovine serum (FBS), 170 nM insulin, 1 mM rosiglitazone, 0.25 mM dexamethasone ,and 2 mM methyl-isobutylxanthine (IBMX). Prepare insulin as a 1,000X concentrated stock solution in 0.01 N HCl and store at 4°C. Prepare rosiglitazone as a 10,000X concentrated stock solution in DMSO and store at −20°C. Prepare dexamethasone as a 10,000X concentrated stock solution in ethanol and store at −20°C. Prepare IBMX as a 500X concentrated stock solution in 0.35 N KOH and store at −20°C (*see* **Note 1**).
3. The postdifferentiation medium is DMEM supplemented with 10% FBS and 170 nM insulin.
4. Maintain differentiated adipocytes in DMEM supplemented with 10% FBS.
5. 10 X Solution phosphate-buffered saline (PBS) is 1.37 M NaCl, 27 mM KCl, 100 mM Na2HPO4, 18 mM KH2PO4, pH 7.2. Prepare 1X working solution by dilution of one part with nine parts water.
6. Medium I: DMEM supplemented with 220 mM bicarbonate and 20 mM HEPES pH 7.4.
7. Medium II: 150 mM NaCl, 1 mM $CaCl_2$, 5 mM KCl, 1 mM $MgCl_2$ and 20 mM HEPES pH 7.2.
8. Fix cells using PBS supplemented with 3.7 % formaldehyde (*see* **Note 2**).

2.2. Antibodies and Staining

1. Purify the anti-HA mouse monoclonal antibody (HA-11) from ascites (Covance) by using a protein G affinity column (Amersham Pharmacia AB; **Note 3**). The concentration required to saturate the HA-epitope of HA-GLUT4-GFP is determined by measuring cell-associated anti-HA antibody after a 10-min pulse at 37°C (*see* **Subheading 3.2.1.1**). The saturating concentration is typically 50 μg/mL. The concentration used for nonsaturating HA staining is 10 μg/mL.
2. Fluorescently labeled secondary antibodies are purchased from Jackson Immuno-Research. Secondary antibodies are typically used at 1.5 μg/mL. The blocking

concentration of the secondary antibodies is typically 15 μg/mL and is determined experimentally following the procedure given **Subheading 3.2.1.1**.
3. Stainings are done using 100 μL of the appropriate dilution of antibody in PBS supplemented with 2% FBS to reduce nonspecific binding. To detect intracellular antigens, the detergent saponin is added to the antibody dilution at a final concentration of 250 μg/mL (*see* **Note 4**).

2.3. Microscopy and Fluorescence Quantification

1. Plate electroporated 3T3L1 adipocytes onto cover slip-bottom dishes. Use the same dishes with living cells for the experiments as well as after fixation of cells for staining and imaging.
2. Collect images using a DMIRB inverted microscope with a 40X oil-immersion objective and a cooled charge-coupled device (CDD) 12-bit camera.
3. Perform fluorescence quantifications using Metamorph image-processing software (Molecular Devices Corp.; *4,5*).

3. Methods

The authors developed a panel of quantitative immunofluorescence assays using HA-GLUT4-GFP expressed in 3T3-L1 adipocytes to study GLUT4 intracellular trafficking. All assays follow a common general procedure: differentiation of 3T3-L1 fibroblasts into adipocytes, electroporation of adipocytes with HA-GLUT4-GFP, specific experiment using HA-GLUT4-GFP expressing adipocytes, immunofluorescence staining, quantification of the fluorescence, data processing, and analysis.

In the first part of this section, we present the procedures common to our quantitative fluorescence assays. We then discuss how these assays are used to study different aspects of GLUT4 trafficking.

3.1. General Procedure

3.1.1. 3T3-L1 Fibroblasts Differentiation into Adipocytes

1. Cultivate 3T3-L1 fibroblasts in a 10-cm dish and passage before reaching 80% confluence. Split cells every other day as follows: Plate one-fourth of the cells into a new 10-cm dish to maintain a culture, and the remaining cells at a high density in another 10-cm dish to allow them to rapidly reach confluence (*see* **Note 5**).
2. Cultivate 100% confluent cells for an additional 2 to 4 d without changing the medium.
3. Switch the cells to differentiation medium for 3 d.

4. Replace the medium with postdifferentiation medium for an additional day. On day 4, use 3T3-L1 adipocytes for electroporation.

3.1.2. 3T3-L1 Adipocytes Electroporation

1. On day 4 of differentiation, harvest adipocytes as follows: Wash cells twice in PBS and incubate in trypsin (Gibco, Invitrogen Corp.), supplement with collagenase (0.5 mg/mL) to prevent cell aggregation for 2 min. Mechanically de-attach cells from the plastic and from each others by pipeting up and down.
2. One confluent 10-cm dish contains approximately 6×10^6 3T3-L1 adipocytes and one well of a 6-well plate contains approximately 1×10^6 3T3-L1 adipocytes. 1×10^6 Cells are used per electroporation cuvette.
3. Wash de-attached cells in PBS and resuspend in 0.5 mL prewarmed DMEM per 10^6 cells.
4. Transfer 500 µL cell suspension ($\sim 10^6$ cells) to a 1-cm electroporation cuvette (Bio-Rad) and add 45 µg cDNA per cuvette.
5. Mix DNA with the cell suspension by gently tapping the cuvette.
6. Electroporat the cells at 0.18 kV, 960 µF (Bio-Rad GenePulser; Bio-Rad).
7. Add 80 µL electroporated cell suspension to a 120-µL drop of complete medium on the cover slip. Allow cells to attach to cover slip for 30 to 60 min at 37°C in 5% CO2 (*see* **Note 6**).
8. Feed the cells by adding 2 mL of complete medium in the dish and incubate at 37°C in 5% CO^2. The experiments are usually done 24 h after electroporation.

3.1.3. Fluorescence Assays

Details of different experiments are given in **Subheading 3.2**.

3.1.4. Imaging and Fluorescence Quantification

1. Acquire images in epifluorescence mode of a DMIRB inverted microscope coupled to a CCD camera.
2. Choose exposure times for each fluorescence channel such that more than 95% of the image pixel intensities are below camera saturation. Importantly, exposure times for each channel should be kept constant within each experiment. Therefore, it is necessary to begin each imaging session by setting the exposure times for each channel using the condition with the maximum signal (specific examples are given in the following sections).
3. Identify HA-GLUT4-GFP electroporated cells based on GFP expression and subsequently take images in each channel of interest. For each cover slip-bottom dish (i.e., for each condition) collect 20 images corresponding to at least 20 individual.
4. To quantify the fluorescence signals, the authors use the Metamorph image-processing software (Molecular Devices Corp.). Manually outline HA-GLUT4-

GFP expressing cells, identified in the GFP channel, and log the average GFP pixel intensity in the cell. Transfer the outlined area (cell) to the corresponding image on the other fluorescence channels and log the average pixel intensities per cell.
5. Using the same method, measure the fluorescence background, except choose cells that do not express GFP. Average these measured pixel intensities from a few GFP-negative cells for each channel. Subtract these background values from specific signals for each individual HA-GLUT4-GFP-expressing cell.
6. Analyze these quantified fluorescence data according to the specific experiment. Detailed examples are given in the following sections.

3.2. Examples of Immunofluorescence-Based Assays

Changes in the surface-to-total ratio of a protein correlate with changes in exocytosis, in endocytosis, or in both. To understand how insulin affects the surface level of GLUT4 in adipocytes, it is necessary to examine both endocytosis and exocytosis. The following section describes three experimental procedures used to measure the effects of insulin on the level of GLUT4 in the plasma membrane of adipocytes; the exocytosis rate of GLUT4; and the endocytosis rate of GLUT4.

3.2.1. Antibodies Saturating/Blocking Concentration Determination

3.2.1.1. HA.11 SATURATING CONCENTRATION DETERMINATION

1. For each batch of purified HA-11 antibody, determine the concentration of antibody required to saturate the HA epitope of HA-GLUT4-GFP.
2. Electroporate one cuvette of 3T3-L1 adipocytes with 45 μg of HA-GLUT4-GFP cDNA and plate the electroporated cells onto five cover slip-bottom dishes.
3. The day after the electroporation, prepare successive dilutions of purified anti-HA (1/10, 1/20, 1/40, 1/80, 1/160) in warm medium I.
4. Add 100 μL of each of the dilution to the HA-GLUT4-GFP-electroporated cells on the cover slip. Incubate cells for 10 min at 37°C, allowing the HA-11 antibody to be taken by HA-GLUT4-GFP-expressing cells (*see* Note 7).
5. Wash cells three times in PBS and fix by a 10-min incubation in PBS/3.7% formaldehyde.
6. Permeabilize and stain the cells for 30 min at 37°C with 1.5 μg/mL Cy3-conjugated goat-anti-mouse secondary antibodies in PBS/2% FBS/250 μg/mL saponin.
7. Wash cells and take and analyze images as described in **Subheading 3.1.4**.
8. Plot the Cy3/GFP ratios as a function of the concentration of anti-HA used for the uptake. The saturated concentration of anti-HA is reached when the ratio Cy3/GFP plateaus. The typical saturating concentration of purified anti-HA is 50 μg/mL.

3.2.1.2. DETERMINATION OF THE CONCENTRATION OF SECONDARY ANTIBODY TO USE FOR SATURATION OF EPITOPES (BLOCKING CONCENTRATION)

1. Electroporate one cuvette of 3T3-L1 adipocytes with 45 µg of HA-GLUT4-GFP cDNA and plate the electroporated cells onto six cover slips.
2. The day after the electroporation, stimulate the cells with 170 nM insulin for 30 min at 37°C: Replace culture medium with 2 mL Medium I supplemented with 170 nM insulin per cover slip-bottom dish.
3. Wash cells with PBS and fix for 10 min in PBS/3.7% formaldehyde.
4. Stain cells with 100 µL per cover slip of 10 µg/mL anti-HA in PBS/2% FBS for 30 min at 37°C.
5. Wash cells three times with PBS to remove unbound antibody.
6. Use increasing concentrations of Cy5-conjugated secondary antibodies to reveal anti-HA staining.
7. Wash the cells three times in PBS to eliminate the unbound antibodies.
8. Stain the cells again using Cy3-conjugated secondary antibodies at a regular concentration (typically 1.5 µg/mL for Jackson Labs goat-anti-mouse secondary antibodies).
9. Image the cells in the GFP, Cy5, and Cy3 channels.
10. Cy5 and Cy3 fluorescence signals are quantified for GFP positive cells. The blocking concentration of Cy5-conjugated antibodies is reached when the second staining with Cy3 antibodies is prevented.

3.2.2. HA-GLUT4-GFP Surface-to-Total Ratio Determination

This assay is used to determine the proportion of HA-GLUT4-GFP in the plasma membrane of cells, and it can be used to compare the amounts of GLUT4 in the plasma membrane of adipocytes in the basal and the insulin-stimulated states or to determine the rate of redistribution of GLUT4 to the plasma membrane upon insulin stimulation *(4,6,7)*. In this assay, the amount of HA-GLUT4-GFP in the plasma membrane is quantified by measuring the amount of HA epitope on the surface of cells. In a cell-by-cell analysis, the amount of anti-HA antibody bound to the plasma membrane (surface HA-GLUT4-GFP) to the amount of GFP expressed in the cell (total HA-GLUT4-GFP) is normalized. This normalization corrects for different levels of total expression of HA-GLUT4-GFP among the electroporated cells (**Fig. 2A**).

3.2.2.1. PREPARATION OF THE EXPERIMENT AND ELECTROPORATION

The goal of the following experiment is to determine the time necessary for GLUT4 to redistribute to the plasma membrane in response to 170 nM insulin *(6)*. In this example, the levels of GLUT4 in the plasma membrane of basal

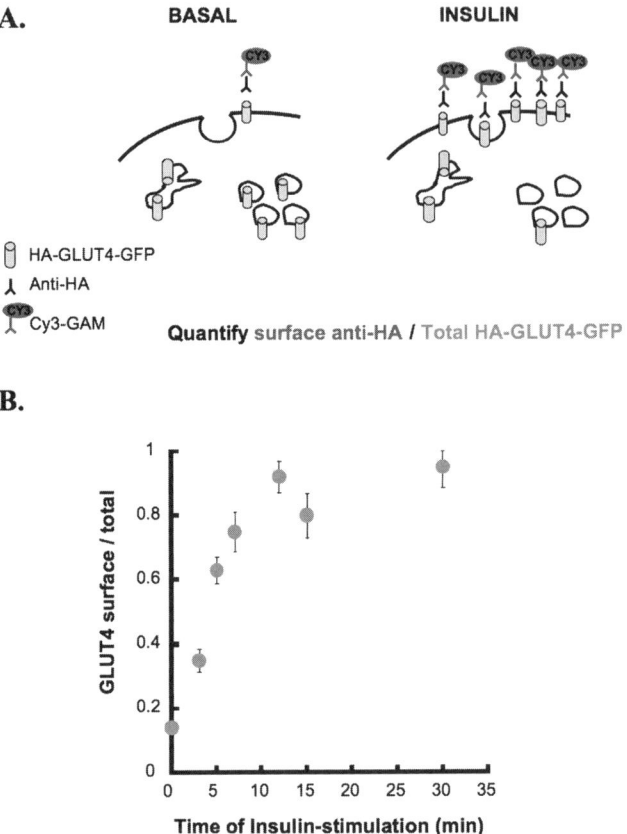

Fig. 2. HA-GLUT4-GFP surface-to-total ratio determination. (**A**) Cartoon of the experiment protocol. HA-GLUT4-GFP electroporated cells are left in their basal unstimulated state (left) or stimulated for 30 min with 170 nM insulin (right). Cells are fixed and stained with anti-HA epitope HA.11 monoclonal antibody and Cy3-conjugated secondary antibodies. Cells are imaged and the fluorescence is quantified. The ratio Cy3 (surface HA-GLUT4-GFP) / GFP (total HA-GLUT4-GFP) is calculated for each condition. (**B**) Plot of the change in GLUT4 surface-to-total. GLUT4 surface-to-total ratio was measure as indicated in A, after increasing times of insulin stimulation. Time 0 corresponds to cells in the basal, nonstimulated state. The results are from a single experiment. Each time point is the averaged value and SEM from at least 20 individual cells.

cells (time 0) and in the plasma membrane of cells stimulated with insulin for increasing times—3, 5, 7, 12, 15, and 30 min—are measured.

The experiment has seven different conditions and thus requires seven cover slips of HA-GLUT4-GFP expressing adipocytes. The 3T3L1 cells are electroporated and plated on seven cover slip-bottom dishes as described in **Subheading 3.1.2**.

3.2.2.2. TIME COURSE OF INSULIN STIMULATION

1. Twenty-four hours after the electroporation, wash the cells once with prewarmed Medium I. Incubate the cells in Medium I for 2 h at 37°C in 5% CO_2.
2. Stimulate the cells with Medium I supplemented with 170 nM insulin. Approximately 2 mL per cover slip-bottom dish should be prepared and warmed to 37°C. Perform cell stimulation by exchanging the Medium I with prewarmed Medium I with insulin.
3. Return the cells to 37°C in 5% CO_2 for the stimulation times (*see* **Note 8**).
4. At the end of the stimulation quickly place the cover slip-bottom dishes on ice-water slurry.
5. Replace the medium with ice-cold Medium II. Wash the cells twice with cold Medium II and fix for 10 min with PBS/3.7 % formaldehyde.
6. Store the fixed cells for a few days at 4°C or use immediately for staining and imaging.

3.2.2.3. IMMUNOFLUORESCENCE STAINING

1. Incubate fixed cells with 100 μL per cover slip anti-HA antibody in PBS/2% FBS for 30 min at 37°C (*see* **Note 7**).
2. Wash the cells three times with PBS to eliminate unbound antibody.
3. Reveal anti-HA staining by incubation with Cy3-labeled goat-anti-mouse secondary antibodies for 30 min at 37°C.
4. Wash the cells three times in PBS and keep them in PBS.

3.2.2.4. IMAGING

HA-GLUT4-GFP in basal adipocytes is mainly concentrated in a perinuclear region and insulin induces a redistribution of the protein between intracellular compartments and the cell surface *(5,6)*. In basal cells, the GFP signal is more concentrated than in insulin-stimulated cells. We therefore use cells at basal to set the maximum GFP exposure time. On the other hand, GLUT4 surface level is higher in the insulin-stimulated state than at basal. Thus, we choose the condition where the cells are exposed to insulin for 30 min to set the maximum exposure time for the Cy3 channel. The maximum exposure times are determined for both the GFP and the Cy3 channels and the images are acquired as described in **Subheading 3.1.4**.

3.2.2.5. DATA PROCESSING AND ANALYSIS

1. Perform quantification of the fluorescence signals as described in **Subheading 3.1.4**. *(8)*.
2. After subtraction of the backgrounds from the specific GFP and Cy3 signals, calculate the ratio Cy3/GFP for each cell (i.e., the amount of surface-exposed HA epitope normalized to the total amount of HA-GLUT4-GFP in each cell).
3. For each condition, basal or insulin, average the background-corrected Cy3/GFP ratios from at least 20 cells and plot the results (**Fig. 2B**).

From this single experiment, it can be deduced that insulin induces an increase of HA-GLUT4-GFP in the surface of adipocytes. GLUT4 redistribution to the plasma membrane occurs within the first minutes of insulin treatment. The new steady-state distribution of HA-GLUT4-GFP is reached after about 10 min of insulin stimulation. In this experiment, maximum GLUT4 translocation upon insulin stimulation was about sevenfold the basal level. The level of insulin-induced GLUT4 translocation in 3T3-L1 adipocytes typically ranges from 7- to 12-fold.

3.2.3. HA-GLUT4-GFP Exocytosis/Cycling Rate Determination

The following assay has been designed to measure the rate of appearance of GLUT4 at the cell surface of adipocytes in different conditions. In this assay, cells are incubated at 37°C in medium containing a saturating concentration of anti-HA antibody. The assay is based on the capacity of the anti-HA antibody to bind the exofacial HA epitope on the HA-GLUT4-GFP construct when it is inserted in the plasma membrane (**Fig. 3A**). Because there is only one HA epitope on the HA-GLUT4-GFP and the HA.11 antibody bound to its epitope does not fall off as HA-GLUT4-GFP moves through intracellular compartments, each HA-GLUT4-GFP can only capture antibody once, the first time it cycles to the plasma membrane *(6,7)*. The amount of cell-associated anti-HA antibody will increase with increasing incubation times at 37°C, as HA-GLUT4-GFP unbound by antibody cycles to the plasma membrane, until it reaches a plateau level, at which time all HA-GLUT4-GFP is bound by HA.11 (that is all HA-GLUT4-GFP has cycled at least once to the plasma membrane). The graph plotting the increase in cell-associated anti-HA over time is fit to a single exponential increase described by the equation:

$$(\text{Cell associated anti-HA/GFP})_t = A - B \times \exp(-k_{ex}t)$$

where A is the plateau reached, B is the amplitude between signal at time 0 and the plateau level, t is the time and k_{ex} is the exocytosis rate constant *(7)*.

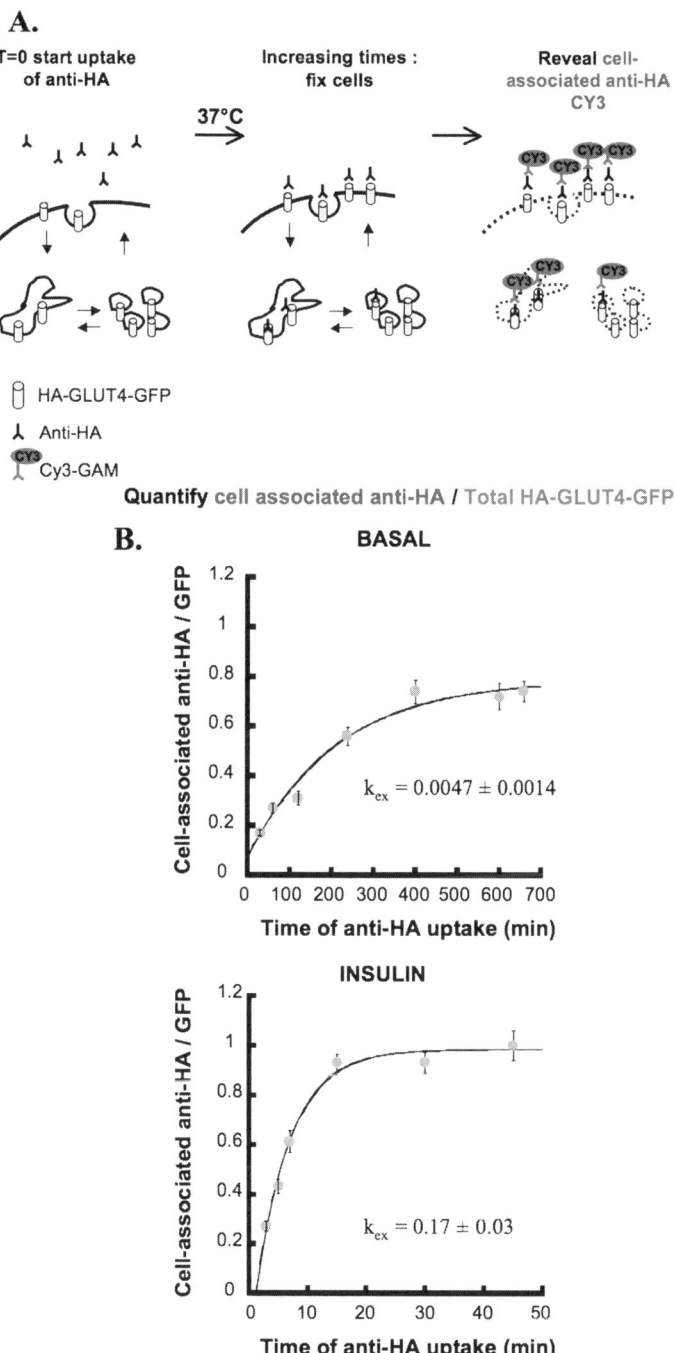

Fig. 3. (Continued)

3.2.3.1. PREPARATION OF THE EXPERIMENT AND ELECTROPORATION

In the following example, the experiment is designed to measure GLUT4 exocytosis in basal and insulin-stimulated adipocytes. The time course of the experiment is determined empirically. There must be sufficient early time points to describe the initial rise of the curve and the time course must be long enough to reach the plateau of anti-HA uptake. To determine GLUT4 exocytosis rate constant in the insulin-stimulated state, anti-HA uptake is measured for 3, 5, 7, 15, 20, 30, and 45 min. In the basal state, 30-, 60-, 120-, 240-, 400-, 600-, and 660-min incubations are used. Thus, 14 cover slip-bottom dishes of cells electroporated with the HA-GLUT4-GFP construct are needed to measure exocytosis in basal and insulin-stimulated cells. Cells are electroporated and plated on cover slip-bottom dishes as previously described.

3.2.3.2. TIME COURSE OF ANTI-HA BINDING

1. Twenty-four hours after electroporation, wash the cells once with prewarmed Medium I and starve from serum in the same medium for 2 h at 37°C in 5% CO_2.
2. Prepare 100 μL of Medium I containing a saturated concentration of anti-HA (usually 50 μg/mL) per cover slip-bottom dish. Divide the solution in half and add insulin to a final concentration of 170 nM to one tube. Warm the antibody solutions to 37°C.

◀──

Fig. 3. (Continued)   HA-GLUT4-GFP exocytosis/cycling rate determination. (**A**) Cartoon of the experiment protocol. Nonstimulated or insulin-stimulated HA-GLUT4-GFP electroporated cells are allowed to take up a saturated concentration of HA.11 anti-HA antibody for increasing times at 37°C. At appropriate time, cells are placed on ice, washed, and fixed. To reveal cell-associated anti-HA, cells are permeabilized (dashed lines) stained with a solution of Cy3-conjugated secondary antibodies supplemented with the detergent saponin. Cells are imaged and fluorescence is quantified. The ratio Cy3 (HA-GLUT4-GFP that cycled at least one to the surface) / GFP (total HA-GLUT4-GFP) over time is determined for each condition. (**B** and **C**) Plots of increase of cell associated anti-HA over time in basal (B) or insulin-stimulated (C) cells. Ratios cell-associated anti-HA/GFP were determined as indicated in A and plotted over time. Plots were fitted to an exponential increase described by the equation:

$$(\text{Cell associated anti-HA/GFP})_t = A - B \text{ xexp}(-k_{ex}t)$$

where A is the plateau reached, B is the amplitude between signal at time 0 and the plateau level, t is the time and k_{ex} is the exocytosis rate constant. Results are from a single experiment and each time point is the averaged value and SEM from at least 20 individual cells.

3. For the insulin condition, prestimulate the cells for 30 min with Medium I supplemented with 170 nM insulin (170 nM insulin is included in the antibody solution used for anti-HA binding; *see* **Note 9**).
4. Set up a plate warmer at 37°C on the bench.
5. For each time point, take a cover slip-bottom dish from the incubator and quickly place it on the plate warmer so that the temperature change is minimal (*see* **Note 8**).
6. Replace the serum-free Medium I with medium containing the anti-HA antibody.
7. Quickly return the cover slip-bottom dish to the CO_2 incubator (*see* **Note 10**).
8. At the end of the incubation period, place the cover slip-bottom dishes on ice-water slurry.
9. Replace the medium with ice-cold Medium II. Wash the cells twice with cold Medium II and fix for 10 min with PBS/3.7% formaldehyde.
10. Store the fixed cells for a few days at 4°C or use immediately for staining and imaging.

3.2.3.3. IMMUNOFLUORESCENCE STAINING

1. To reveal the total amount of cell-associated anti-HA antibody (i.e., the amount on the cell surface and internalized) cells must be permeabilized (**Fig. 3A**). Permeabilize and stain the cells with a solution of Cy3-conjugated goat-anti-mouse antibody supplemented with 250 μg/mL saponin for 30 min at 37°C.
2. Wash the cells three times with PBS to remove unbound antibodies and stored in PBS.

3.2.3.4. IMAGING

1. Use a cover slip-bottom dish of basal cells is used to set the exposure time in the GFP channel (*see* **Subheading 3.2.2.4**). Because exocytosis of GLUT4 is greatly accelerated by insulin, the most intense anti-HA-associated Cy3 staining can be expected in the insulin-stimulated conditions. Thus, a cover slip-bottom dish corresponding to the longest time of anti-HA uptake in the insulin-stimulated condition can be used to set the exposure time in the Cy3 channel.
2. Maximum exposure times are determined for both the GFP and Cy3 channels.
3. The images are then taken in both channels as described in **Subheading 3.1.4**.

3.2.3.5. DATA PROCESSING AND ANALYSIS

1. Perform quantification of the fluorescence signals as described in **Subsection 3.1.4**.
2. After subtraction of the backgrounds from the specific signals, calculate the ratio Cy3/GFP for each cell (i.e., the amount of HA-GLUT4-GFP that cycled at least once to the plasma membrane normalized to the total amount of HA-GLUT4-GFP contained in the cells).

3. For each time point, average and plot the background-corrected Cy3/GFP ratios of at least 20 cells for each basal or insulin-stimulated condition (**Fig. 3B**).
4. Fit plots to single exponential. The rise to the plateau is the GLUT4 exocytosis rate constant.
5. The data presented in **Fig. 3B** indicate that GLUT4 cycles to the plasma membrane in both the basal and the insulin-stimulated conditions. Insulin increases the amount of GLUT4 cycling to the plasma membrane by accelerating GLUT4 exocytosis rate rather than, for example, controlling the size of the cycling pool.

3.2.4. HA-GLUT4-GFP Endocytosis Rate Determination

The following experiment is designed to measure GLUT4 internalization. This internalization assay is a derivative of the INTERNAL/SURFACE (IN/SUR) method (9), where the amount of internalized ligand, normalized to the amount of ligand bound to the cell surface, is measured. In the case of HA-GLUT4-GFP, the ligand is the anti-HA antibody. The plot of the ratio of internalized GLUT4 to surface GLUT4 as a function of time yields a straight line, the slope of which is proportional to the endocytosis rate constant.

3.2.4.1. Preparation of the Experiment and Electroporation

For recycling proteins like GLUT4, the IN/SUR method is only valid for conditions in which GLUT4 internalized from the plasma membrane does not return to the cell surface during the time course of the experiment. This condition is approximated when the internalization measurement times are less than the recycling halftimes. In insulin-stimulated conditions, the halftime for GLUT4 recycling is about 10 min, and in basal conditions the recycling halftime is more than 100 min (**Fig. 3B**). Therefore, in insulin-stimulated cells the internalization measurement times are restricted to less than 10 min, whereas in basal conditions longer internalization pulse times can be used. In basal cells there is about one-tenth the amount of GLUT4 in the plasma membrane as in insulin-stimulated cells. Thus, the longer pulse times in basal conditions allow for sufficient accumulation of internal anti-HA antibody to be accurately measured.

Four time points are sufficient for the fit of a straight line with good statistical significance. In basal cells, uptake is measured for 15, 20, 25, and 30 min. In insulin-stimulated cells, endocytosis is measured on a time scale of 3, 5, 7, and 9 min. The experiment requires a total of eight cover slips of HA-GLUT4-GFP-expressing adipocytes.

1. Electroporated and plate cells on eight cover slip-bottom dishes as previously described.

3.2.4.2. TIME COURSE OF ANTI-HA UPTAKE

1. At 24 h after electroporation, incubate the cells with serum-free medium I for 2 h.
2. Prepare Medium I solutions containing a saturated amount of anti-HA antibody with or without 170 nM insulin. Equilibrate it to 37°C in a water bath, as described previously for the exocytosis assay.
3. Set up a plate warmer to 37°C on the bench.
4. At the end of the serum-free incubation time, prestimulate half the cells, those used to measure insulin-stimulated state GLUT4 endocytosis, for 30 min with 170 nM insulin as described in the previous assay.
5. For each time point, take the appropriate cover slip-bottom dish from the incubator and put it on the plate warmer (*see* **Notes 8** and **9**).
6. Replace the serum-free medium I with the uptake medium and quickly place the cover slip-bottom dish back in the incubator (*see* **Note 10**).
7. At appropriate times, place the cells on ice-water slurry and replace the medium with ice-cold Medium II.
8. Wash the cells twice with cold Medium II and fix them for 10 min with PBS/3.7 % formaldehyde.
9. Store the fixed cells for a few days at 4°C or use them immediately for staining and imaging.

3.2.4.3. IMMUNOFLUORESCENCE STAINING

To reveal the amount of internalized anti-HA, excluding the anti-HA remaining bound to the surface, the cells are in two steps: First, use a blocking concentration of secondary antibody to stain surface-associated anti-HA (for protocol to determine the blocking concentration, *see* **Subheading 3.2.1.2**). Second, permeabilize the cells and stain internalized anti-HA using secondary antibodies conjugated to a different fluorophore (**Fig. 4A**). Typically, Cy5-conjugated goat-anti-mouse is used to block the surface anti-HA and Cy3-conjugated goat-anti mouse to reveal internalized anti-HA, but different combinations can be used (*see* **Notes 11**).

1. Stain the cells with the blocking concentration of Cy5-conjugated goat-anti-mouse for 30 min at 37°C.
2. Wash the cells three times with PBS to remove unbound antibodies.
3. Re-fixed the cells at room temperature for 5 min with PBS/3.7% formaldehyde.
4. After three washes with PBS, permeabilize and stain the cells using the regular 1.5 µg/mL concentration of Cy3-conjugated goat-anti-mouse supplemented with 250 µg/mL saponin for 30 min at 37°C.
5. Wash the cells three times in PBS to remove unbound antibodies and store in PBS.

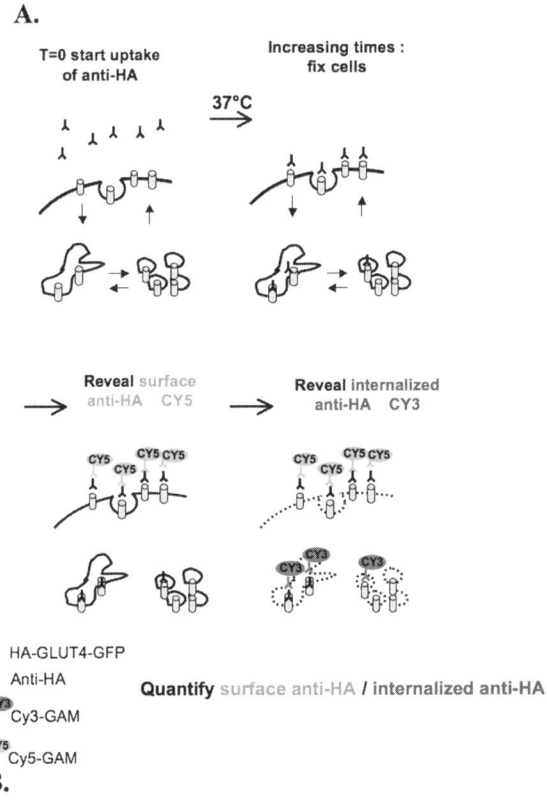

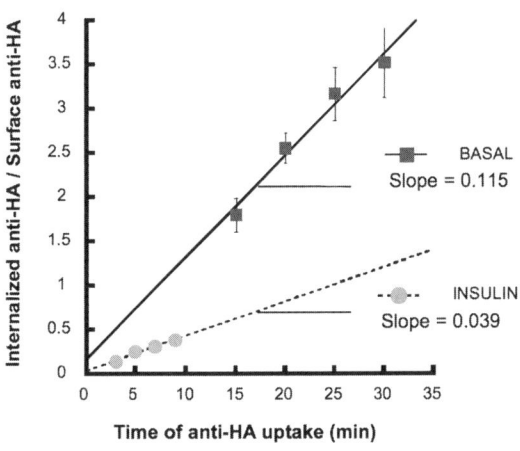

Fig. 4. (Continued)

3.2.4.4. IMAGING

We set up maximum GFP exposure time on basal cells and maximum Cy5 exposure time (surface staining) on cells treated with insulin (**Subheading 3.2.2.4**). For internalized anti-HA (Cy3 channel), we use the longest times of uptake in the different conditions tested and set up the exposure time on the brightest signal. Images are then acquired in the three different channels, as described in **Subheading 3.1.4**.

3.2.4.5. DATA PROCESSING AND ANALYSIS

1. Perform quantification of fluorescence signals as described in **Subheading 3.1.4**.
2. After subtraction of the background from the specific signals, calculate the Cy3/Cy5 ratios (i.e. internalized (IN) GLUT4 normalized to the amount of GLUT4 in the surface (SUR), for each cell).
3. For each time point within each condition, average the ratios IN/SUR of at least 20 cells.
4. Fit the plot of the averaged ratios IN/SUR over time for each condition to a straight line, the slope of which is proportional to GLUT4 internalization rate constant (**Fig. 4B**). It is important to note that because we used two different fluorophores to measure IN GLUT4 and SUR GLUT4, we cannot determine the actual endocytosis rate constant of GLUT4 but an endocytosis parameter that is proportional to the endocytosis rate constant. To adapt this protocol in order to measure the actual endocytosis rate constant, please *see* **Note 12**.
5. Results presented in **Fig. 4B** indicate that insulin-stimulation of adipocytes induces a decrease in GLUT4 internalization *(10)*.

Fig. 4. (Continued) HA-GLUT4-GFP endocytosis rate determination. (**A**) Cartoon of the experiment protocol. Nonstimulated or insulin-stimulated HA-GLUT4-GFP-electroporated cells are incubated in medium with a saturating concentration of anti-HA HA.11 antibody for increasing times at 37°C. At the times noted, the cells are place on ice, washed, and fixed. Surface anti-HA is revealed with a blocking concentration of Cy5-conjugated secondary antibodies and cells are fixed a second time. To reveal internalized anti-HA, cells are permeabilized (dashed lines) stained with Cy3-conjugated secondary antibodies. Cells are imaged and the fluorescence is quantified. The ratio Cy3 (internalized anti-HA) / Cy5 (surface anti-HA) over time is determined for each condition. (**B**) Plots of increase of internalized anti-HA over time in basal (plain line) or insulin-stimulated (dotted line) cells. The ratios of internalized anti-HA/surface anti-HA were determined as indicated in A. The data were fit to straight lines, the slopes of which are HA-GLUT4-GFP basal and insulin-stimulated endocytosis parameters. These endocytosis parameters are proportional to HA-GLUT4-GFP endocytosis rate constant.

3.3. Conclusions

In this chapter, we presented three quantitative immunofluorescence assays developed to measure different parameters of the subcellular trafficking of the GLUT4 glucose transporter. GLUT4 trafficking in adipocytes is one of the main targets of insulin signaling. The first assay we presented demonstrated that the level of GLUT4 in the plasma membrane of adipocytes is acutely increased by insulin signaling. The two other assays showed that insulin acts on GLUT4 surface distribution by both increasing GLUT4 exocytosis and decreasing GLUT4 endocytosis.

Some of the current open questions in the field include the following. What are the downstream effectors of the insulin signaling required to affect the different step of GLUT4 trafficking? What are the ultimate cellular trafficking machineries targeted by insulin signaling? What are the molecular determinants in GLUT4 that regulate its trafficking and the insulin-sensitivity of this trafficking? Quantitative immunofluorescence assays such as those described here can be used to address some of those questions.

Beyond insulin signaling and GLUT4 trafficking, quantitative immunofluorescence can be used to study the trafficking of virtually any membrane protein.

4. Notes

1. To avoid repetitive freeze–thaw, aliquot solutions in small volumes.
2. PBS/3.7% formaldehyde solution must be prepared fresh by 1/10 dilution of commercially available 37% formaldehyde solution.
3. After purification, store the antibody in PBS at 4°C.
4. Prepare 100X concentrated saponin solution in PBS fresh the day of the experiment.
5. It is important to split cells before they are confluent because cell–cell contacts would lead to inhibition of cell growth.
6. Electroporation leads to some cell death and accumulation of foam on top of cell suspension. We recommend manually eliminating the debris before resuspending and plating the cells onto cover slips.
7. Because of the small volume of incubation and in order to prevent its evaporation during long incubations, we recommend covering the cover slips with wet towel paper.
8. The time course can be done in reverse (starting with the longest time point) so that the end of the time course is the same for all time points.
9. The basal and insulin-stimulated conditions can be done either separately or in parallel.

10. For an accurate time course, especially in the insulin-stimulated conditions where the time points are short and closed to each other, it is crucial to act fast and to minimize the temperature changes.
11. We experimentally determined that the concentration of Jackson Labs Cy5-conjugated goat-anti-mouse necessary to block the same Cy3-conjugated goat-anti-mouse antibodies binding is 15 µg/mL. It is important to use same antibodies conjugated to different fluorophores they must have similar affinities for the same epitopes in order for the one to block the binding of the other.
12. To obtain the endocytosis rate constant of GLUT4, we must measure IN GLUT4 and SUR GLUT4 using the same fluorophores. The endocytosis assay must thus be adapted. For each condition, an additional cover slip-bottom dish is used for one of the time points of uptake and the experiment is conducted exactly as described **Subheading 3.2.4**. After fixation, all cover slip-bottom dishes are treated as described previously except the extra one. The extra cover slip-bottom dish is used to measure SUR anti-HA staining with the same fluorophore as the one used to measure IN anti-HA (i.e., Cy3 in this example). Cells are imaged and fluorescence is quantified as described earlier. In each condition we determine the ratio Cy3 (SUR)/GFP (total) from the extra cover slip. For each time point in each condition, we calculate the background-corrected Cy3 (IN)/GFP (total) ratio on a cell-by-cell basis. Finally, we calculate the ratio [Cy3 (IN)/GFP (total)]/ [Cy3 (SUR)/GFP (total)], which is equivalent to a measurement of IN/SUR with the same Cy3 fluorophore. The plot of the IN/SUR ratio over time is fit to a straight line, the slope of which is GLUT4 internalization rate constant.

Acknowledgments

The work was supported by grants from the NIH DK52852 (TEM) and DK69982 (TEM).

References

1. Dugani C.B. and Klip A. (2005) Glucose transporter 4: cycling, compartments and controversies. *EMBO Rep.* **6**, 1137–1142.
2. Ishiki M. and Klip A. (2005) Minireview: recent developments in the regulation of glucose transporter-4 traffic: new signals, locations, and partners. *Endocrinology* **146**, 5071–5078.
3. Dawson K., Aviles-Hernandez A., Cushman S., and Malide D. (2001) Insulin-regulated trafficking of dual-labeled glucose transporter 4 in primary rat adipose cells. *Biochem. Biophys. Res. Commun.* **287**, 445–454.
4. Lampson M.A., Racz A., Cushman S.W., McGraw T.E. (2000) Demonstration of insulin-responsive trafficking of GLUT4 and vpTR in fibroblasts. *J. Cell Sci.* **113**, 4065–4076.

5. Lampson M.A., Schmoranzer J., Zeigerer A., Simon S.M., and McGraw T.E. (2001) Insulin-regulated release from the endosomal recycling compartment is regulated by budding of specialized vesicles. *Mol. Biol. Cell* **12**, 3489–3501.
6. Zeigerer A., Lampson M., Karylowski O., et al. (2002) GLUT4 retention in adipocytes requires two intracellular insulin-regulated transport steps. *Mol. Biol. Cell* **13**, 2421–2435.
7. Karylowski O., Zeigerer A., Cohen A., and McGraw T.E. (2004) GLUT4 is retained by an intracellular cycle of vesicle formation and fusion with endosomes. *Mol. Biol. Cell* **15**, 870–882.
8. Dunn K., Mayor S., Myers J., and Maxfield F. (1994) Applications of ratio fluorescence microscopy in the study of cell physiology. *FASEB J.* **8**, 573–582.
9. Wiley H.S. and Cunningham D.D. (1982) The endocytotic rate constant. A cellular parameter for quantitating receptor-mediated endocytosis. *J. Biol. Chem.* **257**, 4222–4229.
10. Blot V. and McGraw TE. (2006) GLUT4 is internalized by a cholesterol-dependent nystatin-sensitive mechanism inhibited by insulin. *Embo J.* **25**, 5648–5658.

27

Dissecting GLUT4 Traffic Components in L6 Myocytes by Fluorescence-Based, Single-Cell Assays

Costin N. Antonescu, Varinder K. Randhawa, and Amira Klip

Summary

Postprandial blood glucose homeostasis is regulated by an insulin-stimulated increase in glucose transport into muscle and fat tissues via glucose transporter isoform 4 (GLUT4). In the basal state, this constitutively recycling membrane protein predominantly resides intracellularly. In order to achieve the insulin-stimulated increase in glucose flux, GLUT4 increases its cell surface abundance at the expense of preformed intracellular depots. By confocal microscopy of cultured L6 muscle cells stably expressing *myc*-tagged GLUT4 (L6-GLUT4*myc*), we can visualize the two arms of GLUT4 traffic: exocytosis (movement to the cell surface) and endocytosis (internalization from the cell surface).

Key words: Endocytosis; exocytosis; GLUT4 traffic; recycling.

1. Introduction

The regulation of membrane protein traffic has received considerable attention in recent times. Among the best understood of models of this phenomenon is the regulated traffic of the facilitative glucose transporter 4 (GLUT4), the principal insulin-responsive glucose transporter in muscle and fat cells (*1*). GLUT4 constitutively cycles between the plasma membrane and internal membrane compartment(s) and in the basal state, it is localized primarily intracellularly, as a result of rapid endocytosis and slower exocytosis at the plasma membrane

---

C.N.A. and V.K.R. contributed equally to the writing of this manuscript.

combined with idle cycling between intracellular depots (reviewed in **ref. 2**). Upon insulin stimulation, a gain in cell surface GLUT4 occurs at the expense of intracellular GLUT4 *(2)*. In addition to insulin, other physical stimuli also elevate cell surface GLUT4 levels, such as muscle contraction *(3)*, hypoxia *(4)*, hypertonic stress *(5)*, metabolic stress in the form of mitochondrial uncouplers *(6)*, and DNA and protein synthesis inhibitors such as anisomycin *(7)*.

Any gain in cell surface GLUT4 could conceivably be caused by either an increase in exocytosis (movement to the plasma membrane), a decrease in endocytosis (internalization from the plasma membrane) of the transporter, or both. Thus, in order to understand the regulatory mechanism(s) of GLUT4 traffic, it is important to differentiate between these two possible modes of action. It is well known that insulin augments GLUT4 exocytosis (reviewed in **ref. 8**), although a decrease in GLUT4 endocytosis has also been reported *(9)*. However, hypertonicity is thought to increase plasma membrane levels of GLUT4 largely by inhibition of internalization *(5)*. The mechanism of action of other agents that elevate surface GLUT4 content is less well understood, although inhibition of GLUT4 internalization by metabolic stressors has recently been reported *(10)*.

L6 myoblasts in culture expressing *myc*-tagged GLUT4 are ideal to study GLUT4 membrane traffic, through the dynamic availability of the *myc* epitope to the extracellular milieu. Here, we describe single-cell immunofluorescence-based methods to separately determine the exocytic and endocytic rates of GLUT4 in single L6 myoblasts. Although the protocols described herein are for measurement of GLUT4 traffic parameters, these approaches are applicable to any membrane protein with an exofacial epitope that is stably expressed in a cell line. Stable expression ensures that all cells express the epitope to the same extent. If transient transfections are used, then all values gathered must be expressed relative to the expression level of the protein in each cell.

2. Materials

2.1. Cell Culture

1. Alpha Modified Eagle's Medium (α-MEM; Wisent Inc.) with Earle's salts and ribonucleosides, L-glutamine and deoxyribonucleosides supplemented with 10% fetal bovine serum (FBS; HyClone).
2. Solution of 0.5% trypsin with 0.53 mM ethylenediamine-tetraacetic acid (EDTA) from Wisent Inc.
3. Human insulin (Humulin R) 100 U/mL (Lilly) is diluted to 100 nM working solution in α-MEM

2.2. L6 Myocytes Stably Expressing GLUT4myc (L6-GLUT4myc)

L6 myoblasts stably expressing GLUT4 harboring a *myc* epitope in the first exofacial loop were previously created and characterized *(11–13)*. Briefly, the human c-*myc* epitope (AEEQKLISEEDLLK) was inserted into the first ectodomain of rat GLUT4. This cDNA construct was then subcloned into the pCXN2 vector *(14)*, and transfected into L6 myoblasts and stably expressing clones were selected.

2.3. Confocal Immunofluorescence for GLUT4myc

1. Microscope coverslips (18 mm^2 or 25 mm^2 circle) from VWR International.
2. Multiwell 6- or 12-well tissue culture plates from Becton Dickinson.
3. BD Falcon 75-cm^2 tissue culture flasks are from BD Biosciences Canada.
4. Phosphate-buffered saline (PBS; 10 X stock): 1.37 M NaCl, 27 mM KCl, 100 mM Na_2HPO_4, 18 mM KH_2PO_4 (adjust to pH 7.4 with HCl if necessary) and store at room temperature. Prepare a working solution of 1X PBS by dilution of one part 10X PBS with nine parts water. Prepare PBS+ by supplementing 1X PBS with 1 mM $CaCl_2$ and 1 mM $MgCl_2$.
5. 16% Paraformaldehyde (PFA; Electron Microscopy Sciences): 4% (v/v) solution in PBS+ fresh for each experiment.
6. Quench solution: 0.1 M glycine in PBS+.
7. Permeabilization solution: 0.1% (v/v) Triton X-100 in PBS+.
8. Blocking buffer: either 5% (w/v) milk or 5% (v/v) goat serum in PBS+.
9. Primary antibody: monoclonal anti- 9E10 antibody (Santa Cruz Biotechnology Inc.) or polyclonal anti-*myc* polyclonal antibody (Sigma-Aldrich Canada Ltd.). Prepare working dilution of antibody in blocking buffer.
10. Secondary antibody: Anti-mouse or anti-rabbit IgG conjugated to Cy3 (Molecular Probes). Prepare working dilution of antibody in blocking buffer.
11. Mounting medium: DakoCytomation Fluorescent Mounting Medium (Dako Canada Inc.).
12. Parafilm from Pechiney Plastic Packaging.
13. Kimwipes EX-L Delicate Task Wipers from Kimberly-Clarke Corporation.
14. Confocal microscope (e.g., Zeiss LSM510 laser scanning confocal microscope, Carl Zeiss).

3. Methods

Confocal immunofluorescence microscopy allows the visualization of the steady-state levels of GLUT4*myc* transporters at the cell surface under various conditions (**Subheading 3.2**). Additionally, individual components of GLUT4*myc* traffic to and from the cell surface can be resolved: GLUT4*myc* internalization (**Subheading 3.3**) and externalization (**Subheading 3.4 and 3.5**). The methods outlined here allow the measurement of GLUT4*myc* traffic parameters in individual cells. Thus, one can employ additional molecular techniques such as cDNA transfection, even when a small percentage of cells are transfected because transfected cells can be easily discerned from vicinal untransfected cells. As an example, GLUT4*myc* internalization in L6 myoblasts transfected with green fluoresent protein (GFP) cDNA (**Fig. 2A**) is shown, under conditions such that about 30% of cells express GFP.

3.1. Cell Culture

1. L6-GLUT4*myc* cells are thawed into 75-cm^2 tissue culture flasks in α-MEM supplemented with 10% FBS (10% α-MEM) under sterile conditions.
2. On passage (typically when cells reach 50–60% confluence), cells are discarded of 10% α-MEM media and quickly washed once with sterile PBS, which is also discarded.
3. Cells are incubated with the solution of trypsin for about 30 to 60 s (until the cells begin to round up).
4. Cells are resuspended in 10 mL of 10 % α-MEM media, and 1 mL is diluted for passage into 12 mL of fresh 10% α-MEM media.

3.2. Confocal Immunofluorescence for Steady-State Surface GLUT4myc

1. Seed L6-GLUT4*myc* cells under sterile conditions onto 18 mm^2 coverslips in 12-well plates. Sterilize coverslips by holding with tweezers, wetting with 95% ethanol, and passing through the flame of an alcohol lamp.
2. At the time of the experiment (at least 24 h after seeding), serum-starve cells by first rinsing and then incubating with serum-free α-MEM for 3 to 5 h. Ensure that cells are below confluence to allow individual cells to be clearly discerned by immunofluorescence.
3. Stimulate the cells as required. For example, leave cells untreated (basal) or treat with 100 nM insulin at 37°C for the required times, and then rinse rapidly three times with ice-cold PBS+ on ice.
4. Block cells with 5% milk in PBS+ for 10 min.
5. During blocking, prepare ice-cold 2 µg/mL anti-*myc* antibody solution in blocking solution. For each coverslip, spot 200 µL of antibody solution on clean Parafilm placed on an ice-cold surface.
6. Following blocking, label surface GLUT4*myc* in intact cells (*see* **Note 1**). Carefully lift each coverslip with forceps, and remove excess liquid by dabbing the edge of the coverslip on a static-free Kimwipe tissue. Then, place coverslip with the cell-adhered face down on an antibody solution spot (on Parafilm), and incubate for 1 h at 4°C. During this period, retain 12-well plates containing blocking solution on ice.
7. Following anti-*myc* incubation, lift coverslips from Parafilm and replace it face-up in respective blocking solution-containing wells. Wash coverslips five times in PBS+, with aspiration of solution between each wash.
8. Fix the cells with 4% PFA solution for 20 to 30 min on ice. Following fixation, discard the PFA into a hazardous waste container and then wash samples quickly with PBS+.

9. Quench residual PFA by incubation in the 0.1 M glycine solution for 10 min at room temperature, followed by another quick PBS+ wash.
10. Detect the amount of anti-*myc* antibody-bound GLUT4*myc* at the cell surface (*see* **Note 1**). Incubate intact cells in 750 μL of ice-cold blocking solution containing appropriate Cy3-conjugated secondary antibody (1:750 dilution) for 1 h at 4°C. Ensure that samples are kept in the dark.
11. Following incubation with secondary antibody, wash cells six times in ice-cold PBS+ with aspiration of solution between each wash.
12. Mount the samples by inverting each coverslip into a drop of Dako mounting medium on a microscope glass slide (*see* **Note 2**). Samples can be viewed immediately once dry, or be stored in the dark at −20°C for up to 1 month.
13. View the slides by confocal microscopy. Excitation at 543 nm induces Cy3 fluorescence (red emission) for GLUT4*myc*. For each sample, xy-plane images are acquired along the entire z-plane of the cells and then collapsed to generate a single composite xy-projection in LSM Image software. Examples of the cell surface signal obtained for the steady-state gain in GLUT4*myc* over time in intact L6-GLUT4*myc* cells in the basal and insulin-stimulated states are shown in **Fig. 1**.
14. For quantification, export single composite xy-projections as TIFF files. To quantify cell surface GLUT4*myc* staining intensity in each condition, employ ImageJ software (NIH). Following outlining each cell's contour, determine the mean pixel intensity for surface GLUT4*myc* stain. The mean surface GLUT4*myc* measurement of at least 30 cells per condition is required.

3.3. Confocal Immunofluorescence for GLUT4myc Endocytosis (Internalization)

1. Seed L6-GLUT4*myc* myoblasts on 25 mm^2 glass coverslips in 6-well plates. Sterilize coverslips as described in **Subheading 3.2.1**. For each condition in which GLUT4*myc* internalization is to be measured, typically four coverslips are required in order to measure multiple time points of internalization.
2. Twenty-four hours after seeding, transfect cells with appropriate cDNAs as required, using Lipofectamine 2000 transfection reagent, as per manufacturer's instructions.
3. At the time of the experiment (18–48 h after transfection), serum-deprive L6-GLUT4*myc* cells for 3 to 5 h and then stimulate as required.
4. Wash cells three times in ice-cold PBS+, and then block by incubation in 5% GS in PBS+ for 20 min, on ice.
5. During blocking, prepare ice-cold anti-*myc* antibody solution (1:100) in blocking solution and label GLUT4*myc* as described in **Subheadings 3.2.5** to **3.2.7** (*see* **Note 1**).
6. To rewarm cells prior to internalization, quickly wash cells twice in 37°C PBS+.
7. Immediately following rewarming, place cells in prewarmed 37°C α-MEM containing the appropriate stimulus, and allow internalization of GLUT4*myc* by incubation at 37°C for either 0 (no internalization), 2, 5, or 10 min.

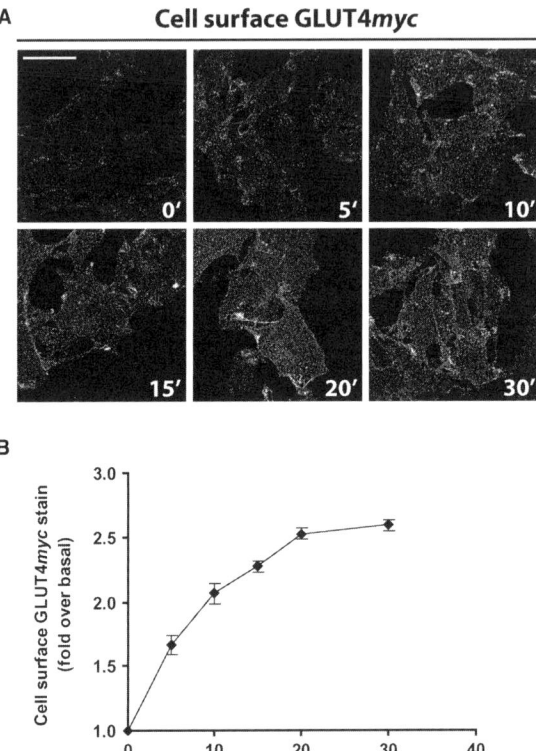

Fig. 1. Insulin-dependent gain in surface GLUT4*myc* in L6 myoblasts as a function of time. (**A**) Intact L6-GLUT4*myc* cells were left untreated (0') or treated with 100 nM insulin over time, and processed for the gain in surface GLUT4*myc*. Shown are micrographs obtained by confocal microscopy, representative of three independent experiments, showing the GLUT4*myc* content at the plasma membrane following stimulation with insulin as a function of time, as indicated. Scale bar, 20 μm. (**B**) Cell surface labeling of GLUT4*myc* at 0, 5, 10, 15, 20, or 30 min of insulin stimulation was quantified as described in **Subheading 3.2**. Shown are the means ± S.E. of 3 independent experiments with cell surface GLUT4*myc* gain expressed as a fold over basal.

8. To halt internalization, wash cells twice with ice-cold PBS+. Block cells in 5% GS in PBS+ for 20 min.
9. Detect the amount of anti-*myc* antibody-bound GLUT4*myc* remaining at the cell surface following internalization (*see* **Note 1**). Incubate intact cells in 750 μL of ice-cold blocking solution containing Cy3-conjugated secondary antibody (1:750) for 1 h at 4°C. Ensure that samples are kept in the dark.
10. Following incubation with secondary antibody, wash cells six times in ice-cold PBS+ with aspiration of solution between each wash.

11. Fix cells in ice-cold 4% PFA in PBS+ for 30 min. Discard the PFA into a hazardous waste container and quickly wash the samples with PBS+.
12. Quench fixative with 0.1 mM glycine in PBS+ for 10 min.
13. Mount the samples by inverting each coverslip into a drop of Dako mounting medium on a microscope glass slide (*see* **Note 2**). Samples can be viewed immediately once dry, or be stored in the dark at –20°C for up to 1 month.
14. View the slides by confocal microscopy. Excitation at 543 nm induces Cy3 fluorescence (red emission) for GLUT4*myc*. Scan whole cell mounts along the z-axis in order to subsequently obtain single composite xy-plane images (sum projection). A typical series of images for GLUT4 internalization in L6 myoblasts transfected with GFP cDNA is shown in **Fig. 2A**.
15. For quantification, export single composite xy-projections as TIFF files. To quantify the cell surface GLUT4*myc* staining intensity at each time of internalization and treatment condition, employ ImageJ software. Following outlining each cell's contour, determine the mean pixel intensity for surface GLUT4*myc* stain.
16. The mean surface GLUT4*myc* measurement of at least 30 cells per time point per treatment condition is expressed as a fraction of that at t = 0 min of internalization (**Fig. 2B**).

3.4. Confocal Immunofluorescence for GLUT4myc Recycling

1. Seed L6-GLUT4*myc* cells under sterile conditions onto 18 mm² coverslips in 12-well dishes. Sterilize coverslips are described in **Subheading 3.2.1**. Determination of total cellular GLUT4*myc* in L6-GLUT4*myc* cells in a distinct but parallel sample (described in **Subheading 3.5**) is also required for each experiment.
2. At the time of the experiment (at least 24 h after seeding), serum-starve cells by first rinsing and then incubating with serum-free α-MEM for 3 to 5 h. Ensure that cells are below confluence to allow individual cells to be clearly discernable by immunofluorescence.
3. Incubate the cells with primary anti-*myc* antibody (1:100) in α-MEM for the appropriate times as required. Then, rapidly rinse cells three times with ice-cold PBS+ on ice to arrest membrane traffic.
4. Fix cells in 4% PFA in PBS+ for 20 to 30 min on ice. Discard the PFA into a hazardous waste container and quickly wash the samples with PBS+.
5. Quench residual PFA by incubation in the 0.1 M glycine solution for 10 min at room temperature, followed by another quick PBS wash.
6. Permeabilize cells by incubation with 0.1% Triton X-100 for 15 to 20 min at room temperature, and then rinse quickly with PBS.
7. Block cells with 5% milk in PBS+ for 10 min.
8. Detect the amount of anti-*myc* antibody-bound GLUT4*myc* within permeabilized cells (see **Note 1**). Incubate cells in 750 µL of ice-cold blocking solution containing Cy3-conjugated secondary antibody (1:750) for 1 h at room temperature. Ensure that samples are kept in the dark.

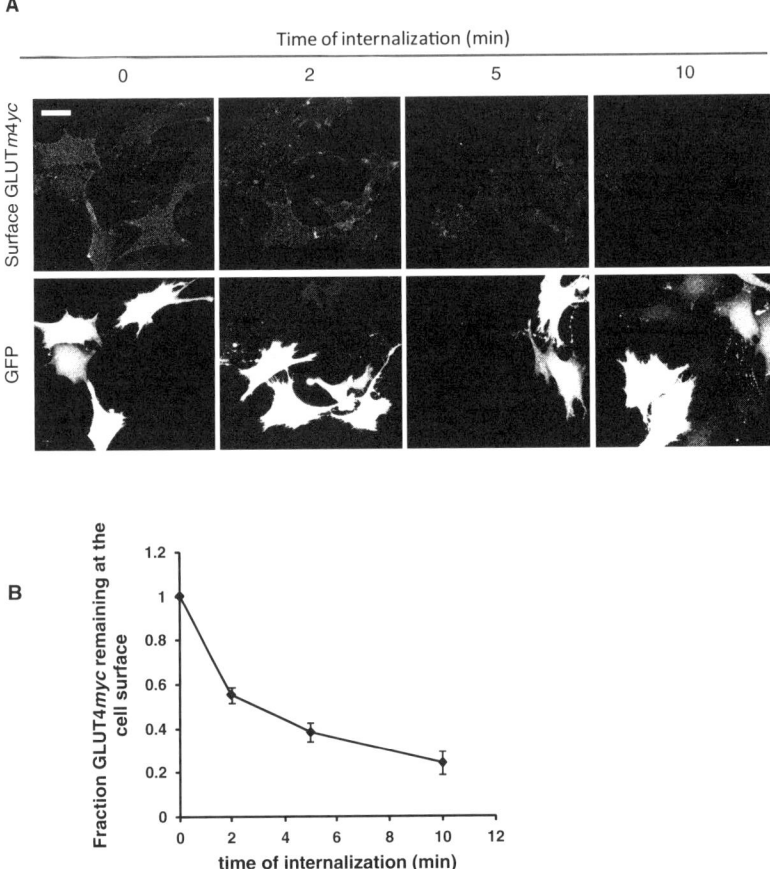

Fig. 2. GLUT4*myc* internalization in L6 myoblasts as a function of time. (**A**) L6-GLUT4*myc* myoblasts were transfected with cDNA encoding green florescent protein (GFP) and subsequently, GLUT4*myc* internalization was determined in the basal, unstimulated state. GFP is used as an example of the use of transfection in conjunction with measurements of GLUT4 traffic. Shown are micrographs obtained by confocal microscopy, representative of three independent experiments, showing the amount of GLUT4*myc* remaining at the cell surface as a function of time of internalization, or GFP fluorescence, as indicated. Scale bar, 20 μm. (**B**) The amount of GLUT4*myc* remaining at the cell surface, following initial labeling, in untransfected cells, at either 0, 2, 5, or 10 min of internalization was quantified as described in **Subheading 3.3**. Shown are the means ± S.E. of 3 independent experiments expressed as a fraction of the GLUT4*myc* remaining at the cell surface.

9. Mount the samples by inverting each coverslip into a drop of Dako mounting medium on a microscope glass slide (see **Note 2**). Samples can be viewed immediately once dry, or be stored in the dark at –20°C for up to 1 month.

10. View the slides by confocal microscopy. Excitation at 543 nm induces Cy3 fluorescence (red emission) for GLUT4*myc*. For each sample, xy-plane images are acquired along the entire z-plane of the cells and then collapsed to generate a single composite xy-projection in LSM Image software. Examples of the amount of GLUT4*myc* recycling over time in L6-GLUT4*myc* cells in the basal state are shown in **Fig. 3**.

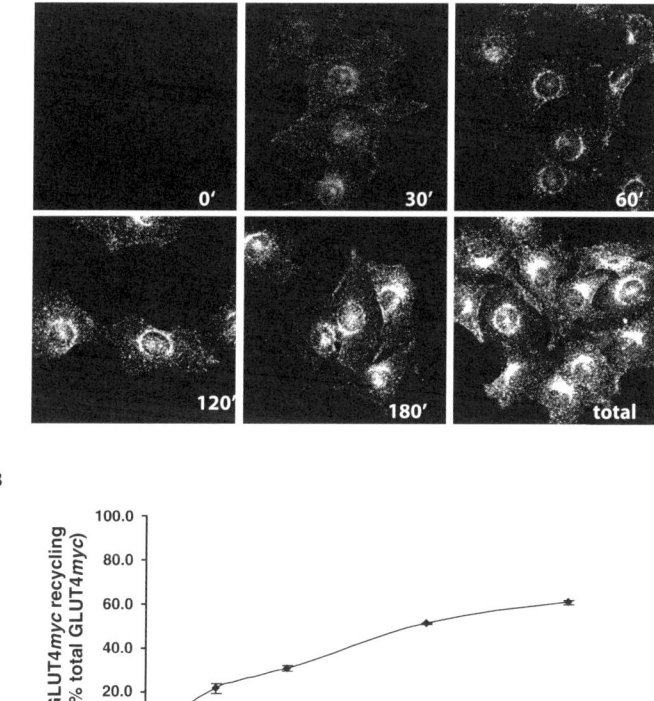

Fig. 3. GLUT4*myc* recycling in L6 myoblasts as a function of time. (**B**) L6-GLUT4*myc* cells were loaded with anti-*myc* antibody at 37°C for various times, and processed for the recycling of GLUT4*myc* in the basal, unstimulated state. Shown are micrographs obtained by confocal microscopy, representative of four independent experiments, showing the rate of GLUT4*myc* recycling to the plasma membrane as a function of time, as indicated. Scale bar, 20 μm. (**B**) GLUT4*myc* recycling at 0, 30, 60, 120, or 180 min and total cellular GLUT4*myc* content was quantified as described in **Subheadings 3.4** and **3.5**, respectively. Shown are the means ± S.E. of 3 independent experiments with recycling expressed as a percent of total GLUT4*myc*.

11. For quantification, export single composite xy-projections as TIFF files. To quantify the GLUT4*myc* staining intensity at each time of recycling and treatment condition, employ ImageJ software. Following outlining each cell's contour, determine the mean pixel intensity for surface GLUT4*myc* stain.
12. The mean GLUT4*myc* staining measurement of at least 30 cells per time point per condition is expressed as a fraction of total cellular GLUT4*myc*, determined in a separate but parallel sample as described in **Subheading 3.5**.

3.5. Confocal Immunofluorescence of Total Cellular GLUT4myc

1. Seed L6-GLUT4*myc* cells under sterile conditions onto 18 mm^2 coverslips in 12-well dishes. Sterilize coverslips are described in **Subheading 3.2.1**.
2. At the time of the experiment (at least 24 h after seeding), serum-starve cells by first rinsing and then incubating with serum-free α-MEM for 3 to 5 h. Ensure that cells are below confluence to allow individual cells to be clearly visible in the immunofluorescence.
3. Rapidly rinse the cells three times with ice-cold PBS+ on ice, and then fix with 4% PFA solution for 20 to 30 min on ice. Discard the PFA into a hazardous waste container and quickly wash the samples with PBS+.
4. Quench residual PFA by incubation in 0.1 M glycine solution for 10 min at room temperature, followed by another quick PBS+ wash.
5. Permeabilize the cells by incubation with 0.1% Triton X-100 for 15 to 20 min at room temperature, and then rinse quickly with PBS+.
6. Block the cells with 5% milk in PBS+ for 10 min.
7. During blocking, prepare ice-cold anti-*myc* antibody solution (1:100) in blocking solution and label GLUT4*myc* as described in **Subheadings 3.2.5 to 3.2.7** (*see* **Note 1**).
8. Detect the amount of anti-*myc* antibody-bound GLUT4*myc* within permeabilized cells. Incubate cells in 750 μL of ice-cold blocking solution containing Cy3-conjugated secondary antibody (1:750) for 1 h at room temperature. Ensure that samples are kept in the dark.
9. Following incubation with secondary antibody, wash cells six times in ice-cold PBS+ with aspiration of solution between each wash.
10. Mount the samples by inverting each coverslip into a drop of Dako mounting medium on a microscope glass slide (*see* **Note 2**). Samples can be viewed immediately once dry, or be stored in the dark at −20°C for up to 1 month.
11. View the slides by confocal microscopy. Excitation at 543 nm induces Cy3 fluorescence (red emission) for GLUT4*myc*. For each sample, xy-plane images are acquired along the entire z-plane of the cells and then collapsed to generate a single composite xy-projection in LSM Image software. An example of total GLUT4*myc* staining in L6-GLUT4*myc* cells is shown in **Fig. 3**.
12. For quantification, export single composite xy-projections as TIFF files. To quantify the total GLUT4*myc* staining intensity, employ ImageJ software. Following outlining each cell's contour, determine the mean pixel intensity for total

GLUT4*myc* stain. The mean GLUT4*myc* staining measurement of at least 30 cells is required.

4. Notes

1. Antibody labeling for immunofluorescence can be done by either inversion of a coverslip with cells faced down onto Parafilm spotted with the antibody solution or by application of the antibody solution directly onto a coverslip while it remains in the well of a tissue culture plate. In the case of the latter, be sure to use an appropriate volume of the antibody solution that will cover the entire surface of the coverslip.
2. To minimize the appearance of air bubbles in the mounted sample, place the coverslip carefully down at one edge of the mounting medium and slowly lower the coverslip onto the microscope slide.

Acknowledgments

The authors would like to thank Professor Y. Ebina for help in the development of the stable L6-GLUT4*myc* cell line and Dr. P. Bilan for the careful reading of this manuscript. This work was supported by a Canadian Institutes of Health Research (CIHR) grant (MT7307) to AK. C.N.A. has received doctoral studentship funding from an Ontario Graduate Scholarship (OGS) and a Doctoral Research Award from the Canadian Diabetes Association. V.K.R. has received doctoral studentship funding from CIHR, Banting and Best Diabetes Centre-Novo Nordisk and University of Toronto, an OGS, a SickKids Research Training Centre-University of Toronto MD/PhD Award, and a Dr. Joseph A. Connolly Award in Cell Biology.

References

1. Zisman, A., Peroni, O. D., Abel, E. D., et al. (2000) *Nat. Med.* **6,** 924–928.
2. Dugani, C. B., and Klip, A. (2005) *EMBO Rep.* **6,** 1137–1142.
3. Brozinick, J. T., Jr., Etgen, G. J., Jr., Yaspelkis, B. B., 3rd, and Ivy, J. L. (1994) *Biochem. J.* **297,** 539–545.
4. Cartee, G. D., Douen, A. G., Ramlal, T., Klip, A., and Holloszy, J. O. (1991) *J. Appl. Physiol.* **70,** 1593–1600.
5. Li, D., Randhawa, V. K., Patel, N., Hayashi, M., and Klip, A. (2001) *J. Biol. Chem.* **276,** 22883–22891.
6. Patel, N., Khayat, Z. A., Ruderman, N. B., and Klip, A. (2001) *Biochem. Biophys. Res. Commun.* **285,** 1066–1070.
7. Hayes, N., Biswas, C., Strout, H. V., and Berger, J. (1993) *Biochem. Biophys. Res. Commun.* **190,** 881–887.

8. Ishiki, M., and Klip, A. (2005) *Endocrinology* **146,** 5071–5078.
9. Blot, V., and McGraw, T. E. (2006) *Embo J.* **25,** 5648–5658.
10. Yang, J., and Holman, G. D. (2005) *J. Biol. Chem.* **280,** 4070–4078.
11. Wang, Q., Khayat, Z., Kishi, K., Ebina, Y., and Klip, A. (1998) *FEBS Lett.* **427,** 193–197.
12. Ueyama, A., Yaworsky, K. L., Wang, Q., Ebina, Y., and Klip, A. (1999) *Am. J. Physiol.* **277,** E572–578.
13. Foster, L. J., Li, D., Randhawa, V. K., and Klip, A. (2001) *J. Biol. Chem.* **276,** 44212–44221.
14. Kanai, F., Nishioka, Y., Hayashi, H., Kamohara, S., Todaka, M., and Ebina, Y. (1993) *J. Biol. Chem.* **268,** 14523–14526.

28

Functional Genetic Analysis of the Mammalian Mitochondrial DNA Encoded Peptides

A Mutagenesis Approach

María Pilar Bayona-Bafaluy, Nieves Movilla, Acisclo Pérez-Martos, Patricio Fernández-Silva, and José Antonio Enriquez

Summary

Animal mitochondria are refractory to transformation. This fact has hampered the study of the oxidative phosphorylation system biogenesis by genetic manipulation of the mitochondrial DNA (mtDNA). In humans, a larger variety of mutants have been obtained from patients with mitochondrial diseases, but still we lack a great portion of the range of potential mutants and there is a major obstacle: Animal models cannot be derived from human mtDNA mutants. Until now the only source of mtDNA mutants in mouse was restricted to some drug-resistant-specific cell lines in which a given mtDNA mutation provided growth advantage in the presence of the inhibitor for a specific complex. To overcome these limitations, the authors have developed a protocol that allows the systematic generation of cells harboring mutations in their mtDNA affecting all types of mitochondrial genes. Chemical mutagenesis followed by mtDNA copy number reduction and the use of large-scale negative selection in duplicate cultures, are the key steps of the strategy used.

Key words: Functional genetics; mammalian; mitochondria; mtDNA mutagenesis; OXPHOS iogenesis.

1. Introduction

Mammalian oxidative phosphorylation system (OXPHOS) biogenesis, within the inner mitochondrial membrane, relies on both nuclear and mitochondrially encoded proteins, which are assembled together to form the respiratory complexes. Complexes I, III, IV, and V contain 7, 1, 3, and 2 mitochondrial DNA (mtDNA)-encoded subunits respectively, that are essential for their function. OXPHOS subunits encoded by mtDNA genes are transcribed inside the mitochondrial matrix and translated by mitochondrial ribosomes, which contain two ribosomal RNAs (12S and 16S) and use a tRNA set also encoded in mtDNA.

mtDNA is located in the matrix side of the inner mitochondrial membrane and it is present in high copy number (1,000–10,000 copies per cell). Until now, it has not been possible to transform mammalian mitochondria and therefore site-directed mutagenesis could not be used to generate mitochondrial gene mutants at convenience *(1)*. Chemical random mutagenesis has been successfully used to isolate mutants in other organisms *(2)*, but the mtDNA high copy number makes it very difficult for a random-generated mutation to become predominant and induce phenotypic consequences. To overcome the dilution factor we have taken advantage of the reversible depletion of the cellular mtDNA obtained after treatment of cell cultures with low doses of ethidium bromide (EthBr), a DNA-intercalating agent that inhibits replication *(3)*. Thus, after mutagenesis the cells are depleted of mtDNA by culturing them in the presence of EthBr, ideally to reduce the copy number to one molecule per cell. Then, the intercalating drug is removed, allowing the cells to recover their normal mtDNA copy number. The cells are plated at a density of one cell per well in order to generate homogeneous clones, and duplicated micro-cultures are made to test their OXPHOS performance by growing in two different conditions. One set of clones is grown in high glucose to guarantee the survival of all the cells, whereas the other set is grown in the presence of galactose as the main energy source (**Fig. 1**).

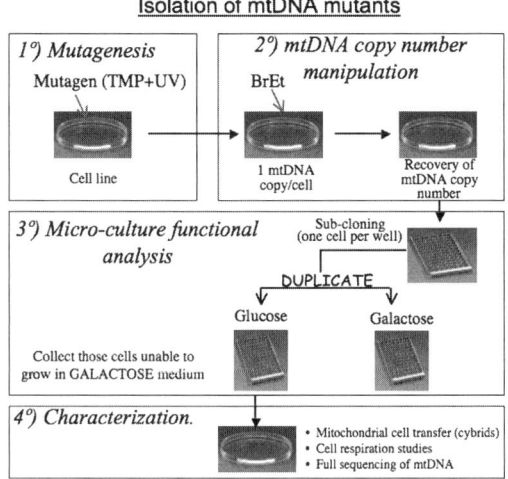

Fig 1. Scheme of the strategy used to isolate mitochondrial DNA mutants.

Functional Genetic Analysis

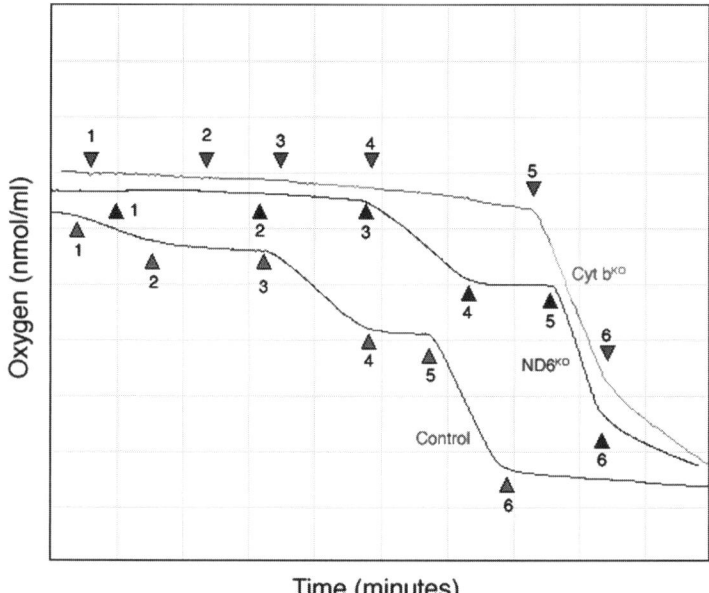

Fig. 2. Examples of oxygen consumption traces obtained by polarography for the control cell line, and for two different mtDNA mutants, one with complex I affected (Nd6 ^{KO}) and other with complex III affected (Cyt b^{KO}). The numbers indicate addition of substrates and inhibitors, (1- glutamate and malate ; 2- rotenone; 3- succinate and glycerol 3-phospahte; 4- Aantimycin A; 5- TMPD; and 6- KCN).

The clones unable to grow or with severely delayed growth in the galactose medium are potential candidates harboring mtDNA mutations. Oxygen consumption measurements by polarography allow the evaluation of the respiratory capacity of the cells (**Fig. 2**). Because mitochondrial mutants are generally respiratory deficient, the polarographic data can indicate the presence or absence of an OXPHOS mutation *(4,5)*. Still, because the mutagenesis can also affect nuclear genes, the mtDNA cause of the defect has to be determined. For this purpose, transmitocondrial cybrids are generated by polyethylene glycol (PEG) promoted membrane fusion between cytoplasts (the enucleated mitochondrial donor cells) and a ρ^0 cell line (lacking mtDNA), and the cotransference of the respiratory-deficient phenotype is in that way determined *(6,7)*.

2. Materials
2.1. Cell Lines

Election of the cell line is a key issue for this protocol. Immortalized cells, especially when derived from tumors, may harbor variations from the reference mitochondrial sequence. It is convenient to sequence the mitochondrial genome of the selected cell line in order to exclude acquired variants, which have been described in mouse *(8)* and in human *(9)* cells. In the authors' assays they have performed the mutagenesis in L929 and NIH3T3 fibroblasts obtained from two different strains of mice: L929 C3H/An and NIH3T3. L929 cell-line contains mutations acquired during the years in culture *(10)*, therefore they previously transferred C57 mitochondria to the L929 nucleus in order to perform the mutagenesis in a defined mitochondrial background.

2.2. Cell Culture Media

1. Complete medium: Dulbecco's Modified Eagle's Medium 4.5 g/L gucose (high-glucose DMEM) (Gibco/BRL; *see* **Note 1**), supplemented with fetal bovine serum (FBS, Gibco/BRL), 0.11 mg/mL sodium pyruvate (Sigma-Aldrich), 50 µg/mL uridine (Sigma-Aldrich), 100 U/mL penicillin G and 100 µg/mL streptomycin sulfate (Gibco/BRL).
2. Galactose medium: DMEM (without glucose; Gibco/BRL), supplemented with dialyzed FBS) dialyzed; *see* protocol of dialysis **Subheading 3.5.3.**), 0.9 g/L galactose (Sigma-Aldrich), 0.11 mg/mL sodium pyruvate (Sigma-Aldrich), 100 U/mL penicillin G and 100 µg/mL streptomycin sulfate (Gibco/BRL).
3. Solution of trypsin: trypsin 0.25%, ethylenediamine-tetraacetic acid (EDTA) 1 mM (Gibco/BRL).
4. Phosphate-buffered saline solution (PBS; Gibco/BRL).
5. Freezing medium: DMEM supplemented with 30% FBS and 10% dimethyl sulfoxide (DMSO; Sigma-Aldrich).
6. Stock solutions of galactose 90 mg/mL; sodium pyruvate 50 mg/mL, and uridine 50 mg/mL should be sterilized by filtration through a 0.2 µm cellulose nitrate filter (Nalgene).

2.3. Chemical Mutagenesis of mtDNA in Mammalian Cell Cultures

1. Stock solution of 4, 5′,8 trimethylpsoralen (TMP) 3 mg/mL.
2. Stock solution of 5 bromo-2′deoxiuridine (5BrdU) 6 mg/mL.

2.4. Reduction of mtDNA Copy Number With Ethidium Bromide

1. Stock solution of EthBr (Sigma-Aldrich) 5 mg/mL.

2.5. Oxygen Consumption Measurements

1. Respiration buffer: 20 mM HEPES pH 7.1, 250 mM sucrose, 10 mM $MgCl_2$, 2 mM potassium phosphate, 1 mM ADP. ADP is added before use from a 0.5 M stock solution.
2. Stock solution of digitonin 10% in DMSO (*see* **Note 8**).

Functional Genetic Analysis

3. Solutions of substrates in water: sodium malate, sodium glutamate, sodium succinate and glycerol 3-phospahte all 1 M; and solution of N,N,N',N'-tetramethyl-p-phenylenediamine (TMPD) 200 mM.
4. Solutions of inhibitors: rotenone 2 mM in ethanol, antimycin A 4 mM in ethanol, and KCN 200 mM in water.
5. Solution of bovine serum albumin (BSA) 10%.

All reactants, substrates, and inhibitors are from Sigma-Aldrich and all solutions are stored at –20°C.

2.6. Cybrid Fusion Reagents and Culture Media

1. Cytochalasin B solution (Sigma-Aldrich) at 1 mg/mL in 95% ethanol; store at –20°C.
2. Polyethylene glycol solution (PEG; Serva Electrophoresis) 50% (w/v).
3. Solution of 10% DMSO in DMEM medium without additions.
4. Selection medium: DMEM 4.5 g/L gucose (high-glucose DMEM; Gibco/BRL) supplemented with 5% dialyzed fetal bovine serum (dFB; *see* protocol of dialysis in **Subheading 3.5., step 3**), 100 U/mL penicillin G and 100 μg/mL streptomycin sulfate (Gibco/BRL), and 500 μg/mL geneticin (G418; Gibco/BRL).
5. Stock solution of rhodamine 6G (Sigma-Aldrich) 2.5 mg/mL in ethanol.

3. Methods

Mitochondrial DNA is less protected by proteins than nuclear DNA, furthermore, mitochondria lack certain DNA repair mechanisms *(11)*. Lesions that are removed by base excision repair are efficiently repaired in mitochondria, whereas some bulky lesions that are removed by nucleotide excision repair seem not to be repaired in these organelles *(12)*. This suggests that mtDNA may be a biologically important target for some drugs. Thus, it has been established that mitochondrial genome is a preferential target of some mutagens such as trimethylpsoralen (TMP) *(13)*. The authors have mainly used this compound in their protocol (**Fig. 1**). However, mutagenic agents differ in efficiency and specificity, and when different kinds of mutations (transitions, transversions, and small deletions) are desired, different mutagens should be used. TMP produces mainly transitions and small deletions, while other mutagenic compounds like 5BrdU induce two-direction transitions or ethylnitrosourea (ENU) yields a wide variety of transitions and transversions *(2)*. Therefore, the authors have also used 5BrdU and are currently testing ENU and other mutagens in their protocols.

3.1. Chemical Mutagenesis of Mammalian Cell Cultures

1. TMP mutagenesis: Culture mouse cells in complete medium in the presence of 30 μg/mL of TMP for 2 d. Trypsinize cells, placed in suspension in a 10-cm

diameter Petri plate, and irradiate with ultraviolet light (440,000 µJ/cm^2). Following the mutagenesis, keep the plate in a cell culture incubator. Two to 4 h later, when cells have adhered to the surface, replace the medium with fresh complete medium (*see* **Note 2**).
2. 5BrdU mutagenesis: Culture L929 cells in complete medium in the presence of 100 µg/mL of 5 BrdU for 7 d, and 2 d more in 200 µg/mL of 5 BrdU.

3.2. Reduction of mtDNA Copy Number in Mutagenized Cells

1. Following the mutagenesis protocol, grow cells in complete medium for 10 d to allow recovery (*see* **Note 3**).
2. Subsequently treat the cultures with 5 µg/mL EthBr in complete medium for 12 d. During the treatment with this DNA intercalating agent, is progressively reduce the mtDNA copy number to very few molecules per cell. Recover the normal mtDNA copy number upon removal of the EthBr. This process is key in the protocol because it allows a redistribution of the mutant mtDNA molecules within the cell population making possible for a molecule of mtDNA with a single newly generated mutation to repopulate the entire mtDNA complement of a cell by its clonal expansion when the EthBr is removed from the culture medium (*see* **Note 4**).

3.3. Negative Selection in Galactose of mtDNA Mutants and Expansion of the Clones of Interest

After EthBr treatment, cells growing on a 10-cm Petri dish are harvested, diluted, and plated at a density of 0.8 cells per well in four 96-well round-bottom plates (*see* **Note 5**). The adequate dilution is 5.333 cells per milliliter of complete medium (0.8 cells in 150 µL) and 59 mL are needed to prepare four 96-well plates. Place 150 µL of the cell dilution in each well using a multichannel pipet (*see* **Note 6**).

In 12 d, cellular clones should be visible under the microscope in most of the wells. Each clone is washed with 100 µL PBS and tripsinized in 60 µL of trypsin-EDTA. Seed 25 µL in replicated 96-well plates, one containing complete medium and the second containing galactose-selective medium (**Fig. 1**).

Those cells that after 4 or 5 d either die or show very low growth in galactose medium versus complete medium (Gal$^-$ clones), are considered potential OXPHOS mutants. They are picked up from the glucose containing complete medium, plated, and expanded for further analysis. Twenty-four-well plates, 6-well plates, and 10-cm Petri dishes are sequentially used for the expansion. Several vials of each clone are stored frozen in liquid nitrogen.

3.4. Characterization of the Selected Clones

1. To have a fast first screening, measure total cell respiration for each Gal$^-$ clone. Harvest and resuspend cells at 6×10^6 cells/mL in respiration buffer. Place 500 μL in the chamber of a calibrated Clark oxygen electrode (Hansatech Instruments) and determine endogenous oxygen consumption as fmol O_2/cell per minute. For Gal$^-$, clones showing oxygen consumption rate significantly lower than the control, a more detailed respiration profile should be obtained to establish the respiratory complex or complexes affected.
2. To determine the specific respiratory complex deficiency, harvest and resuspend cells at 5×10^6 cells per mililiter in respiration buffer (*see* **Note 7**). Place 1 mL of the suspension in the chamber of a calibrated Clark oxygen electrode and permeabilize cells with 1 μg digitonin (*see* **Note 8**), per 1×10^6 cells; add 5 μL from a 1% solution in respiration buffer freshly made form the stock 10% in DMSO to the chamber (*see* **Note 9**). After 10 min of equilibration, close he chamber and record oxygen consumption while adding substrates and inhibitors for each respiratory complex using Hamilton syringes (*see* **Note 10**): sodium malate and sodium glutamate, 5 mM each as complex I substrates, rotenone 100 nM as specific complex I inhibitor; sodium succinate and glycerol 3-phospahte 5 mM as complex II+III substrates, atimycin A 20 nM as inhibitor; TMPD 0.2 mM as complex IV substrate and KCN 1 mM as specific inhibitor of this complex (**Fig. 2**).

It is very important to rinse the electrode chamber between two measurements first with water, then three times with 10% BSA and again with water, in order to remove all traces of inhibitors.

The rate of oxygen consumption after inhibition is subtracted from the rate obtained with the specific substrates for each complex. At least three measurements are needed for each complex to have statistic values. The comparison of the data with those for the control cell line allows to detect single complex defects.

3.5. Confirmation of the Cytoplasmic Nature of the Defect

Mitochondria from Gal$^-$ clones are transferred to a diffcrent nuclear environment, (generation of transmitochondrial cybrids), in order to confirm the cytoplasmic (mtDNA) nature of the defect.

1. For physical enucleation, plate 0.3×10^6 cells in a 35-mm dish 24 h before cnucleation to promote maximal attachment to the growth surface. A confluence of 80% in the dish is adequate. Add 32 mL of enucleation media (30 mL of DMEM, 2 mL of FBS and 0,32 mL of cytochalasin B 1 mg/mL) to an autoclave-sterilized 250 mL wide-mouth centrifuge bottle (*see* **Note 11**). Remove the culture medium from the 35-mm dish, and place it upside-down within the centrifuge bottle with

the help of previously autoclaved forceps (*see* **Note 12**). A small pocket of air is usually trapped within the dish but it is not necessary to eliminate it. Prewarm a fixed-angle centrifuge to 37°C. Place the centrifuge bottle with the culture dish in the rotor and centrifuge for 20 min at 8.000*g*, (*see* **Note 13**). After centrifugation, and once inside the hood, remove the culture plate from the bottle with forceps. Wipe the outside of the dish with 70% ethanol and place a sterile top onto the dish. Add complete medium to the plate and keep the dish in a CO_2 incubator until use (*see* **Note 14**). The authors use the cytoplasts for fusion right away (*see* **Note 15**).

2. Perform fusion in monolayer. The authors use the line ρ^oL929, which was obtained in their lab from line L929 *(14)*, for cytoplasm transfer studies. The resulting ρ^o line (lacking mtDNA) was rendered geneticin-resistant by transfection with the neocassette containing plasmid pcDNA3.1 (Invitrogen). Remove the medium from the 35-mm Petri dish containing donor cytoplast and place 1×10^6 ρ^oL929neo^r cells, previously typsinized, covering the growth surface, and allow to attach for 1 h. Wash the mixture of cytoplasts and cells in the 35-mm culture dish three times with DMEM (without any additions). After the last DMEM wash, completely remove the medium to avoid dilution of the PEG 50% (w/v), because PEG concentration is critical for efficient cell fusion. Add 2 mL of PEG 50% (w/v) to the dish for exactly 60 s. During approx 20 s, gently swirl the dish to facilitate complete coverage of the bottom of the dish. PEG is highly toxic to cells and therefore the time of exposure is very critical. Thus, quickly remove it after the 60 s and add 4 mL of DMEM with 10% DMSO to the plate to dilute the remaining PEG. Wash the plate twice with the same medium, and once with complete medium. Add 3 mL of complete medium to the dish and keep it in CO_2 incubator prior to trypsinization and culture in selective medium.

3. Selection of cybrid cell lines is carried out in selective medium: high-glucose DMEM supplemented with 5% dialyzed FBS, 100 U/mL penicillin G and 100 μg/mL streptomycin, and 500 μg/mL of geneticin (G418). Prepare dialyzed serum by dialysis of the FBS in dialysis tubing (3,000 molecular-weight cutoff) against PBS, at 4°C. Perform dialysis with buffer to serum ratios of 10:1. Change the buffer every 6 h during daytime and after 12 h overnight, and perform at least 10 buffer changes. Following dialyisis, sterilize the serum by filtration through a 0.2 μm cellulose nitrate filter and store at −20°C. Plate the cells from one 35-mm dish in a 100-mm dish in selective medium 24 h after fusion. When cells start dying, change medium daily to remove death cells. Typically, the authors grow cells for 3 wk in selective medium to be sure that all of them are transmitocondrial cells (*see* **Note 16**).

In case of transference of mutations that fully abolish complex III or IV activities, the mutants are, as ρ^o lines, auxotroph for uridine. Dialyzed serum cannot be used in the selection medium. Instead, ρ^oL929neo^r cells are pretreated with a lethal dose of rhodamine 6G, 2.5 μg/mL, for 1 wk before fusion. This treatment commits the cells to death unless new fresh mitochondria are introduced in their cytoplasm *(8)*.

Table 1
Examples of mtDNA Mutants Generated

Complex	Gen	Mutation	aa change	Reference
Complex I	ND1	G3429A	E224 K	Unpublished
	ND4	A12127	Frame shift	Unpublished
	ND4	AC11453	Frame shift	Unpublished
	ND5	G12275A	Knock-out	Unpublished
	ND6	C13890	Frame shift	*20*
Complex III	Cyt b	G15263A	E373 K	*21*
Complex V	A6	A8414G	N163S	Unpublished
Prot. synthesis	tRNAIle	G3739A		Unpublished

mtDNA, mitochondrial DNA

In order to unequivocally determine that the respiratory defect has been transmitted with mitochondria, the rates of oxygen consumption for respiratory complexes is determined in the cybrids by polarography as in **Subheading 3.4**. and compared with the rates obtained before the transference.

3.6. Determination of the mtDNA Mutation by Sequencing of the mtDNA Genome

To determine the particular mtDNA mutation in each mutant cell line isolated, the mitochondrial genome should be fully sequenced and compared with the reference mouse mtDNA *(10)*. With the use of TMP as mutagenic agent the authors have obtained both G→A and T→C transitions as well as small deletions (A, C, AC). Treatment with 5BrdU has generated a G→A transition (**Table 1**).

The sequencing also allows one to determine the existence of hetereoplasmy (wild-type and mutant mtDNA coexisting inside the cell) versus homoplasmy (only one type of mtDNA). Ideally, the isolated mutation should be in homoplasmy. After successive cell-replication cycles, the heteroplasmy may lead to the reduction of the mutation content and recovery of the OXPHOS wild-type phenotype. A second EthBr treatment can be used to modify the levels of heteroplasmy in order to obtain homoplasmic mutant clones.

4. Notes

1. Respiratory deficient cells have been found to require high-glucose medium and, if they lack complex III or complex IV activity, as is the case for mtDNA-depleted cells (rho 0), they also require pyruvate and uridine supplementation for survival *(15–18)*.

2. TMP photoreacts with DNA and induces interstrand crosslinks, therefore the irradiation step is essential to increase the rate of newly generated mutations.
3. Unless stated otherwise, culture medium is replaced with fresh medium every 2 or 3 d, and cell cultures are trypsinized and diluted 1/5 when confluent.
4. 5µg/mL is the appropriate concentration of EthBr in the culture medium in order to generate mtDNA-less murine cell lines. The time of treatment with EthBr should never last more than 20 d, which could produce a very high mortality in the cell population, and could make the depletion of mtDNA irreversible.
5. Round-bottom 96-well plates are more suitable for this purpose, because cells tend to grow in the center of the well and are more easily found in the well under the microscope objective providing 400 times enlargement.
6. Typically cells are first harvested in 5 mL complete medium and several (three to five) serial dilutions 1/10 are performed to reach a concentration of 5.33 cells per milliliter reducing the pipeting error.
7. It is convenient to grow cells in 150-cm Petri dishes in order to have sufficient number of cells because 70 to 80% confluence should not be exceeded when cells are harvested to measure respiration.
8. Since commercial digitonin (Sigma-Aldrich) contains 30 to 50% impurities, the following method of purification is recommended: dissolve 0.6 g of commercial digitonin in 15 mL of 75% ethanol. Precipitate digitonin by chilling the solution at 0° (ice) for 20 min and separate by centrifugation at 4°C. Repeat this procedure twice. It will result in 60% recovery of the originally used digitonin. Once dried, dissove the pellet to 10% in DMSO.
9. If a cell line other than L929 is going to be used, determine the appropriate concentration of digitonin per cell by checking in trypan blue the number of permeabilized cells. The best concentration is the lower for which almost all of the cells become blue.
10. Use separate Hamilton syringes for the addition of substrates and inhibitors, and especially for addition of antimycin A because it is very difficult to remove.
11. Autoclave the centrifuge bottle with water inside and the cap unscrewed to avoid damage, and wrap it in aluminum foil to keep it sterile until it is open in the laminar flow hood.
12. Wipe the outside surface of the 35-mm plate with 70% ethanol to sterilize it, and autoclave the forceps wrapped in aluminum foil.
13. Establish the optimal time of centrifugation for each cell type.
14. An alternative chemical enucleation protocol may be used instead *(19)*.
15. After a good enucleation there are no cells remaining, only cytoplasts, unable to grow, are attached to the dish.
16. The cell line L929 does not survive more than a 1 wk growing in the presence of 500 µg/mL of geneticin, and the ρ^o line, ρ^oL929 is not able to grow in dialyzed serum (absence of uridine) more than 10 d, therefore, selection of transmitocondrial cell lines should last for more than 10 d. Routinely, the authors grow them for 3 or 4 wk and collect a pool of selected cells.

Acknowledgments

The authors would like to thank Rebeca Acín-Pérez, Raquel Moreno-Loshuertos, Gustavo Ferrín-Sánchez and Esther Perales-Clemente for their collaboration at different steps of the development of the method; and to Santiago Morales for his technical assistance. The authors' work was supported by the Spanish Ministry of Education (SAF2006-00428), the EU (EUMITOCOMBAT-LSHM-CT-2004-503116), by the Diputación General de Aragon Group of Excellence (B55) and PIP083/2005 grants. MPB-B is supported by a Ramon y Cajal contract from the Spanish Ministry of Education.

References

1. Bayona-Bafaluy, M. P., Fernandez-Silva, P., and Enriquez, J. A. (2002) The thankless task of playing genetics with mammalian mitochondrial DNA: a 30-year review *Mitochondrion* **2,** 3–25.
2. Anderson, P. (1995) Mutagenesis *Methods Cell Biol* **48,** 31–58.
3. Wiseman, A., and Attardi, G. (1978) Reversible tenfod reduction in mitochondria DNA content of human cells treated with ethidium bromide *Mol. Gen. Genet.* **167,** 51–63.
4. Hofhaus, G., Shakeley, R. M., and Attardi, G. (1996) Use of polarography to detect respiration defects in cell cultures *Methods Enzymol.* **264,** 476–483.
5. Villani, G., and Attardi, G. (1997) In vivo control of respiration by cytochrome c oxidase in wild-type and mitochondrial DNA mutation-carrying human cells *Proc. Natl. Acad. Sci. U S A.* **94,** 1166–1171.
6. Chomyn, A., Meola, G., Bresolin, N., Lai, S. T., Scarlato, G., and Attardi, G. (1991) In vitro genetic transfer of protein synthesis and respiration defects to mitochondrial DNA-less cells with myopathy-patient mitochondria *Mol. Cell. Biol.* **11,** 2236–2244.
7. King, M. P., and Attardi, G. (1996) Mitochondria-mediated transformation of human rho0 cells *Methods Enzymol.* **264,** 313–334.
8. Acin-Perez, R., Bayona-Bafaluy, M. P., Bueno, M., et al. (2003) An intragenic suppressor in the cytochrome c oxidase I gene of mouse mitochondrial DNA *Hum. Mol. Genet.* **12,** 329–339.
9. Gallardo, M. E., Moreno-Loshuertos, R., Lopez, C., et al. (2006) m. 6267G>A: a recurrent mutation in the human mitochondrial DNA that reduces cytochrome c oxidase activity and is associated with tumors *Hum. Mutat.* **27,** 575–582.
10. Bayona-Bafaluy, M. P., Acin-Perez, R., Mullikin, J. C., et al. (2003) Revisiting the mouse mitochondrial DNA sequence *Nucleic Acids Res.* **31,** 5349–5355.
11. Stuart, J. A., and Brown, M. F. (2006) Mitochondrial DNA maintenance and bioenergetics *Biochim. Biophys. Acta.* **1757,** 79–89.
12. Larsen, N. B., Rasmussen, M., and Rasmussen, L. J. (2005) Nuclear and mitochondrial DNA repair: similar pathways? *Mitochondrion.* **5,** 89–108.
13. Cullinane, C., and Bohr, V. A. (1998) DNA interstrand cross-links induced by psoralen are not repaired in mammalian mitochondria *Cancer Res.* **58,** 1400–1404.

14. Tiranti, V., Hoertnagel, K., Carrozzo, R et al. (1998) Mutations of SURF-1 in Leigh disease associated with cytochrome c oxidase deficiency *Am. J. Hum. Genet.* **63,** 1609–1621.
15. King, M. P., and Attardi, G. (1989) Human cells lacking mtDNA: repopulation with exogenous mitochondria by complementation *Science* **246,** 500–503.
16. Inoue, K., Takai, D., Hosaka, H., et al. (1997) Isolation and characterization of mitochondrial DNA-less lines from various mammalian cell lines by application of an anticancer drug, ditercalinium *Biochem. Biophys. Res. Commun.* **239,** 257–260.
17. Desjardins, P., de Muys, J. M., and Morais, R. (1986) An established avian fibroblast cell line without mitochondrial DNA *Somat. Cell Mol. Genet.* **12,** 133–139.
18. King, M. P., and Attardi, G. (1996) Isolation of human cell lines lacking mitochondrial DNA *Methods Enzymol.* **264,** 304–313.
19. Bayona-Bafaluy, M. P., Manfredi, G., and Moraes, C. T. (2003) A chemical enucleation method for the transfer of mitochondrial DNA to rho(o) cells *Nucleic Acids Res.* **31,** e98.
20. Moreno-Loshuertos, R., Acin-Perez, R., Fernandez-Silva, P., et al. (2006) Differences in reactive oxygen species production explain the phenotypes associated with common mouse mitochondrial DNA variants *Nat. Genet.* **38,** 1261–1268.
21. Acin-Perez, R., Bayona-Bafaluy, M. P., Fernandez-Silva, P., et al. (2004) Respiratory complex III is required to maintain complex I in mammalian mitochondria *Mol. Cell.* **13,** 805–815.

Index

A

Acid stripping, 310
Acrylamide, unpolymerized, 285
Adaptor proteins (AP)-1, *in-vitro* recruitment to peptidoliposome, 224–234
Adherens junctions, 263
Agarose sheet method, 154–155
Akt kinase, 211–212
    phosphorylation sites, 213, 214, 215–216
    role in ER to Golgi transport of the SREBP/SCAP complex, 213–220
        materials used, 213–214
        methodology, 214–219
        precautions and suggestions, 219–220
    See also Receptor tyrosine kinases (RTK)
Alexa fluor 488, 306, 307–309, 318, 322–324
Alkaline phosphatase (ALP)
    accumulation, transportation and glycosylation analysis of, 14–17
    immunoblotting for, 17, 20–21
Anterograde secretory pathway, 4
Anti-HA antibody, 318, 322–324
    uptake during GLUT4
        endocytosis, 357, 358
        exocytosis, 354–355
Arf1, myristoylated, 225–226, 228–229
Atp proteins, 108
Automated image analysis, 190, 191, 193–194
    data evaluation, 191–192
    for fluorescence-based assays, 194
    for morphology-based assays, 194–196
Autoradiography
    for cell-free system in reconstituting protein transport, 43
Auxotrophic markers, 108

B

BCA™ Protein Assay Kit, 172
Bead Beater, 53
Benzyl-guaninefluorescein (BG-FL), 60
BiFC, *see* Bimolecular fluorescence complementation (BiFC)
Bimolecular fluorescence complementation (BiFC), 164
    for *in vivo* protein interaction study in yeast, 164–165
        materials used, 165–167
        methodology, 167–171
        precautions and suggestions, 171–172
BioRad semi-dry transfer apparatus, 283–284
Biotin conjugated E-Cadherin
    endocytosis assay, 266, 267–268
    recycling assay, 268
    See also E-Cadherins
Block face trimming
    in correlative microscopy, 206, 208
Blue Native PAGE (BN-PAGE)
    for *in-vitro* analysis of mitochondrial preprotein import, 61, 65, 75–78
    pipeting scheme for, 78
Bovine serum, 310
BoxShade 3.2, 219
Budding yeast, *see Saccharomyces cerevisiae*

C

Cadherins, 263–264
  See also E-Cadherins
Calcium phosphate precipitation, 333–334, 335, 339
Calcofluor white assay, 28
  genome-wide screen for membrane transport analysis
    detecting Chs3 deletion mutants, 31
    materials used, 30
    methodology, 33
Carbonate extraction of peripheral membrane proteins, 96, 101–102
Carboxypeptidase Y (CPY)
  biosynthesis and processing in yeast spheroplasts, 45
  genome-wide secretion assay for membrane transport analysis, 28, 29
    materials used, 30
    methodology, 31, 32, 34
Carboxypeptidase yscS (CPS)
  accumulation, transportation and glycosylation analysis of, 14–17
  immunoblotting for, 17, 20–21
Cell density, 309
Cell-free system for reconstitution of intercompartmental protein transport
  materials, 40–43
  methodology, 43–52
    assays, 49
    cell lysis via extrusion through polycarbonate filters, 46–48
    cytosol preparation, 48
    immunoprecipitation, 49–51
    precautions and suggestions, 52–54
    preparation of acceptor membranes from yeast spheroplasts, 45
  radiolabeling yeast spheroplasts, 43–45
  SDS-PAGE, 51–52
Cerebellar granule neurons (CGN), 340
  styryl dye-based synaptic vesicles recycling assay in cultured, 330–340
CGN, see Cerebellar granule neurons (CGN)
Charged coupled device (CCD) camera, 179–180
Chemiluminiscence, 145
Chinese Hamster Ovary (CHO) cells, 214, 219
Chitin synthase (Chs3) deletion mutants, 28
  calcofluor white assay for detecting, 31
Chitosomes, 28
Cholesterol
  homeostasis in mammals, 212
  regulatory components
    ER to Golgi transport, 212–220
    scansite results for human's, 216
Chromatic aberrations, 207
Chs3, see Chitin synthase (Chs3)
Cisternae, golgi, 176, 177
Clathrin-coated vesicles (CCV), 224–225, 227–228
cMet RTK, 301–302
  internalization analysis using flow cytometry, 302–311
Coat proteins, 4, 224
  recruitment to peptidoliposomes, 223–234
  role in transport vesicles formation, 5
Collateral damage, 207
Co-localization in fluorescence microscopy, 183–184
Compactin, 215
ConA-coated slides, 170
Confocal microscopy
  to Alexa[488]-labeling, 306–307

Index 393

immunofluorescence for GLUT4myc
  traffic in L6 myocytes, 365
  endocytosis, 367–369, 370
  recycling, 369–372
  steady-state surface, 366–367
  total cellular, 372–373
  See also Microscopy
Conservation of sequences throughout
  evolution, 213, 216, 217
Coomassie blue staining of SDS gels, 97,
  106
Correlative microscopy, 199, 200
  based on laser-inscribing and
    laser-etching, 200–208
  materials used, 200–201
  methodology, 201–206
  precautions and suggestions,
    206–207
  See also Microscopy
Cox1p, 126
Cox proteins, 108
CPY, see Carboxypeptidase Y (CPY)
Cycloheximide, 120, 135
Cytochrome b, 108
Cytochrome b2-DHFR fusion proteins,
  60–61
  mitochondrial import of, 63–64, 72–74
    in MTX presence, 75–78
  preparation from E.coli, 62–63, 69–72

D

Deconvolution, 176, 180–183
Densitometry of digital images, 33
384-density arrays, 31–32
Density gradient centrifugation
  for lipid droplet analysis, 153, 156
Detector device, 179–180
Detergents, 135
  dilution method, 245
DHFR, see Dihydrofolate reductase
  (DHFR)
Diauxic shift, 13
Dichroic mirrors, 183
Differential centrifugation analysis, 143
  for isolating mitochondria

  materials used, 95
  methodology, 99
Differential interference contrast, 155,
  156
Digital imaging, 176
  densitometry, 33
Digitonin, 75, 80, 384
Dihydrofolate reductase (DHFR) in
    cytochrome b2 fusion
    proteins, 60–61
  mitochondrial import of, 63–64, 72–74
    in MTX presence, 75–78
  preparation from E.coli, 62–63,
    69–72
Dithiothreitol (DTT), 277, 285

E

E-Cadherins
  biotin conjugated, see Biotin
    conjugated E-Cadherin
  transport tracking to and from plasma
    membrane, 263–273
    materials used, 264–266
    methodology, 266–272
    precautions and suggestions, 273
EDTA-PBS, 310
Electron microscopy, 199–200
  correlation of light and, 200–208
  imaging
    of golgi apparatus, 177–179
    setting object bound-
      aries/thresholds, 181–182
  See also Microscopy
Electrophoretic Mobility Shift Assay
    (EMSA)
  for nucleocytoplasmic shuttling
    of NFκB in leukocytes,
    278–279, 284
Electroporation
  of 3T3-L1 adipocytes, 347, 349–351
Endocytosis, 4
  assay, 316–318, 319–322
    for E-Cadherin trafficking, 266,
      267–268

Endocytosis (*cont.*)
    kinetic analysis of, 318, 321–322, 326
    imaging with pHluorin-based probes, 290, 291, 295–297
    rate determination of GLUT4, 356–359, 361
    in L6 myocytes, confocal immunofluorescence, 367–369, 370
Endoplasmic reticulum (ER)
    to golgi transport, role of signaling in, 211–220
Enhanced chemiluminescence (ECL) reagents, 98
    detection, 167
Enhanced yellow fluorescent protein (EYFP), 164, 166, 167
Myc epitope
    for GLUT4 traffic analysis, 364–373
ER, *see* Endoplasmic reticulum (ER)
Escherichia coli
    preparation of recombinant mitochondrial precursor proteins from, 59, 60
        materials used, 62–63
        methodology, 69–72
Ethidium bromide (EthBr), 376, 378
Evolution, conservation of sequences throughout, 213, 216, 217
Exocytosis
    imaging with pHluorin-based probes, 290, 291, 295–297
    rate determination of GLUT4, 352–356
EZview Red Streptavidin affinity gel, 273

F
FACS DIVA software, 305
Fermentative growth
    effect of heat lysis on, 16
Fission yeast, *see* Schizosaccharomyces pombe

Floatation assay
    for coat proteins recruitment to peptidoliposomes, 226, 230–231
Flow cytometry, 302
    based recycling assay, 318, 322–326, 327
    kinetic analysis of, 324–326
    RTK internalization analysis using, 301–311
Fluorescence-based assays
    automated data evaluation for, 194
Fluorescence microscopy
    of Nile Red-labeled yeast cells, 155–158
    quantitative, *see* Quantitative fluorescence microscopy
    See also Microscopy
Fluorescence recovery after photobleaching (FRAP), 184–185
Fluorescence resonance energy transfer (FRET), 164, 237–238
Fluorescence spectrometry
    for measuring mitochondrial membrane potential, 62, 67–68
    for mitochondrial import analysis of recombinant precursors, 64, 75
Fluorescent imaging, in correlative microscopy, 202, 203
FM4-64 dye
    for synaptic vesicles recycling assay in cultured neurons, 330–340
    See also Styryl dye
FuGene 6, 219
Functional assays, 190–191
Functional genetic analysis
    of the mammalian mtDNA encoded peptides, 375–384
        materials used, 377–379
        methodology, 379–383
        precautions and suggestions, 383–384

Fungi, vesicular transport in, 5
Fusion assay and analysis, 242–244

G

Galactose negative mtDNA mutants, 380–381
Gaussian beam profile, 207
Genegnome imaging station, 145
Geneticin™, 35
Genome-wide analysis
  of membrane transport using yeast arrays, 28–36
GFP, *see* Green fluorescence protein (GFP)
Glass etching, 203–205, 206–207, 208
Glass projections, 208
Glucose transporter 4 (GLUT4)
  in L6 myocytes, single-cell immunofluorescence-based assays for, 363–373
  trafficking insulin regulation, quantitative immunofluorescence microscopy for, 343–361
    materials used, 345–346
    methodology, 346–359
    precautions and suggestions, 360–361
Glusulase, 52
GLUT4, *see* Glucose transporter 4 (GLUT4)
Glycosylation of zymogen precursors, 16, 17
Glycosyltransferases, golgi, 181
Golgi apparatus, 176, 177
  assay for integrity of, 193
  automated data evaluation, 194–196
  correlation of light and electron microscopy images of, 200–208
  electron microscopic imaging, 177–179
  glycosyltransferases, 181
  quantitative flourescence microscopic analysis
    confocal, 177–179

  contrasting organization in control and ZW10 depleted cells, 184–185
  line scan methodology, 185–186
  object boundary/threshold setting using, 181–183
  wide-field, 181–182, 183
  signaling's role in transport from ER to, 211–220
Green fluorescence protein (GFP), 7, 151–152, 154
  fluorescence microscopy, 156–158
  technology, 164
Green fluorescent protein (GFP)-SCAP, 214, 215
Green fluorescent protein (GFP)-tagged single gene library, 12

H

H89, 211
HA-11 antibody
  saturating concentration determination, 348
HA-GLUT4-GFP chimera, 344, 345–359
  endocytosis rate determination, 356–359, 361
  exocytosis/cycling rate determination, 352–356
  surface-to-total ratio determination, 349–352
Hemifusion determination assay, 244
Hepatocyte growth factor (HGF), 304, 305
  conjugation to Alexa[488], 306, 307
High-throughput protein extraction from single-gene deletion library
  materials used, 12–13
  methodology, 13–16
    cell culture and extraction, 16–18
    immunoblotting, 17, 20–21
    SDS-PAGE, 18–20
  precautions and suggestions, 21–25
Hippocampal synapses
  imaging of pHluorin-based probes at, 289–298

Hippocampal synapses (cont.)
   analysis of signals, 295–297
   primary culture of neurons, 291, 293–294
   transfection of neurons, 291–292, 294, 298
Hoechst 33342, 196
Homo sapiens
   gene maps of mtDNA genomes, 113
Hypertonic shock, 308

I
IαBα, 276
   analysis of nucleocytoplasmic shuttling of, 276–286
Igepal CA-630, 285
IgorPro, 295
Iγ phosphatidylinositol 4-phosphate 5-kinase (PIPKIγ), 264
   role in E-Cadherin transport to and from plasma membrane, 266, 267–268, 270–271, 272
ImageQuant software, 63, 75
Immobilization of cells, 154–155
Immunoblotting of high-throughput protein extracts
   materials required for, 13
   methodology, 15, 17, 20–21
   precautions and suggestions, 21–22
Immunodecoration, 97–98, 107
Immunofluorescence-based assays
   single-cell, *see* Single-cell immunofluorescence-based assays
Immunofluorescence microscopy
   quantitative, *see* Quantitative immunofluorescence microscopy
   for subcellular proteins localization, 250, 252–253, 254, 255
   See also Microscopy
Immunofluorescent staining of MDCK cells for E-Cadherin, 265–266, 271–272

Immunoprecipitation
   for cell-free system in reconstituting protein transport
      materials used, 42
      methodology, 49–51
   in ER-to-Golgi SREBP/SCAP transport analysis, 213, 216–218
PreSu9-DHFR import into yeast mitochondria, 88
In organello analysis
   labeling of mitochondrial translational products, 94
   materials used, 95–96
   methodology, 99–101
   protein synthesis, 124, 127–128, 131–133
Insulin
   regulation of GLUT4 trafficking, 343–361
Internalin B (InlB), 301
   internalization of Alexa[488]-labelled, 306, 307–309
   recombinant, 303
   to study cMet endocytosis, 304
INTERNAL/SURFACE (IN/SUR) method, 356, 361
Intracellular protein trafficking
   schematic representation of, 3
In-vitro study
   generation of TOM-TIM supercomplex, 65, 75–76
   of mitochondrial preprotein import, 59–87
      Blue Native PAGE (BN-PAGE) for, 61, 65, 75–78
   recruitment to peptidoliposome of adaptor proteins (AP)-1, 224–234
In vivo study
   insertion of [^{35}S]-methionine, 115
   labeling of mitochondrial translational products, 94

Index

materials used, 97
methodology, 103–105
of mitochondrial translation analysis, 114–119
protein interaction in yeast
bimolecular fluorescence complementation (BiFC) for, 164–172
Iodination, 317, 319–320, 326
Iodixanol velocity sedimentation gradient, 257–258, 259
Iodogen, 316–317, 319, 326
IPLab software, 295

J
"Journal," 194–195

K
Kan^R trafficking mutants, 35
Kinases, 211
See also Akt kinase; Receptor tyrosine kinases (RTK)

L
Labeling technique for correlative microscopy, 200–208
See also Correlative microscopy
LabTek chamber slides, 196
Laser
cutter systems, 207
etching, 203–205, 206–207, 208
inscription, 201–203, 207
scanning confocal microscope (LSCM), 176, 179, 180, 186
Leptomycin B, 276, 284–285
to study nucleocytoplasmic shuttling of IαBα in U-937 cells, 276–286
Leukocytes
nucleocytoplasmic shuttling of NFκB in, 275–286
Light microscopy, 199–200
correlation of electron and, 200–208
Line scan methodologies, 185–186

Lipid droplets metabolism and dynamics
microscopic analysis, 149–150
materials used, 150–151
methodology, 151–152
cell growth, 152
fixation of cells, 154
immobilization, 154–155
microscopy, 155–158
Nile Red staining, 153–154
quiescent cells separation, 153
precautions and suggestions, 158–160
Lipid-mixing assay, 237
principles of, 238
for studying fusogenic properties of SNARE, 239–246
Listeria monocytogenes, 302
Lithium acetate
for yeast transformation in protein-protein interactions, 127, 130–131
L6 myocytes, 363
single-cell immunofluorescence-based assays for GLUT4 traffic in, 363–373
Low-density lipoprotein (LDL) receptor-related protein (LRP), 316
endocytosis and recycling of, 316–327
materials used, 316–318
methodology, 319–326
precautions and suggestions, 326–327
LRP, see Low-density lipoprotein (LDL) receptor-related protein (LRP)
LSCM, see Laser scanning confocal microscope (LSCM)
LY294002, 211–212
Lysate, 89, 140
in BiFC, 165, 168–169

M
Madin-Darby Canine Kidney (MDCK)
culture, 264, 271–272

Mammalian mtDNA, 376
  functional genetic analysis of peptides encoded by, 376–384
  isolation of mutants, 376–377
  See also Mitochondria DNA (mtDNA)
Manual pinning, 29–30, 31–32
MATa single-gene deletion library
  for high-throughput protein extraction and immunoblotting analysis, 11–21
    materials used, 12–13
    methodology, 13–21
    precautions and suggestions, 21–25
McIlwain tissue chopper, 339
MDCK, see Madin-Darby Canine Kidney (MDCK)
Membrane flotation analysis, 96, 103
Membrane potential of mitochondria
  fluorescence spectrometry for measuring, 62, 67–68
Membrane trafficking
  genome-wide analysis using yeast arrays, 28–36
    materials used, 29–30
    methodology, 30–34
    precautions and suggestions, 34–36
    quantitative fluorescence image analysis, 181–183
  secretory, see Secretory membrane traffic measurement
MetaMorph software, 194–195, 197
Methotrexate (MTX), 58, 61
  cytochrome b2-DHFR fusion protein mitochondrial import in presence of, 75–78
Microscopy
  in BiFC, 167, 170–171
  for lipid droplets metabolism and dynamics, 155–158
  See also Confocal microscopy; Correlative microscopy; Electron microscopy; Fluorescence microscopy; Immunofluorescence microscopy; Light microscopy
Mirus TransIT CHO kit, 321, 326
Mitochondria
  biogenesis of, 93–94
  DNA (mtDNA), 375–376
    gene maps of, 113
    proteins encoded by, see Mitochondrial DNA-encoded proteins study
    See also Mammalian mtDNA
  proteome, 6
  translation, see Mitochondrial translation
  See also Mitochondrial preproteins import in-vitro analysis
Mitochondrial DNA-encoded proteins study, 94
  materials used, 94–98
  methodology, 98–108
    analysis, 107–108
    carbonate extraction of peripheral membrane proteins, 101–102
    growth of yeast, 98
    membrane flotation analysis, 103
    mitochondrial isolation by differential centrifugation, 99
    in organello labeling, 99–101
    outer membrane disruption, 101
    protein transfer and immunodecoration, 106–108
    SDS-PAGE, 105–106
    in vivo labeling, 103–105
  precautions and suggestions, 108–109
  protein–protein interactions analysis in, 124–136
  See also Mitochondrial translation
Mitochondrial preproteins import in-vitro analysis, 59–80, 83–85
  materials used, 61–66, 85–86
  import, 86
  synthesis, 85

methodology, 66–79, 86–87
   analysis of import reaction, 74–75
   fluorescence spectrometry for measuring membrane potential, 67–68
   import, 88–89
   import of recombinant precursor proteins, 72–74
   inward translocation force assessment, 78–79
   isolation from yeast, 66–67
   preparation of precursor proteins from *E.coli*, 69–72
   synthesis, 87
   translocation intermediate dissection on BN-PAGE, 75–78
  precautions and suggestions, 79–80
  principles of, 57–59
Mitochondrial translation
  in organello analysis, 124, 127–128, 131–133
  in vivo analysis of, 114–115
    materials used, 115–117
    methodology, 117–119
    precautions and suggestions, 120–121
  See also Mitochondrial DNA-encoded proteins study
  mLRP4, 316, 322
Morphology-based assays
  automated data evaluation for, 194–196
Mss51p, 126
mtDNA, *see* Mitochondrial DNA (mtDNA)
mtHsp70, 65–66
Multivesicular body (MVB), 28, 29
Mutagenesis approach
  for functional genetic analysis of mammalian mtDNA peptides, 375–384
Vps mutants, 29
Myristoylated Arf1, 225–226, 228–229

N

N-acetylgalactosyltransferase 2 (GalNAcT2), 177–178, 181
Neuronal survival, 297
NFκB in leukocytes, 276
  nucleocytoplasmic shuttling of, 275–286
NIH 3T3 cells, 196
Nile Red labeling, 149, 151–152, 153–154
  fluorescence microscopy, 155–158
  photobleaching, 160
  precautions and suggestions, 158–160
Nitrobenzoxadiazol (NBD)
  fluorescence dequenching, 243–244
  to rhodamine energy transfer, 237–238
Nitrocellulose filter
  for CPY secretion assay, 31, 32, 34
Nucleocytoplasmic shuttling of NFκB in leukocytes, 275–286
  materials used, 277–279
  methodology, 279–284
  precautions and suggestions, 284–285
Nyquist criterion, 179–180

O

Oligomycin, 89
Olympus IX81 epifluorescence microscope, 332
Osmotic swelling, 96, 101
Oxa1 mitochondria, 100
Oxidative phosphorylation system (OXPHOS) complex, 94, 112, 123
  biogenesis, 375–376
Oxygen consumption in mitochondria, 378–379
  measurements by polarography, 377, 383

P

Parafilm "M," 270
PC12 cells, *see* Rat pheochromocytoma (PC12) cells

PCR amplification
    use in protein-protein interactions study, 129–130
Peptidoliposomes
    coat proteins recruitment to, 223–234
        materials used, 224–227
        methodology, 227–233
        precautions and suggestions, 233–234
"PET duet," 245
Phenotypic screening arrays
    materials used, 29–30
    methodology, 32–34
Photobleaching, 176
    correction in single vesicle imaging experiments, 295–297
    of Nile Red, 160
Pinning robot, 31, 32–33
PIPKIγ, *see* Iγ phosphatidylinositol 4-phosphate 5-kinase (PIPKIγ)
Plasmid maps, 166, 167
Plastic embedding, 206, 208
Polarography measurements
    for oxygen consumption in mitochondria, 377, 383
Polycarbonate filters
    yeast spheroplasts lysis via extrusion through, 46–48, 53
Polyethylene glycol solution (PEG)
    for yeast transformation in protein-protein interactions, 127, 130–131
Preproteins mitochondrial import *in-vitro* analysis, 59–80, 83–85
    materials used, 85–86
        import, 86
        synthesis, 85
    methodology, 86–87
        import, 88–89
        synthesis, 87
    precautions and suggestions, 89–91
    preparation for import of, 63–64, 72–74

preparation from Escherichia coli, 62–63, 69–72
Presequences, 84
Presequence-translocase-associated motor (PAM), 58, 61
Prevacuolar compartments
    cell-free system for reconstitution of protein transport between, 39–54
Procollagen I (PC I), 192, 197
    secretion assay, 193
Propidium iodide (PI), 311
Protease resistance of translocation intermediates, 78
Proteinase K digestion, 96, 101
Protein–protein interactions analysis
    involving mitochondrial-encoded proteins, 124–136
        materials used, 126–129
        methodology, 124–125, 129–134
    in organello analysis, 124, 127–128, 131–133
    pull-down assay, 133–134
    tagged protein strains, 126–127, 129–131
        precautions and suggestions, 134–136
    in vivo study of BiFC in yeast, 164–172
Proteins
    high-throughput extraction from single-gene deletion library, 12–25
    localization, polarized PC12 cells for monitoring, 249–260
    sorting, 2–4
        secretion, 4–6
        translocation, 4
Proteoliposomes, 239–240, 241–242
Pull-down assay, 128, 133–134

Q

Quantitative fluorescence microscopy
    future perspectives, 186–187
    principles of, 175–187
        co-localization, 183–184

diffusion measurements by FRAP, 184–185
full detector resolution, 179–180
line scan methodologies, 185–186
setting object boundaries/thresholds, 180–184
submicroscopic analysis, 177–179
in voxel space, 184
wide-field imaging, 183
See also Fluorescence microscopy
Quantitative immunofluorescence microscopy
to study insulin regulation of GLUT4 trafficking, 343–361
See also Immunofluorescence microscopy
Quenching antibody, 318, 324–326
Quiescent yeast cells purification, 153

R
Radiolabeling
for *in-vivo* analysis of mitochondrial translation, 114, 116–117, 118–119
yeast spheroplasts for cell-free system in reconstituting protein transport
materials used, 42
methodology, 43–45
Rat pheochromocytoma (PC12) cells, 250
polarized for protein localization monitoring, 250–260
materials used, 251–254
methodology, 254–260
cell fractionation, 253–254, 254–255, 256–258
growth and differentiation, 252, 254–255
immunofluorescence microscopy, 252–253, 254, 255
precautions and suggestions, 258–260
Receptor tyrosine kinases (RTK), 301

internalization using flow cytometry, 301–311
materials used, 302–303
methodology, 303–309
precautions and suggestions, 309–310
See also Akt kinase
Recombinant precursor proteins
for *in-vitro* study of mitochondrial preprotein import, 60–61
preparation for import of, 63–64, 72–74
preparation from Escherichia coli, 62–63, 69–72
Recycling assay
for E-Cadherin trafficking, 268
flow cytometry-based, 318, 322–326, 327
kinetic analysis of, 324–326
Redigrad™ density gradient centrifugation, 158
Respiratory deficient cells, 383
Retrograde secretory pathway, 4
Rhodamine-nitrobenzoxadiazol (NBD) energy transfer, 237–238
Robotic pinning, 31, 32–33
Rosetta Escherichia coli, 239
RTK, *see* Receptor tyrosine kinases (RTK)

S
Saccharomyces cerevisiae
cell-free system for reconstitution of intercompartmental protein transport in, 39–54
examples of commercial genomic libraries of, 21
gene maps of mtDNA genomes, 113
high-throughput protein extraction and immunoblotting analysis in, 11–25
in-vivo analysis of mitochondrial translation, 114–121
culture conditions for growth, 115–116, 118

Saccharomyces cerevisiae (*cont.*)
    radiolabeling, 114, 116–117, 118–119
    mitochondrial genome-encoded proteins in, 112–114
    mitochondrial genome of, 94
    mitochondrial translation analysis in, 94–109
    as a model for mitochondrial biogenesis, 114
    protein-protein interactions analysis involving mitochondrial-encoded proteins, 124–136
    use of BiFC in, 163–172
    See also Yeast
Scan R system, 191
Scansite, 216
SCAP, *see* SREBP cleavage activating protein (SCAP)
Schekman, Randy, 5
Schizosaccharomyces pombe
    gene maps of mtDNA genomes, 113
    in-vivo analysis of mitochondrial translation, 114–121
        culture conditions for growth, 116, 118
        radiolabeling, 116–117, 118–119
        mitochondrial genome-encoded proteins in, 112–114
        as model for mitochondrial biogenesis, 114
    See also Yeast
Scientific Volume Imaging Inc., Hilversum, Netherlands, 184
SDS-PAGE, *see* Sodium Dodecyl Sulfate-Polyacrylamide Gel Electrophoresis (SDS-PAGE)
Secretory membrane traffic measurement, 189–197
    materials used, 190–192
    methodology, 192–196
    precautions and suggestions, 196–197
Sec translocase, 6
Sedimentation assay
    for coat proteins recruitment to peptidoliposomes, 227, 231–233
Semi-dry protein transfer, 97, 106–107
PH-sensitive green fluorescent protein (pHluorin), 290
    imaging at hippocampal synapses, 289–298
        materials used, 291–292
        methodology, 293–297
        precautions and suggestions, 297–298
Serine/threonine kinase, *see* Akt kinase
Serum starvation, 309–310
SEY6210, 45, 46, 50
Shiga toxin B transport, 182–183
Single-cell immunofluorescence-based assays
    for GLUT4 traffic in L6 myocytes, 363–373
        materials used, 364–365
        methodology, 365–373
        precautions and suggestions, 373
Single-gene deletion library
    high-throughput protein extraction from, 12–25
"SiRNA", 7
[^{35}S]-methionine *in vivo* insertion, 115
SNAP isoform of SNARE, 238
SNAP-tag
    in cytochrome b2-DHFR fusion proteins, 63, 80
    mitochondrial import of, 64, 74
    preparation from *E.coli*, 70, 71–72
Sodium Dodecyl Sulfate-Polyacrylamide Gel Electrophoresis (SDS-PAGE)
    in BiFC, 165–166, 169
    for cell-free system in reconstituting protein transport
        materials used, 42–43
        methodology, 51–52, 53–54

in ER-to-Golgi SREBP/SCAP
transport analysis, 214, 218
for high-throughput protein extracts
materials used, 12–13
methodology, 16, 18–20
for *in-vivo* analysis of mitochondrial
translation, 117, 119
for mitochondrial-encoded proteins
analysis, 97, 105–106
for nucleocytoplasmic shuttling of
NFκB in leukocytes, 278,
282–283
for protein-protein interactions
analysis, 128–129, 134
in tracking E-Cadherin transport,
264–265
Sodium salicylate solution, 54
Soluble NSF attachment receptor
(SNARE), 5–6, 238–239
lipid-mixing assay for analysing
fusogenic properties of,
239–246
materials used, 239–240
methodology, 240–245
precautions and suggestions,
245–246
Sorting Nexin, 316
Spheroplasts, yeast
for cell-free system in reconstituting
protein transport, 41
lysis via extrusion through
polycarbonate filters, 46–48
radiolabelling, 42, 43–45
SREBP, *see* Sterol regulatory element-
binding protein (SREBP)
SREBP cleavage activating protein
(SCAP), 212
alignment of conservation sequences,
217
role of Akt kinase in ER to Golgi
transport of SREBP and,
213–220
Sterol regulatory element-binding protein
(SREBP), 212

role of Akt kinase in ER to Golgi
transport of SCAP and,
213–220
Styryl dye
for synaptic vesicles recycling in
cultured neurons, 329–340
materials used, 330–332
methodology, 332–338
precautions and suggestions,
338–339
Sucrose density gradient centrifugation
for purification of yeast
membranes and organelles,
139–145
composition of, 143
materials used, 140
methodology, 141–145
analysis of fractions, 143–145
cell fractionation, 142–143
cell growth and lysis, 141–142
precautions and suggestions, 145
Sulfo-N-hydroxysulfosuccinimide (NHS)
SS-biotin, 266, 273
Synapses, hippocampal
imaging of pHluorin-based probes at,
289–298
Synaptic vesicles
endocytosis, 329
styryl dye-based recycling assay in
cultured neurons, 329–340
Synaptophysin (sypHy), 290, 291,
295–297, 298
Syntaxins, 238
sypHy, *see* Synaptophysin (sypHy)

T
Tagged proteins
use in protein-protein analysis,
126–127, 129–131
Tandem-affinity purification (TAP) tagged
single gene library, 12
TIM22 complex, 84
"Time-lapse" protocol, 295
3T3L1 cells, 345, 346–347

TOM-TIM supercomplex, *see* Translocase of outer membrane (TOM)- translocase of inner membrane (TIM) supercomplex
Transfection
　of CGN cells by calcium phosphate precipitation, 333–334, 335, 339, 340
　of hippocampal neurons, 291–292, 294, 298
Translocase of outer membrane (TOM)- translocase of inner membrane (TIM) supercomplex, 58, 59, 61
　in-vitro generation of, 65, 75–76
Translocase of the inner membrane (TIM)23 complex, 84
Translocase of the outer membrane (TOM) complex, 84
Translocases, 2–4
Translocation, 4
　future directions in, 6–7
Transmission microscopy, 155
Trichloroacetic acid precipitation, 317, 320
Tris-buffered saline Tween-20 (TBST), 167
Trypsin-EDTA, 310
ts-O45-G protein, 192, 196, 197
　secretion assay, 192–193

U

U-937 cells, 276
　leptomycin B to study nucleocytoplasmic shuttling of IαBα in, 276–286
Urea-denatured precursor proteins
　in-vitro mitochondrial import of, 64, 74

V

Valinomycin, 67, 68, 89
Velocity gradient centrifugation
　for polarized PC12 cells in protein localization, 253–254, 257–258, 259–260
Vesicular transport, 4–6
Voxel space imaging, 184

W

Watershed approach, 183
Western blotting
　in BiFC, 166–167, 169–170
　for cell fraction analysis from sucrose gradient, 144–145
　in ER-to-Golgi SREBP/SCAP transport analysis, 214, 218–219
　for nucleocytoplasmic shuttling of NFκB in leukocytes, 278, 283–284
　signals for mitochondrial import analysis of recombinant precursors, 64, 74
　for SNARE-mediated fusion of liposomes, 244–245
　in tracking E-Cadherin transport, 265, 268–271
Whole-genome DNA extraction, 127
Wickner, Bill, 6
Wide-field image flourescence microscopy, 181–182, 183
Wide-field screening microscope automated imaging, 193–194

Y

Yeast
　extract/peptone/dextrose (YPD), 29–30
　knockout genes, 6
　　arrays for genome-wide analysis of membrane transport, 28–36
　membranes and organelles purification by sucrose density gradient centrifugation, 139–145

microscopic analysis of lipid droplet metabolism and dynamics in, 149–160

mitochondria, preSu9-DHFR import into, 88

mitochondrial isolation from, 61–62, 66–67

spheroplasts, *see* Spheroplasts, yeast

transformation in protein-protein interactions, 127, 130–131

See also Saccharomyces cerevisiae; Schizosaccharomyces pombe

Yeast nitrogen base (YNB), 52

Z

ZW10, 184, 185

Zymogen precursors
    effects of growth phases on accumulation of, 14, 15
    immunoblotting for, 17, 20–21
    transport and glycosylation, 15–17

Zymolyase, 135

100T, 41

Printed in the United States of America